Collins

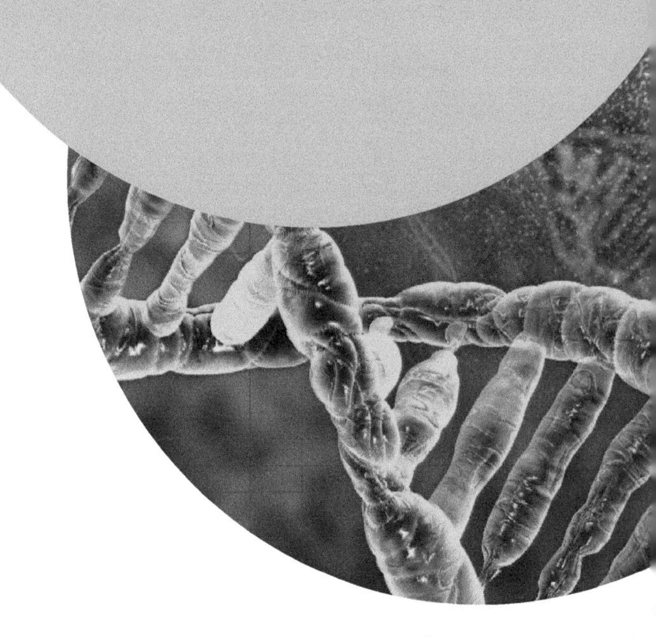

AQA GCSE (9-1)

Life and Environmental Sciences for Combined Science: Synergy

Student Book

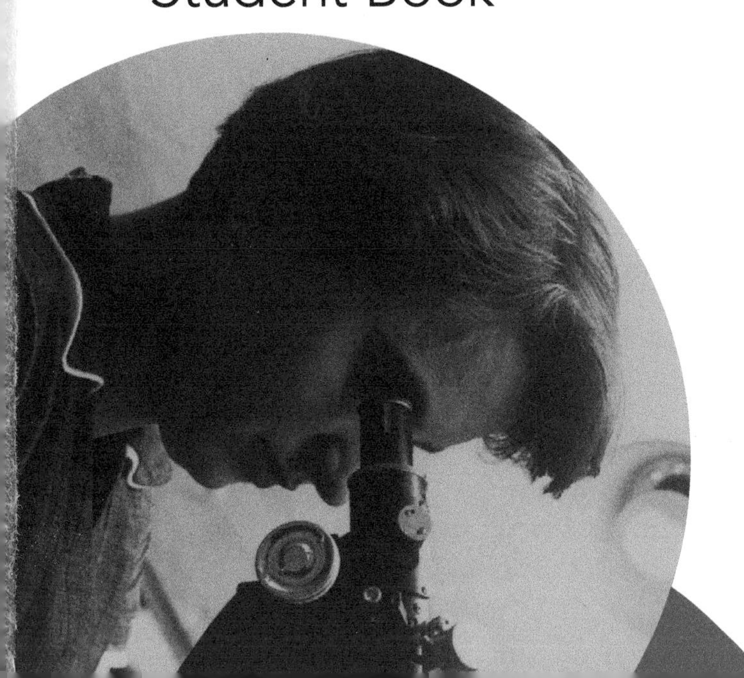

Katy Bloom
Shaista Shirazi
Gina Walker
Series editor: Ed Walsh

William Collins' dream of knowledge for all began with the publication of his first book in 1819.

A self-educated mill worker, he not only enriched millions of lives, but also founded a flourishing publishing house. Today, staying true to this spirit, Collins books are packed with inspiration, innovation and practical expertise. They place you at the centre of a world of possibility and give you exactly what you need to explore it.

Collins. Freedom to teach

HarperCollins Publishers
1 London Bridge Street
London SE1 9GF

Browse the complete Collins catalogue at www.collins.co.uk

First edition 2016

10 9 8 7 6 5 4 3 2 1

ISBN 978-0-00-817495-8

www.collins.co.uk

A catalogue record for this book is available from the British Library

Commissioned by Gillian Lindsey and Katie Sergeant
Edited by Leah Willey and Sarah Thomas
Project managed by Amanda Harman
Copy edited by Life Lines Editorial Services
Development edited by Jane Glendening
Proofread by Write Communications Ltd
Production by Rachel Weaver
Typeset by Jouve India
Cover design by We Are Laura and Jouve India
Printed by CPI Group (UK) Ltd, Croydon, CR0 4YY
Cover images © World History Archive / Alamy Stock Photo; Shutterstock/vitstudio

With thanks to John Beeby, Ann Daniels, Charles Golabek, Sandra Mitchell and Anne Pilling

Approval message from AQA

This textbook has been approved by AQA for use with our qualification. This means that we have checked that it broadly covers the specification and we are satisfied with the overall quality. Full details of our approval process can be found on our website.

We approve textbooks because we know how important it is for teachers and students to have the right resources to support their teaching and learning. However, the publisher is ultimately responsible for the editorial control and quality of this book.

Please note that when teaching the GCSE Combined Science: Synergy course, you must refer to AQA's specification as your definitive source of information. While this book has been written to match the specification, it cannot provide complete coverage of every aspect of the course.

A wide range of other useful resources can be found on the relevant subject pages of our website: aqa.org.uk

Contents

How to use this book

Remember! To cover the content of the AQA GCSE Combined Science: Synergy Specification you should study the text and attempt the End of chapter questions.

These tell you what you will be learning about in the lesson and are linked to the AQA specification.

This introduces the topic and answers the linking question at the bottom of the previous page.

Each topic is divided into three sections. The level of challenge gets harder with each section.

Each section has level-appropriate questions, so you can check and apply your knowledge.

The link shows how the ideas in the book fit together.

Waves

1.4g Uses of electromagnetic waves

Learning objectives:

- give examples of practical uses of electromagnetic waves
- show that the uses of electromagnetic waves illustrate the transfer of energy from source to absorber
- recall that radio waves can be produced by, or can induce, oscillations in electrical circuits.

KEY WORDS

gamma rays
infrared
microwaves
radio waves
ultraviolet
visible light
X-rays

Electromagnetic radiation has a huge number of uses in medicine, communications and everyday life. These uses illustrate the transfer of energy from a source (something that emits electromagnetic radiation) to an absorber (something that takes in electromagnetic radiation). An absorber is often a detector or a receiver.

Gamma rays and X-rays

Gamma rays have the shortest wavelengths in the electromagnetic spectrum and can be harmful to living cells. In controlled doses, gamma rays are used in radiotherapy to kill cancer cells, and to sterilise surgical instruments.

X-rays pass through soft tissues in the body but are absorbed by bone. This means X-rays can be used to check for broken bones, and also in computerised tomography (CT) scans – multiple images taken at different angles build up a detailed picture of inside a patient's body (Figure 1.4.18).

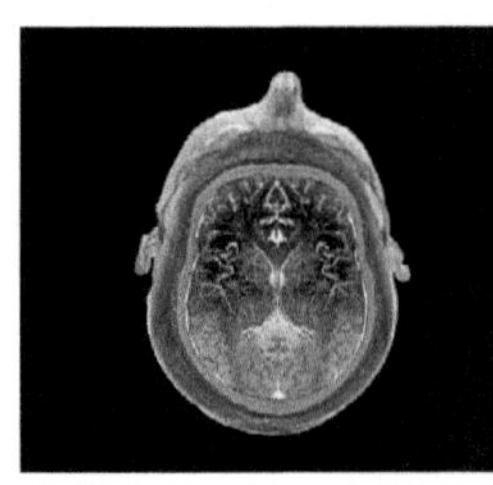

Figure 1.4.18 A CT scan can give good pictures of soft tissue regions but exposes a patient to a much higher radiation dose than a single X-ray

1. **What are the similarities and differences between gamma rays and X-rays?**
2. **Should X-rays used in radiotherapy have longer or shorter wavelengths than those used for medical diagnosis? Explain your answer.**

DID YOU KNOW?

X-ray machines called pedoscopes were used in shoe shops in the 1930s to measure people's feet. Although children loved to watch their toe bones wriggle, people did not realise how dangerous the energy transferred by X-rays can be. Some shop assistants developed cancer because of using them.

Ultraviolet, visible and infrared radiation

Ultraviolet rays have higher frequencies and shorter wavelengths than visible light. One use of ultraviolet is in fluorescent lighting, which is more energy efficient than traditional filament light bulbs.

In small doses, ultraviolet light from sunlight is good for us, as it produces vitamin D in our skin, but larger doses can be harmful to our eyes and may cause skin cancer.

Visible light from lasers is used to send digital data down fibre-optic cables at huge speeds with little loss of signal quality. This has transformed global communications over recent decades.

106 AQA GCSE Life and Environmental Sciences for Combined Science: Synergy Student Book Link: 1.4g → 1.4h

Important scientific vocabulary is highlighted. You can check the meanings in the Glossary at the end of the book.

1.4g

Infrared radiation has longer wavelengths than visible red light. Electrical heaters and traditional cookers (sources) transfer energy by infrared radiation. Thermal imaging cameras (absorbers) detect low levels of infrared radiation from warm objects.

3. **Suggest why you can't get sunburned through a window, even though you can still see the sunlight.**
4. **Suggest how a thermal image of your house might be useful.**

Microwaves and radio waves

Microwaves are radio waves with short wavelengths, between 1 mm and 30 cm. Microwaves are used for cooking in microwave ovens, and also to transmit mobile phone signals. Transmitters are placed on high buildings or masts to give better communication over large distances. Sometimes signals are sent from a transmitter (a source) to a receiver (an absorber) via a satellite.

Microwaves used for communication have a longer wavelength than those used for cooking.

Terrestrial radio and TV signals are sent by **radio waves**, travelling at the speed of light. A radio telescope forms images of objects in space by detecting the radio waves they emit. Radio waves have much longer wavelengths than visible light, and a radio telescope must be very large to have the resolution of an optical telescope.

5. **Suggest why satellite TV dishes are placed on the walls or roofs of houses.**
6. **Like many uses of electromagnetic radiation, a radio telescope illustrates the transfer of energy from a source to an absorber. In this example, identify the source, the form in which energy is transferred, and the absorber.**

HIGHER TIER ONLY

Radio waves are produced by oscillations in electrical circuits. When a current flows through a wire it creates electric and magnetic fields around the wire. When the current changes, fields change, which produces electromagnetic waves of radio frequencies. This is how radio transmitters work.

Radio waves can induce oscillations in an electrical circuit, giving rise to an alternating current with the same frequency as the radio wave itself.

7. **When radio waves are absorbed by an electrical conductor they create an alternating current with the same frequency as the radio wave itself. Suggest how this might be used in a radio receiver.**

Figure 1.4.19 A thermal imaging camera detects infrared radiation given off by an object or person

DID YOU KNOW?

Halogen hobs use ring-shaped halogen lamps beneath a glass top. Although you see a bright red light, the glowing filament radiates mostly infrared radiation.

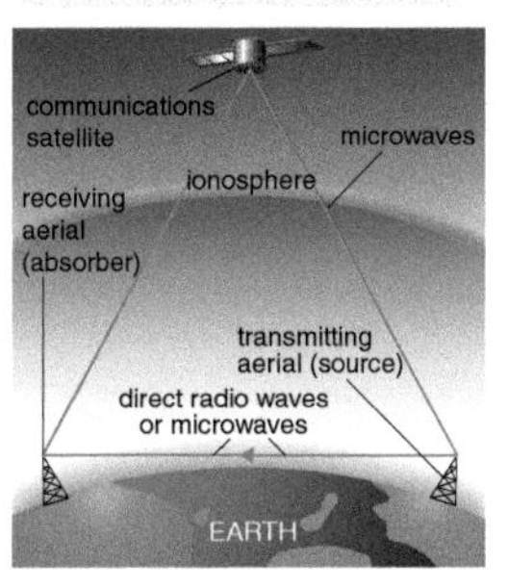

Figure 1.4.20 Direct and satellite wave communication. Microwave satellite communications are used for satellite phones and for satellite TV

DID YOU KNOW?

Bluetooth is a wireless technology for exchanging data over short distances using microwaves.

Qu: How do different surfaces emit and absorb radiation? → 107

Each topic has some fascinating additional background information.

For Foundation tier, you do not need to understand the content in the Higher tier only boxes. For Higher tier you should aim to understand the other sections, as well as the content in the Higher tier only boxes.

The linking question at the bottom of the page prompts you to turn the page to see how the scientific idea develops.

Introduction

At the start of each new Topic, a double-page spread sets out how the ideas in the chapters fit together.

It highlights the increased emphasis on Working Scientifically in the Topic.

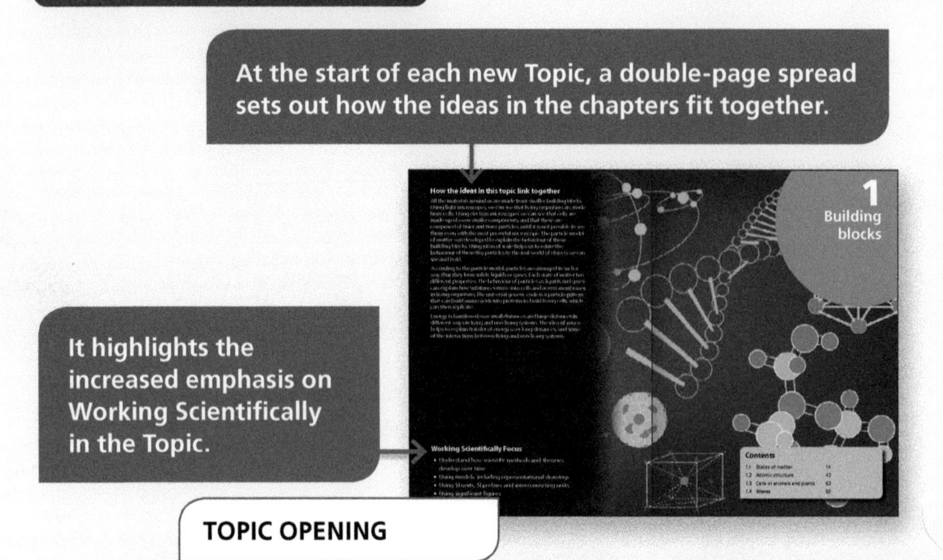

TOPIC OPENING

The first page of a chapter has links to ideas you have met before, which you can now build on.

This page gives a summary of the exciting new ideas you will be learning about in the chapter.

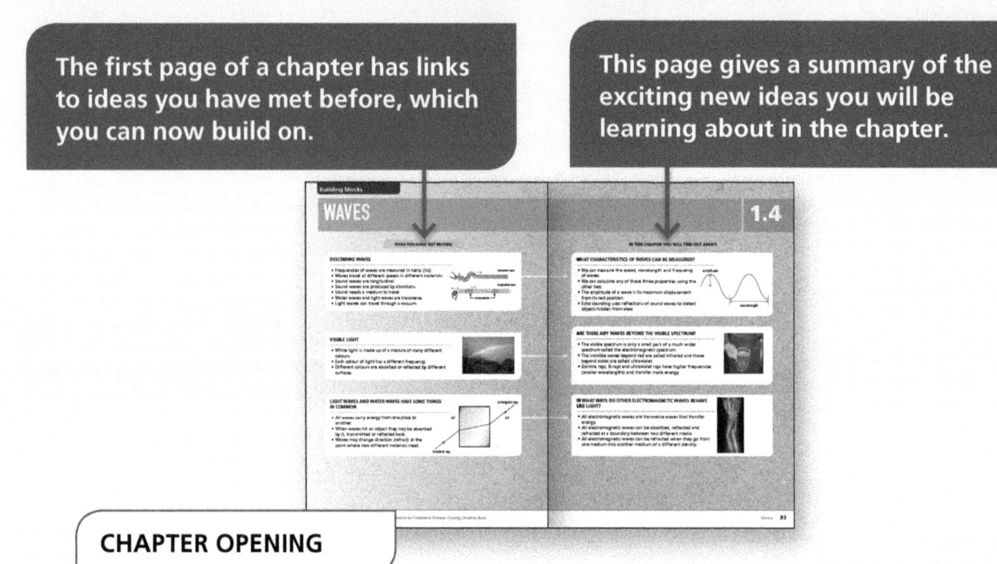

CHAPTER OPENING

The Key Concept pages focus on core ideas. Once you have understood the key concept in a chapter, it should develop your understanding of the whole topic.

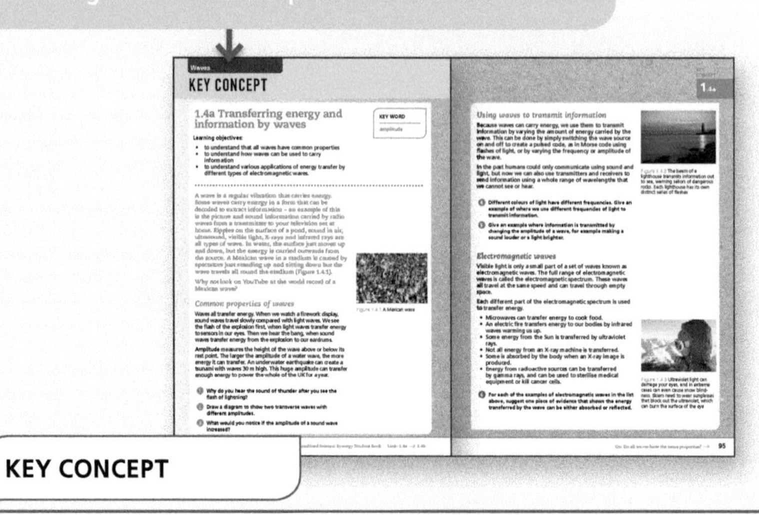

KEY CONCEPT

There is a dedicated page for every Required Practical in the AQA specification. They help you to analyse the practical and to answer questions about it.

The tasks – which get a bit more difficult as you go through – challenge you to apply your science skills and knowledge to the new context.

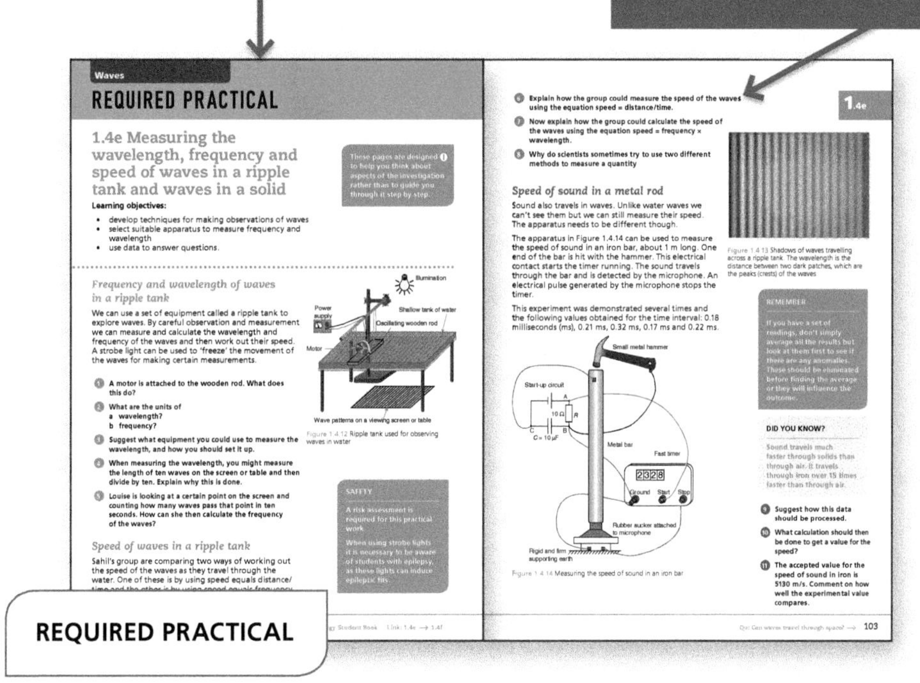
Waves

REQUIRED PRACTICAL

1.4e Measuring the wavelength, frequency and speed of waves in a ripple tank and waves in a solid

Learning objectives:

- develop techniques for making observations of waves
- select suitable apparatus to measure frequency and wavelength
- use data to answer questions.

These pages are designed to help you think about aspects of the investigation rather than to guide you through it step by step.

Frequency and wavelength of waves in a ripple tank

We can use a set of equipment called a ripple tank to explore waves. By careful observation and measurement we can measure and calculate the wavelength and frequency of the waves and then work out their speed. A strobe light can be used to 'freeze' the movement of the waves for making certain measurements.

1. A motor is attached to the wooden rod. What does this do?
2. What are the units of
   a wavelength?
   b frequency?
3. Suggest what equipment you could use to measure the wavelength, and how you should set it up.
4. When measuring the wavelength, you might measure the length of ten waves on the screen or table and then divide by ten. Explain why this is done.
5. Louise is looking at a certain point on the screen and counting how many waves pass that point in ten seconds. How can she then calculate the frequency of the waves?

Figure 1.4.12 Ripple tank used for observing waves in water

SAFETY
A risk assessment is required for this practical work.
When using strobe lights it is necessary to be aware of students with epilepsy, as these lights can induce epileptic fits.

Speed of waves in a ripple tank

Sahil's group are comparing two ways of working out the speed of the waves as they travel through the water. One of these is by using speed equals distance/time and the other is by using speed equals frequency ...

6. Explain how the group could measure the speed of the waves using the equation speed = distance/time.
7. Now explain how the group could calculate the speed of the waves using the equation speed = frequency × wavelength.
8. Why do scientists sometimes try to use two different methods to measure a quantity

1.4e

Figure 1.4.13 Shadows of waves travelling across a ripple tank. The wavelength is the distance between two dark patches, which are the peaks (crests) of the waves

Speed of sound in a metal rod

Sound also travels in waves. Unlike water waves we can't see them but we can still measure their speed. The apparatus needs to be different though.

The apparatus in Figure 1.4.14 can be used to measure the speed of sound in an iron bar, about 1 m long. One end of the bar is hit with the hammer. This electrical contact starts the timer running. The sound travels through the bar and is detected by the microphone. An electrical pulse generated by the microphone stops the timer.

This experiment was demonstrated several times and the following values obtained for the time interval: 0.18 milliseconds (ms), 0.21 ms, 0.32 ms, 0.17 ms and 0.22 ms.

Figure 1.4.14 Measuring the speed of sound in an iron bar

REMEMBER
If you have a set of readings, don't simply average all the results but look at them first to see if there are any anomalies. These should be eliminated before finding the average or they will influence the outcome.

DID YOU KNOW?
Sound travels much faster through solids than through air. It travels through iron over 15 times faster than through air.

9. Suggest how this data should be processed.
10. What calculation should then be done to get a value for the speed?
11. The accepted value for the speed of sound in iron is 5130 m/s. Comment on how well the experimental value compares.

Link: 1.4e → 1.4f

Q: Can waves travel through space? → 103

REQUIRED PRACTICAL

The Maths Skills pages focus on the maths requirements in the AQA specification, explaining concepts and providing opportunities to practise.

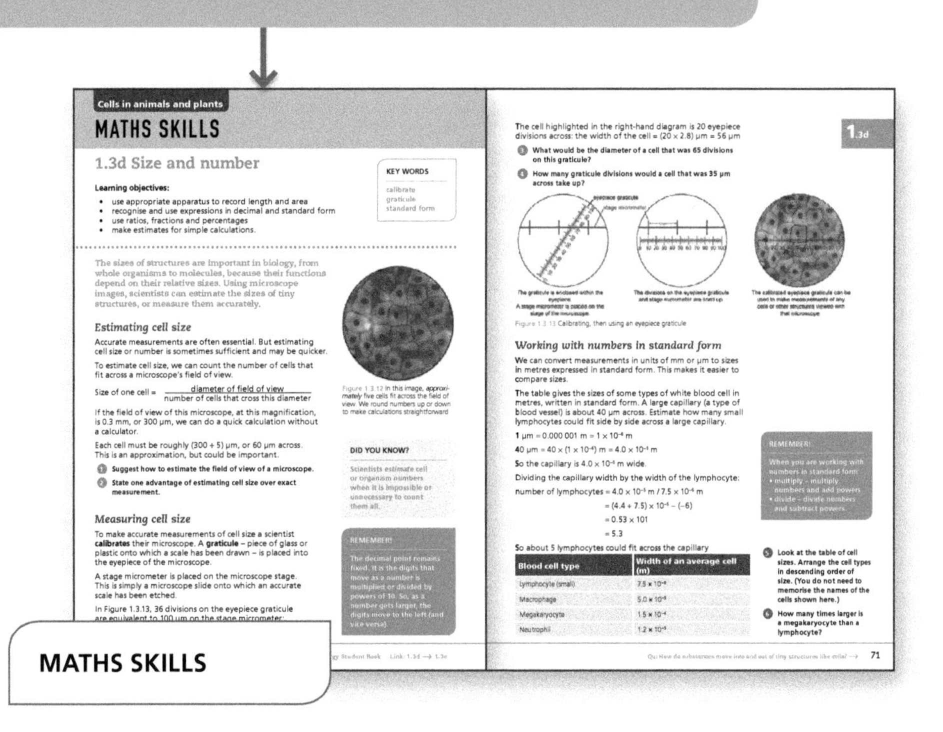
Cells in animals and plants

MATHS SKILLS

1.3d Size and number

Learning objectives:

- use appropriate apparatus to record length and area
- recognise and use expressions in decimal and standard form
- use ratios, fractions and percentages
- make estimates for simple calculations.

KEY WORDS
calibrate
graticule
standard form

The sizes of structures are important in biology, from whole organisms to molecules, because their functions depend on their relative sizes. Using microscope images, scientists can estimate the sizes of tiny structures, or measure them accurately.

Estimating cell size

Accurate measurements are often essential. But estimating cell size or number is sometimes sufficient and may be quicker.

To estimate cell size, we can count the number of cells that fit across a microscope's field of view.

$$\text{Size of one cell} = \frac{\text{diameter of field of view}}{\text{number of cells that cross this diameter}}$$

If the field of view of this microscope, at this magnification, is 0.3 mm, or 300 μm, we can do a quick calculation without a calculator.

Each cell must be roughly (300 ÷ 5) μm, or 60 μm across. This is an approximation, but could be important.

1. Suggest how to estimate the field of view of a microscope.
2. State one advantage of estimating cell size over exact measurement.

Figure 1.3.12 In this image, approximately five cells fit across the field of view. We round numbers up or down to make calculations straightforward

DID YOU KNOW?
Scientists estimate cell or organism numbers when it is impossible or unnecessary to count them all.

Measuring cell size

To make accurate measurements of cell size a scientist **calibrates** their microscope. A **graticule** – piece of glass or plastic onto which a scale has been drawn – is placed into the eyepiece of the microscope.

A stage micrometer is placed on the microscope stage. This is simply a microscope slide onto which an accurate scale has been etched.

In Figure 1.3.13, 36 divisions on the eyepiece graticule are equivalent to 100 μm on the stage micrometer.

REMEMBER!
The decimal point remains fixed. It is the digits that move as a number is multiplied or divided by powers of 10. So, as a number gets larger, the digits move to the left (and vice versa).

Link: 1.3d → 1.3e

1.3d

The cell highlighted in the right-hand diagram is 20 eyepiece divisions across: the width of the cell = (20 × 2.8) μm = 56 μm

3. What would be the diameter of a cell that was 65 divisions on this graticule?
4. How many graticule divisions would a cell that was 35 μm across take up?

Figure 1.3.13 Calibrating, then using an eyepiece graticule

Working with numbers in standard form

We can convert measurements in units of mm or μm to sizes in metres expressed in standard form. This makes it easier to compare sizes.

The table gives the sizes of some types of white blood cell in metres, written in standard form. A large capillary (a type of blood vessel) is about 40 μm across. Estimate how many small lymphocytes could fit side by side across a large capillary.

1 μm = 0.000 001 m = 1×10^{-6} m

40 μm = 40 × (1×10^{-6}) m = 4.0×10^{-5} m

So the capillary is 4.0×10^{-5} m wide.

Dividing the capillary width by the width of the lymphocyte:

number of lymphocytes = 4.0×10^{-5} m / 7.5×10^{-6} m

= (4.4 ÷ 7.5) × $10^{-5-(-6)}$

= 0.53 × 101

= 5.3

So about 5 lymphocytes could fit across the capillary

Blood cell type	Width of an average cell (m)
lymphocyte (small)	7.5×10^{-6}
Macrophage	5.0×10^{-5}
Megakaryocyte	1.5×10^{-4}
Neutrophil	1.2×10^{-5}

REMEMBER!
When you are working with numbers in standard form:
• multiply – multiply numbers and add powers
• divide – divide numbers and subtract powers.

5. Look at the table of cell sizes. Arrange the cell types in descending order of size. (You do not need to memorise the names of the cells shown here.)
6. How many times larger is a megakaryocyte than a lymphocyte?

Q: How do substances move into and out of tiny structures like cells? → 71

MATHS SKILLS

These lists at the end of a chapter act as a checklist of the key ideas of the chapter. In each row, the green box gives the ideas or skills that you should master first. Then you can aim to master the ideas and skills in the blue box. Once you have achieved those you can move on to those in the red box.

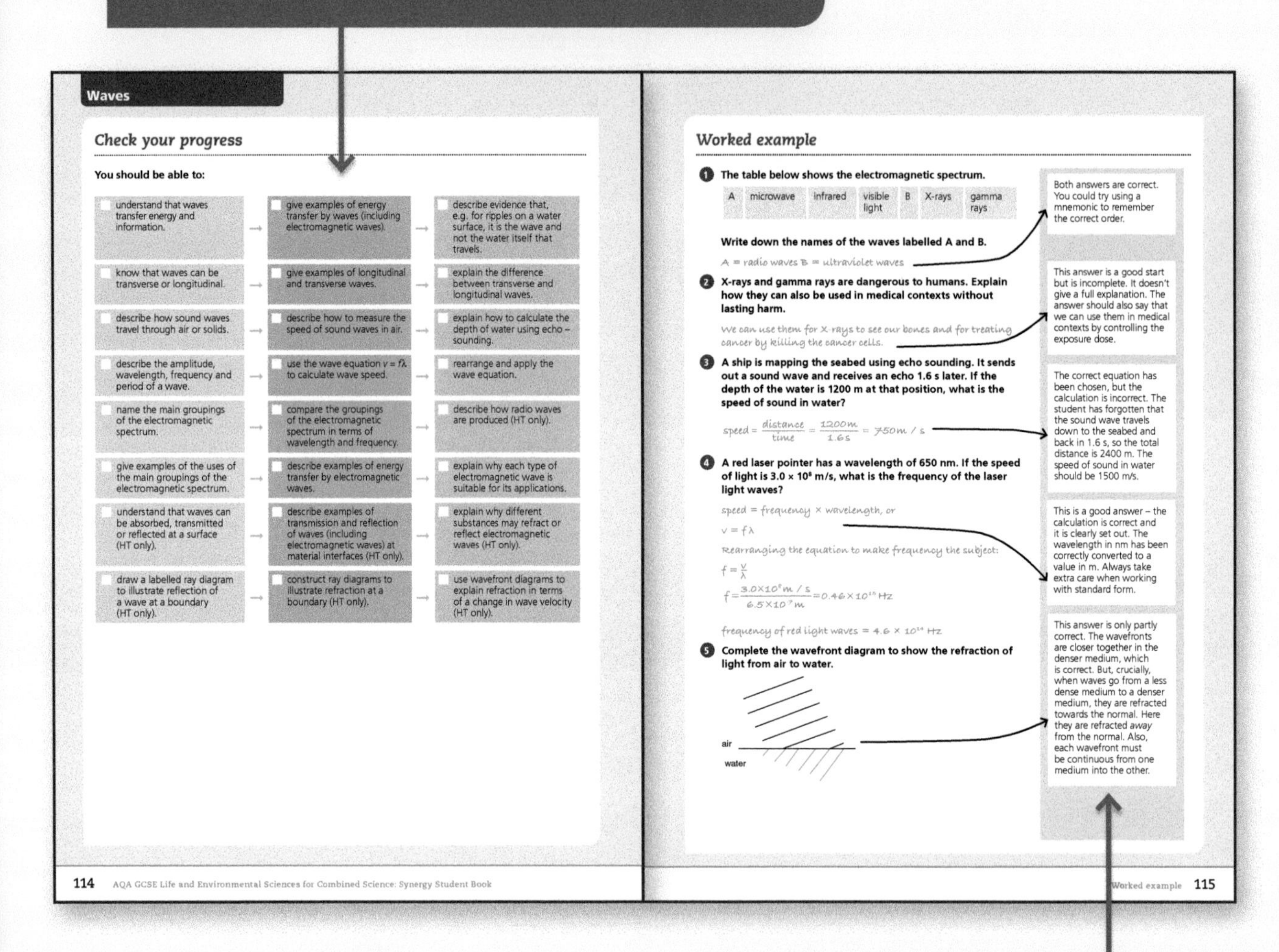

Use the comments to help you understand how to answer questions. Read each question and answer. Try to decide if, and how, the answer can be improved. Finally, read the comments and try to answer the questions yourself.

END OF CHAPTER

The End of chapter questions allow you and your teacher to check that you have understood the ideas in the chapter, can apply these to new situations, and can explain new science using the skills and knowledge you have gained. The questions start off easier and get harder. If you are taking Foundation tier, try to answer all the questions in the Getting started and Going further sections. If you are taking Higher tier, try to answer all the questions in the Going further, More challenging and Most demanding sections.

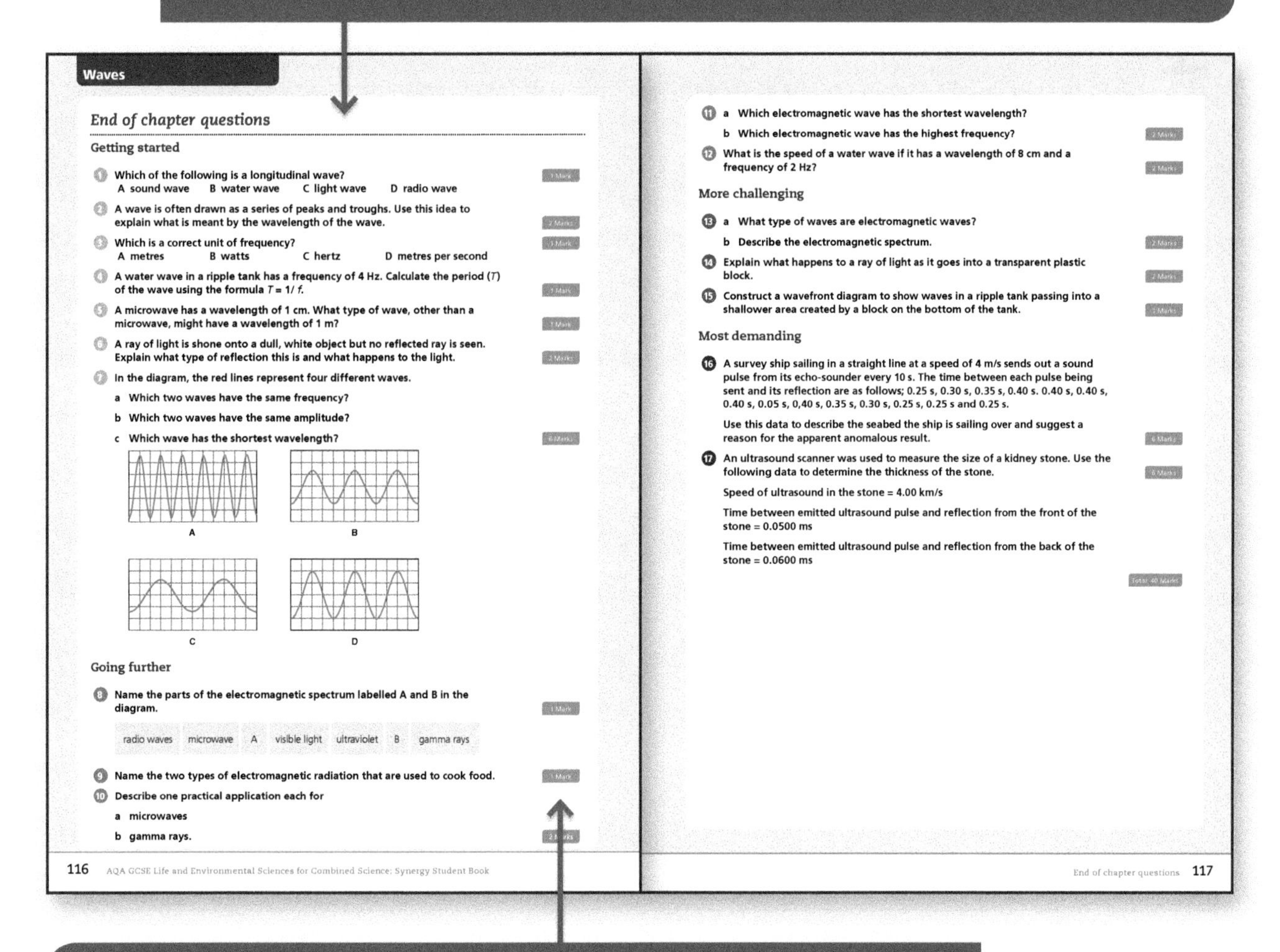

Waves

End of chapter questions

Getting started

1. Which of the following is a longitudinal wave? — 1 Mark
   A sound wave B water wave C light wave D radio wave
2. A wave is often drawn as a series of peaks and troughs. Use this idea to explain what is meant by the wavelength of the wave. — 2 Marks
3. Which is a correct unit of frequency? — 1 Mark
   A metres B watts C hertz D metres per second
4. A water wave in a ripple tank has a frequency of 4 Hz. Calculate the period (T) of the wave using the formula $T = 1/f$. — 1 Mark
5. A microwave has a wavelength of 1 cm. What type of wave, other than a microwave, might have a wavelength of 1 m? — 1 Mark
6. A ray of light is shone onto a dull, white object but no reflected ray is seen. Explain what type of reflection this is and what happens to the light. — 2 Marks
7. In the diagram, the red lines represent four different waves.
   a Which two waves have the same frequency?
   b Which two waves have the same amplitude?
   c Which wave has the shortest wavelength? — 6 Marks

   A B C D

Going further

8. Name the parts of the electromagnetic spectrum labelled A and B in the diagram. — 1 Mark

   radio waves | microwave | A | visible light | ultraviolet | B | gamma rays

9. Name the two types of electromagnetic radiation that are used to cook food. — 1 Mark
10. Describe one practical application each for
    a microwaves
    b gamma rays. — 2 Marks

116 AQA GCSE Life and Environmental Sciences for Combined Science: Synergy Student Book

11. a Which electromagnetic wave has the shortest wavelength?
    b Which electromagnetic wave has the highest frequency? — 2 Marks
12. What is the speed of a water wave if it has a wavelength of 8 cm and a frequency of 2 Hz? — 2 Marks

More challenging

13. a What type of waves are electromagnetic waves?
    b Describe the electromagnetic spectrum. — 2 Marks
14. Explain what happens to a ray of light as it goes into a transparent plastic block. — 2 Marks
15. Construct a wavefront diagram to show waves in a ripple tank passing into a shallower area created by a block on the bottom of the tank.

Most demanding

16. A survey ship sailing in a straight line at a speed of 4 m/s sends out a sound pulse from its echo-sounder every 10 s. The time between each pulse being sent and its reflection are as follows; 0.25 s, 0.30 s, 0.35 s, 0.40 s. 0.40 s, 0.40 s, 0.40 s, 0.05 s, 0,40 s, 0.35 s, 0.30 s, 0.25 s, 0.25 s and 0.25 s.
    Use this data to describe the seabed the ship is sailing over and suggest a reason for the apparent anomalous result. — 6 Marks
17. An ultrasound scanner was used to measure the size of a kidney stone. Use the following data to determine the thickness of the stone. — 6 Marks
    Speed of ultrasound in the stone = 4.00 km/s
    Time between emitted ultrasound pulse and reflection from the front of the stone = 0.0500 ms
    Time between emitted ultrasound pulse and reflection from the back of the stone = 0.0600 ms

Total: 40 Marks

End of chapter questions 117

There are questions for each assessment object (AO). These will help you develop the thinking skills you need to answer each type of question:

AO1 – to answer these questions you should aim to **demonstrate** your knowledge and understanding of scientific ideas, techniques and procedures.

AO2 – to answer these questions you should aim to **apply** your knowledge and understanding of scientific ideas and scientific enquiry, techniques and procedures.

AO3 – to answer these questions you should aim to **analyse** information and ideas to: interpret and evaluate, make judgements and draw conclusions, develop and improve experimental procedures.

How the ideas in this topic link together

All the materials around us are made from smaller building blocks. Using light microscopes, we can see that living organisms are made from cells. Using electron microscopes we can see that cells are made up of even smaller components, and that these are composed of tinier and tinier particles, until it is not possible to see them even with the most powerful microscope. The particle model of matter was developed to explain the behaviour of these building blocks. Using ideas of scale helps us to relate the behaviour of these tiny particles to the real world of objects we can see and hold.

According to the particle model, particles are arranged in such a way that they form solids, liquids or gases. Each state of matter has different properties. The behaviour of particles as liquids and gases can explain how substances move into cells and across membranes in living organisms. The universal genetic code is a particle pattern that can build amino acids into proteins to build living cells, which can then replicate.

Energy is transferred over small distances and large distances in different ways in living and non-living systems. The idea of waves helps to explain transfer of energy over long distances, and some of the interactions between living and non-living systems.

Working Scientifically Focus

- Understand how scientific methods and theories develop over time
- Using models, including representational drawings
- Using SI units, SI prefixes and interconverting units
- Using significant figures

1 Building blocks

Contents

STATES OF MATTER

IDEAS YOU HAVE MET BEFORE:

PARTICLE MODEL AND STATES OF MATTER

- All matter is made up of particles.
- The particle model represents the particles of matter as solid spheres.
- Ice and other solids can turn to liquids and gases.
- Solids melt into liquids at the melting point.
- Liquids turn into gases at the boiling point.

DENSITY

- Density is equal to mass/volume. The unit is kg/m^3.
- Solids usually have a larger density than liquids and gases. The particles are closer together.
- In a gas the particles are far apart and move freely.
- The particles in a gas fill the space available.

TEMPERATURE AND ENERGY TRANSFER

- When there is a difference in temperature between two objects, energy is transferred from the hotter object to the colder one.
- Energy is measured in joules (J) or kilojoules (kJ).
- The amount of energy involved in a temperature change can be calculated.

PARTICLE MODEL AND PRESSURE OF A GAS

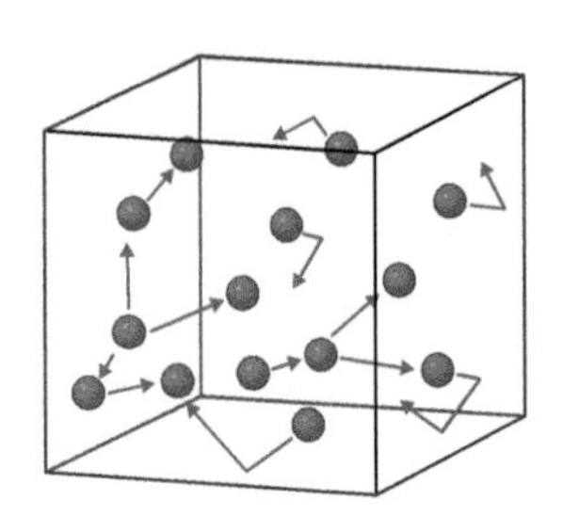

- Every material is made of tiny moving particles.
- In a gas the particles are far apart and move freely.
- The higher the temperature the more kinetic energy the particles have and the faster they move.
- The pressure in a container of gas is caused by the particles colliding with the walls of the container.

1.1

IN THIS CHAPTER YOU WILL FIND OUT ABOUT:

WHAT HAPPENS TO PARTICLES AS SUBSTANCES CHANGE STATE?

- The particles in a system store energy, called internal energy.
- In solids and liquids there are forces between particles that hold them together.
- An increase in internal energy can increase the temperature of the system or change its state.
- The amount of energy needed to change state from solid to liquid or liquid to gas depends on how strong the forces are between the particles.

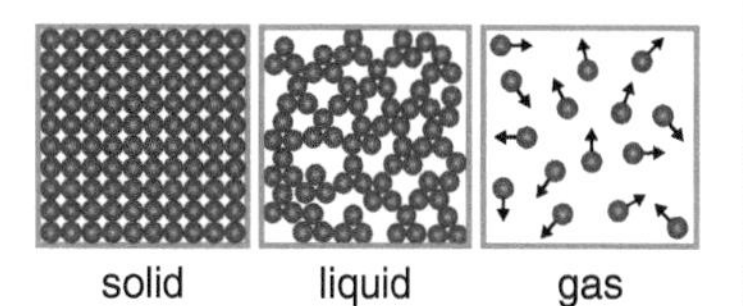

HOW CAN WE FIND THE DENSITY OF LIQUIDS AND SOLIDS?

- The density of a liquid can be found by weighing a known volume.
- The density of a solid can be found from measurements of its mass and measured volume.
- The volume of a solid with a regular shape can be found by accurate measurement.
- The volume of an irregular solid can be found by a displacement method.

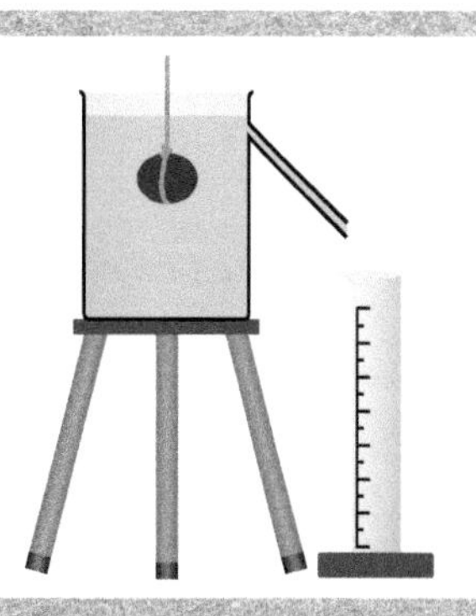

HOW DOES ENERGY TRANSFER BY HEATING CHANGE THE TEMPERATURE OR STATE OF AN OBJECT?

- The increase in temperature of a system depends on the mass of the substance, the type of material and the energy input.
- The specific heat capacity, *c*, is the energy required to raise the temperature of 1 kg of an object by 1 °C. The unit for *c* is J/kg °C.
- The specific latent heat of fusion is the energy required to change 1 kg of an object from a solid to a liquid without a change in temperature.
- The specific latent heat of vaporisation is the energy required to change 1 kg of an object from a liquid to a gas without a change in temperature.

WHAT HAPPENS TO THE PRESSURE OF A GAS WHEN IT IS HEATED, KEEPING THE VOLUME CONSTANT?

- An increase in temperature increases the kinetic energy of the gas particles.
- The particles move faster, colliding more often and with greater force on the walls of their container. The pressure of the gas increases.

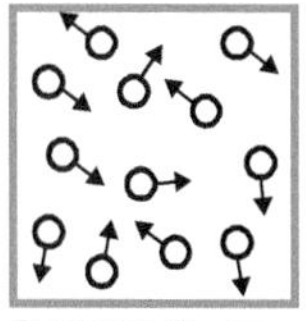

Cool gas, fewer and less energetic collisions

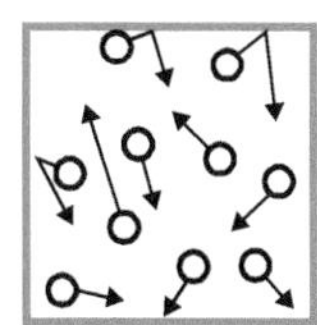

Hot gas, more and more energetic collisions

1.1a The particle model

Learning objectives:

- describe and explain the properties of solids, liquids and gases using the particle model
- relate the size and scale of atoms to objects in the physical world
- identify the strengths and limitations of the particle model.

KEY WORDS

model
particle model
physical change

In science we often use models to help us think about situations in which what we are looking at is too big or too small to see. The particle model is one of the most important concepts in science, because we use it to explain so many different things.

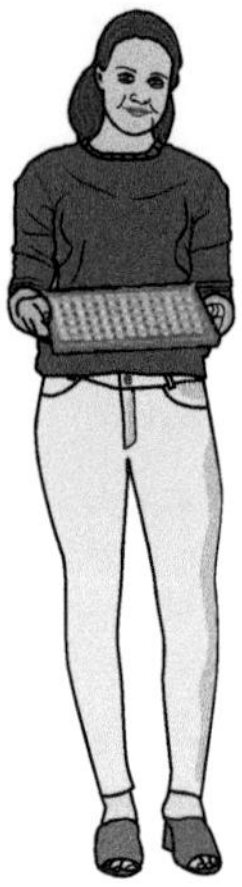

Figure 1.1.1 Modelling a solid by gently shaking a tray of marbles and allowing the marbles to settle

The states of matter

The particle model says that all matter is made from tiny particles. This simple model uses small solid spheres to represent the particles.

There are three states of matter: solid, liquid and gas.

Solid	Liquid	Gas
• particles very close together in an ordered pattern • particles vibrate to and fro but cannot change their positions • fixed shape and volume	• particles close together but not in a regular pattern • particles slide past each other • fixed volume but not a fixed shape	• particles spaced far apart • particles move fast in all directions • no fixed shape or volume

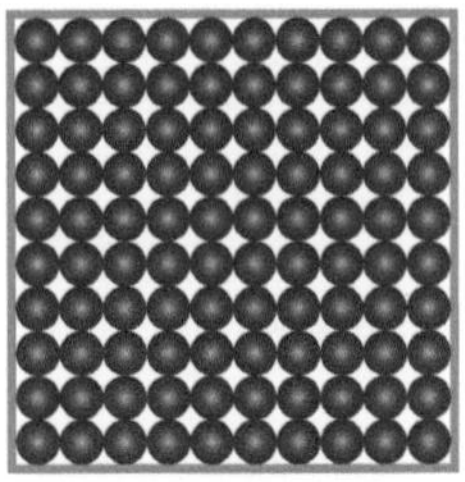

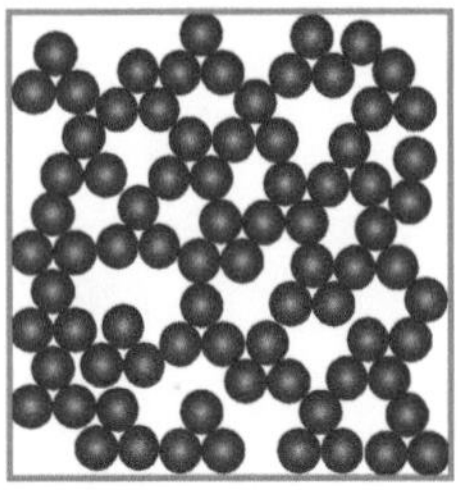

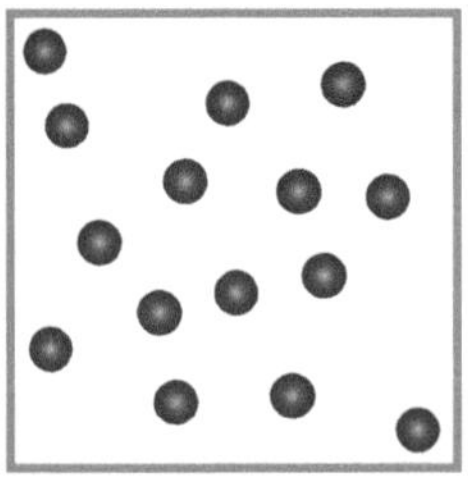

Figure 1.1.2 The three states of matter

DID YOU KNOW?

Atoms are so small that they cannot be seen, even with the most powerful light microscope. If you imagine an atom to be the size of a golf ball, then an actual golf ball would represent the size of the Earth.

1 Which state does each of the following describe?

a fixed shape and volume

b particles move around freely at high speed

c fixed volume but no fixed shape

2. **Explain why it is possible to squash a gas but not a solid or a liquid.**

Changes of state

When substances change state the arrangement and movement of the particles changes.

- Melting and freezing both take place at the melting point.
- Boiling and condensing both take place at the boiling point.

Changes of state are **physical changes**. The change is reversible, as the substance does not change its composition.

Substance	Melting point (°C)	Boiling point (°C)	State at room temperature
W	−18	42	liquid
X	150	875	
Y	−190	−84	
Z	−56	16	

Substance W in the table will have melted at −18°C but will not have boiled at room temperature, so it is a liquid.

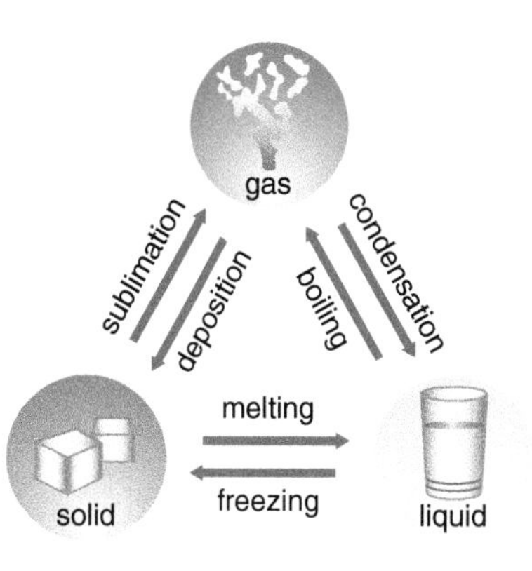

Figure 1.1.3 Energy must be supplied to change state from solid to liquid, and from liquid to gas. Energy is given out when the change happens in the reverse direction

3. **Look at the table above. What are the states of substances X, Y and Z at room temperature?**

HIGHER TIER ONLY

Limitations of models

This simple particle model of solid spheres has limitations, as:

- there are no forces of attraction between the spheres
- the particles are all solid spheres of the same size.

This means that interactions between particles as they get closer together or further apart cannot be represented fully. We also cannot properly model the distance between atoms, molecules and ions.

4. **State with reasons one way in which a tray of marbles is a good model for a solid and one way in which it is not.**
5. **Water vapour condenses to form liquid water. Explain how the limitations of the model relate to the process of condensation.**

REMEMBER!

Particle diagrams are a model, not the real situation. In the model of a solid the individual spheres represent particles that are themselves not solid. It is only when lots of these particles are arranged closely in a regular pattern that they represent a solid.

Qu: Which has the greater density – a shoebox of marbles or a shoebox of feathers? ⟶

1.1b Density

Learning objectives:

- define density
- explain the differences in density between different states of matter using the particle model
- calculate densities of different materials.

KEY WORDS

density
gas
liquid
solid

Identical volumes of marbles and feathers will not have the same mass. A box of marbles has more mass for the same volume (or *per volume*) than the feathers. The marbles are more dense.

Density

When people say 'lead is heavier than iron', what they mean is that a piece of lead is heavier than a piece of iron of the same volume.

Density compares the mass of materials with the same volume. It is defined as the mass of unit volume of a substance:

$$\text{density (in kg/m}^3\text{)} = \text{mass (in kg)} \div \text{volume (in m}^3\text{)}$$

Gold has a density of 19 000 kg/m^3. That means a volume of 1 m^3 of gold has a mass of 19 000 kg. The densities of a range of materials are shown in the table.

Figure 1.1.4 A standard gold bar has a mass of 12.4 kg. A gold bar the size of a brick would have a mass of 20.9 kg. Could you lift that with one hand?

Substance	Physical state at room temperature	Density in kg/m^3
gold	solid	19 000
iron	solid	8000
lead	solid	11 000
cork	solid	200
mercury	liquid	13 600
water	liquid	1000
petrol	liquid	800
air	gas	1.3

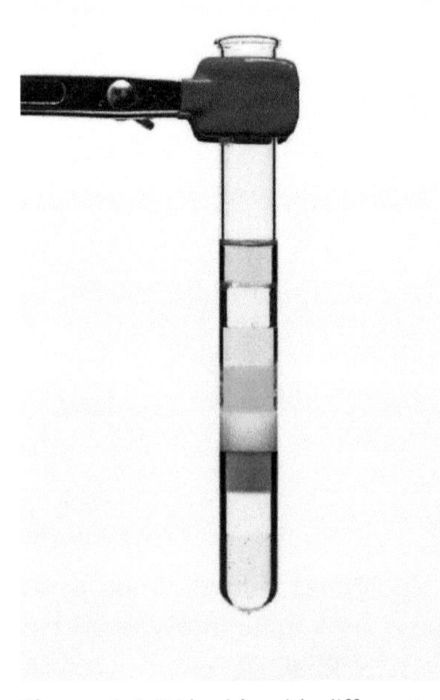

Figure 1.1.5 Liquids with different densities will form layers, but only if the liquids do not mix freely

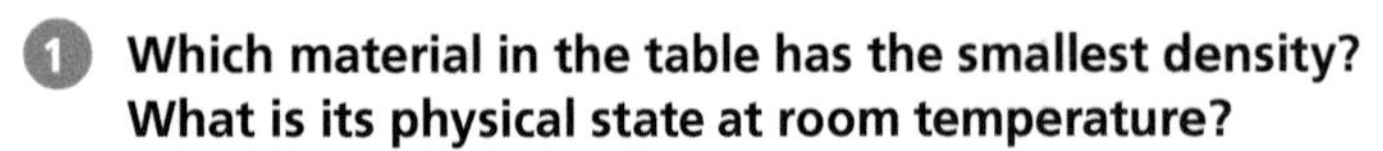

1. **Which material in the table has the smallest density? What is its physical state at room temperature?**
2. **What state of matter has the greatest density?**

Density and the particle model

1.1b

The particle model can help to explain the different densities of solids, liquids and gases.

In a **solid** the particles are very close together. When a solid melts it becomes a **liquid**. The particles are still very close together, but they are not quite as tightly packed as in a solid. So the density of a liquid is usually less than the solid. For example, the density of solid aluminium is 2720 kg/m³ and the density of liquid aluminium is 2380 kg/m³.

When a liquid boils it becomes a **gas**. The particles are very far apart so the density is small.

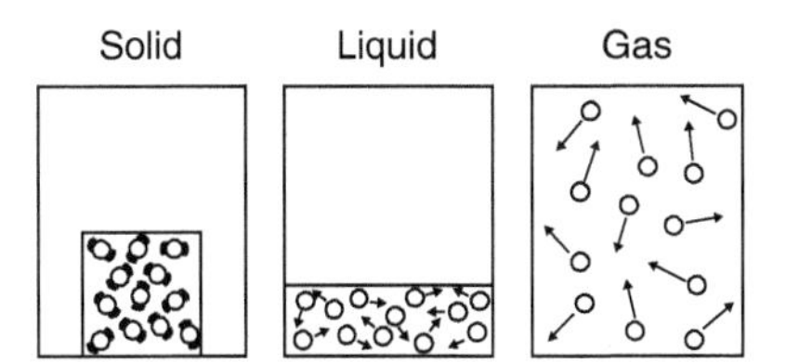

Figure 1.1.6 Particles in a solid, liquid and gas

3 **Use the particle model to explain why solids usually have a higher density than liquids and gases.**

4 **Draw a particle model of a gas, and annotate it to show why gases have low density.**

DID YOU KNOW?

Water vapour has a density of about 0.8 kg/m³. Boiling away 100 g (100 cm³) of water would produce about 100 000 cm³ (100 litres) of water vapour.

Calculations with density

Another way of writing the equation for density is with symbols:

$$\rho = m \div V$$

where ρ is density in kg/m³, m is mass in kg and V is volume in m³.

Density is also given in g/cm³, where mass is in g and volume is in cm³.

$$1 \text{ g/cm}^3 = 1000 \text{ kg/m}^3$$

Example:

Calculate the mass of 3 m³ of water. The density of water is 1000 kg/m³.

$$\rho = m \div V$$

Rearrange the equation to give

$$m = \rho V$$

$$= 1000 \text{ kg/m}^3 \times 3 \text{ m}^3$$

$$= 3000 \text{ kg}$$

Figure 1.1.7 Materials used in aeroplanes have to be light but strong

DID YOU KNOW?

Osmium is the densest substance found on Earth. Its density is 22 600 kg/m³.

5 a **Calculate the density of a 5400 kg block of aluminium with a volume of 2 m³.**

b **Calculate the mass of steel having the same volume as the aluminium block. Density of steel = 7700 kg/m³.**

c **Explain why aluminium is used to build aeroplanes rather than steel (Figure 1.1.7).**

6 **Calculate the mass of air in a room 5 m by 4 m by 3 m. The density of air is 1.3 kg/m³.**

7 **Suggest why cork floats in water, but iron sinks.**

REQUIRED PRACTICAL

1.1c To investigate the densities of regular and irregular solid objects and liquids

KEY WORD

displacement method
significant figures

Learning objectives:

- interpret observations and data
- use spatial models to solve problems
- plan experiments and devise procedures
- use an appropriate number of significant figures in measurements and calculations.

Density is worked out knowing the mass of an object and the volume it occupies. Measuring the volume is easy enough if the object is a regular shape such as a cube but what if it was, say, a crown?

These pages are designed to help you think about aspects of the investigation rather than to guide you through it step by step.

This problem was given to Archimedes, a clever man who lived thousands of years ago in Greece. The king had ordered a new crown to be made but suspected the craftsman had mixed a cheaper metal with the gold. Measuring the density would reveal if the gold was pure, but how could the volume be found? Archimedes realised that by carefully immersing the crown in a full can of water, the water that overflowed would have the same volume as the crown.

Apparently he shouted "Eureka!" ("I have found it!") and the cans used in practical investigations are known as Eureka cans in recognition of this.

Figure 1.1.8 Archimedes making his discovery

Measuring the density of a liquid

The density of any substance can be worked out using the formula:

density = mass/volume

If mass is measured in grams (g) and volume in cubic centimetres (cm^3) then the density will be in g/cm^3.

Measuring the density of a liquid can be done by pouring the liquid into a measuring cylinder to find its volume and using a balance to find the mass. The results might look rather like those in the table.

Liquid	Mass (g)	Volume (cm^3)
coconut oil	18.5	20
acetone	19.6	25
sea water	51.3	50

1. **Explain why you would not get the mass of the liquid by just putting the measuring cylinder with the liquid on the balance and recording the reading.**

2. **Suggest what you could do to get the mass of the liquid.**
3. **Calculate the density of each of the liquids in the table on the previous page.**

Measuring the density of a regular solid

Solids, of course, have their own shape. If that shape is a regular one, we can calculate the volume by measuring dimensions and then using the correct formula. For example, if the solid was a cuboid, the volume would be length multiplied by width multiplied by height. The mass would be divided by this to get the density.

For example, if a 2.0 cm cube of soft rubber had a mass of 8.82 g, its volume would be 2.0 cm × 2.0 cm × 2.0 cm, which is 8.0 cm^3 and the density would be 8.82 g/8.0 cm^3 which is 1.1 g/cm^3.

The answer can only have the same number of **significant figures** as the component with the least number of significant figures (8.0 – it has two significant figures).

Figure 1.1.9 Assortment of cubes of different materials

Material	Mass (g)	Length (cm)	Width (cm)	Height (cm)
cork	3	2.0	2.0	3.0
oak	17	2.0	3.0	4.0
tin	364	2.5	2.5	8.0

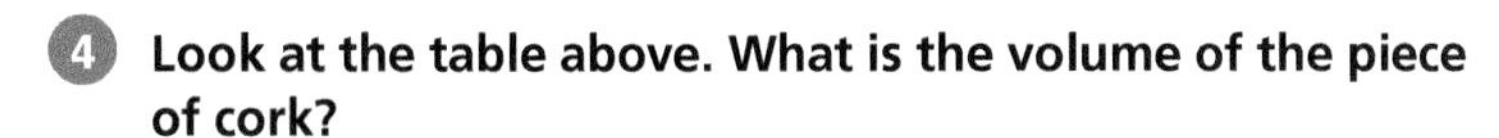

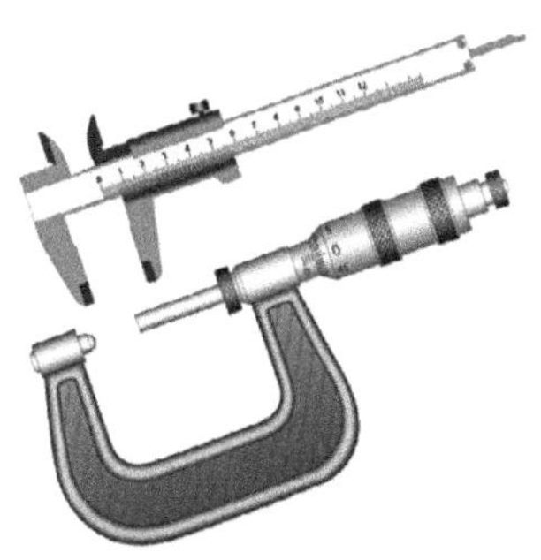

Figure 1.1.10 Precise measuring instruments. Vernier calipers (top) measure to 0.1 mm and micrometers (below) measure to 0.01 mm. Which is more precise?

4. **Look at the table above. What is the volume of the piece of cork?**
5. **What is its density?**
6. **What are the densities of the other two materials?**
7. **The dimensions of the oak were measured using Vernier calipers. Why is it incorrect to use 0.68106 g/cm^3 as the answer for the density of oak?**

Measuring the density of an irregular solid

This is, of course, the problem that Archimedes was trying to solve and his solution (apparently inspired by getting into a bath tub that was too full) was to use the idea of **displacement**. The solid will displace the same volume of water as its own volume. If there's room in the container, it rises. If not, it overflows.

Imagine trying to see if a gold necklace was pure. Knowing that the density of gold is 19.29 g/cm^3, being able to dangle the necklace on a thread and having a glass of water full to the brim, you could carry out a simple experiment.

Figure 1.1.11 How could you find out if this is pure gold?

8. **What procedure would you follow?**
9. **What measurements would you need to take?**
10. **What calculation would you then perform?**
11. **Why might this experiment probably not be very accurate?**

KEY CONCEPT

1.1d Particle theory

Learning objectives:

- use the particle model to explain states of matter
- use ideas about energy and bonds to explain changes of state
- explain the relationship between temperature and energy.

KEY WORDS

specific latent heat of fusion
specific latent heat of vaporisation

The physical state of a substance is determined by the way in which its particles are arranged. Using the idea of forces between particles, the particle theory can explain why substances have different melting and boiling points, and what happens when a substance changes state.

Solids, liquids and gases

All matter is made up from atoms and molecules, but it is the way these atoms and molecules are held together that determines whether a substance is solid, liquid or gas. In a solid the atoms are held tightly together forming strong regular shapes. In a liquid the atoms are held much more loosely, which is why liquids can flow. In a gas the atoms and molecules are free to move about on their own.

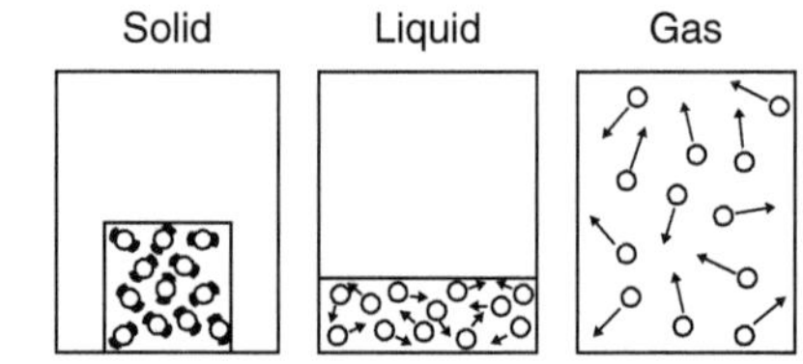

Figure 1.1.12 Movement of particles in the three states of matter

What determines the state of matter is the strength with which the atoms and molecules are held together by their bonds. All atoms also vibrate. Even in a solid, where the atoms are held together in a tight structure, every single atom is moving backwards and forwards next to its neighbour.

To get a solid to become a liquid, energy has to be added to the substance. Energy increases the vibrations of the atoms. As the vibrations increase, the bonds that hold the atoms are stretched and the atoms are pushed further apart from each other (Figures 1.1.12 and 1.1.13).

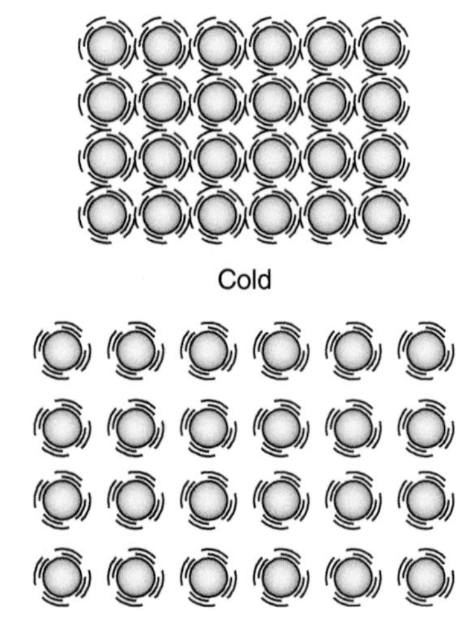

Figure 1.1.13 This particle diagram is exaggerated to show the increased vibration of particles at a higher temperature. Note that the particles do not get bigger

Adding more energy to the solid increases the vibrations even more until comes a point where the bonds can no longer hold together. Many bonds break and the atoms are free to move – the solid has become a liquid. Adding yet more energy breaks the bonds completely and the atoms escape into the atmosphere. The liquid becomes a gas. The amount of energy needed to change state from solid to liquid and from liquid to gas depends on the strength of the forces between the particles of the substance.

Link: 1.1d ⟶ 1.1e

1. **State the difference between a solid and a liquid.**
2. **Explain why solids expand as they are heated.**

Change of state and change in temperature

The state of matter depends on the amount of energy inside it. Heating a substance increases the amount of internal energy and this either raises the temperature of the system or produces a change of state. The process also works in reverse. If you take energy out of a substance it will either decrease the temperature of the system or change state from a gas back to a liquid and then to a solid (Figure 1.1.14).

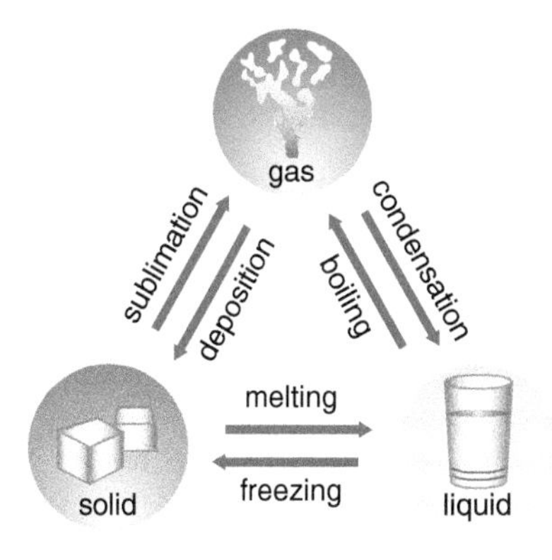

Figure 1.1.14 Changing between states of matter

When you heat a substance there are stages where you just get a rise in temperature and then a stage where you get a change of state.

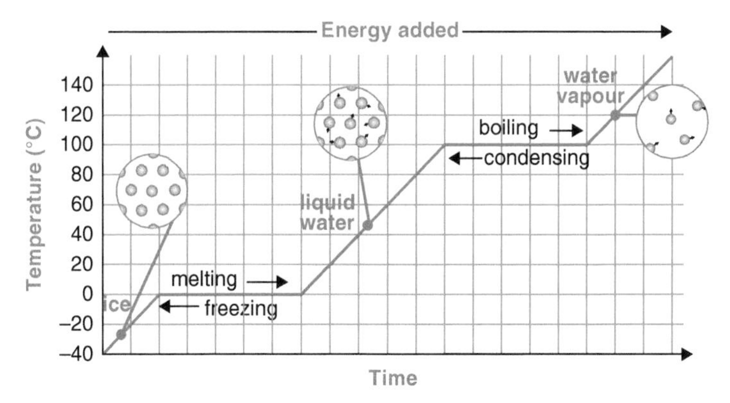

Figure 1.1.15 Temperature–time graph for heating water

Follow the steps on the graph in Figure 1.1.15, which shows how the temperature changes as a block of ice is heated continuously over a period of time.

The ice is initially at –40 °C. As it is heated, the ice gets warmer, but it is still a solid. The upward slope of the graph shows that the temperature is increasing. When the temperature reaches 0 °C energy is still being put into the ice, but there is no increase in temperature. The graph shows a level straight line. This is where the energy is being absorbed by the ice to change it into liquid water. The amount of energy needed to change state from solid to liquid is called the **specific latent heat of fusion**.

Once the ice has melted into water the temperature begins to rise again. The line on the graph is sloping upwards. When the temperature reaches 100 °C we get the second change in state. Energy is absorbed but there is no rise in temperature. The amount of energy required to change the state from liquid to water vapour is called the **specific latent heat of vaporisation**.

3. **Ethanol freezes at –114 °C and boils at 78 °C. Draw a temperature–time diagram for heating ethanol from –120 °C to 100 °C.**
4. **Explain what a line sloping down from left to right on a temperature–time graph means.**

1.1e Gas pressure

Learning objectives:

- use the particle model to relate the temperature of a gas to the average kinetic energy of the particles
- explain how a gas has a pressure.

KEY WORD

randomly

Inside an iron, the hot plate heats up the water, changing it from a liquid to a gas. Gas particles are fast-moving, and collide with the walls of the reservoir to create a force. The steam then escapes from the iron as a jet of gas under pressure.

Figure 1.1.16 Steam escaping under pressure from a steam iron

Temperature and pressure of a gas

The particles in a gas move around **randomly**. These particles have kinetic energy (the energy of moving objects). They move about freely at high speed.

The higher the temperature the faster the particles move and the more kinetic energy they have. The temperature of a gas is related to the average kinetic energy of the molecules. As the temperature increases, the particles move faster. They gain more kinetic energy.

1. **What happens to the molecules of a gas when the gas is heated?**
2. **How does the particle model explain temperature?**

The molecules of a gas collide with each other as well as with the walls of their container (Figure 1.1.17). When they hit a wall there is a force on the wall. Pressure is equal to the force on the wall divided by the area over which the force acts. The total force exerted by all the molecules inside the container that strike a unit area of the wall is the gas pressure.

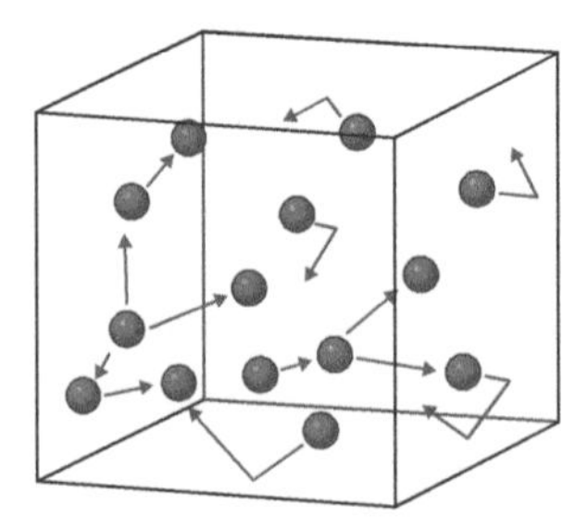

Figure 1.1.17 An expanded model view showing that when gas particles hit a wall of their container there is a force on the wall. This force is the pressure of the gas

3. **When you press on a balloon the pressure inside it increases. How does the particle model explain this?**
4. **Describe, in terms of particles, how a gas exerts a pressure.**

Changing the temperature of a gas

Air is sealed in a container (Figure 1.1.18). If we keep the mass and volume of the air constant, an increase in the temperature will increase the pressure of the gas.

There are the same number of particles because the container is sealed, and there is the same mass of gas. As the container is sealed the volume of gas is also constant. As the particles have more energy, they move faster, hitting the walls more often and with greater force, increasing the pressure.

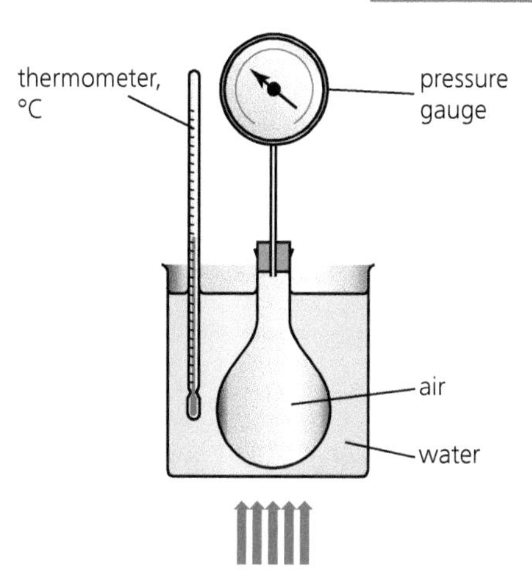

Figure 1.1.18 Heating air in a sealed container

5. **How do we know that the mass of gas is constant?**
6. **Explain, using ideas about energy, why the pressure of gas increases if it gets hotter.**

Linking pressure and volume, at constant temperature

For a fixed amount of gas kept at constant temperature, increasing the volume of the container results in a decrease in pressure. If the volume is doubled, with the same number of particles, there will be fewer collisions between the particles and the walls of the container. The pressure will be halved.

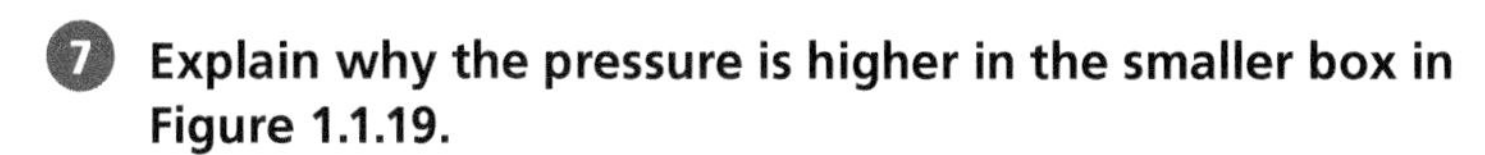

7. **Explain why the pressure is higher in the smaller box in Figure 1.1.19.**

Decreasing the volume of the container results in an increase in pressure. If the volume is halved the pressure will double (figure 1.1.19).

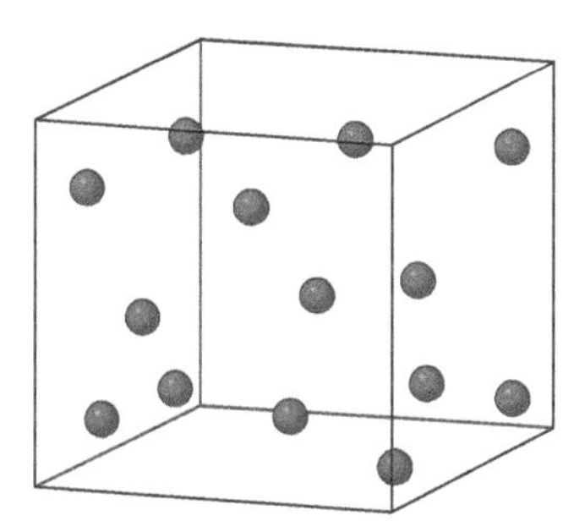

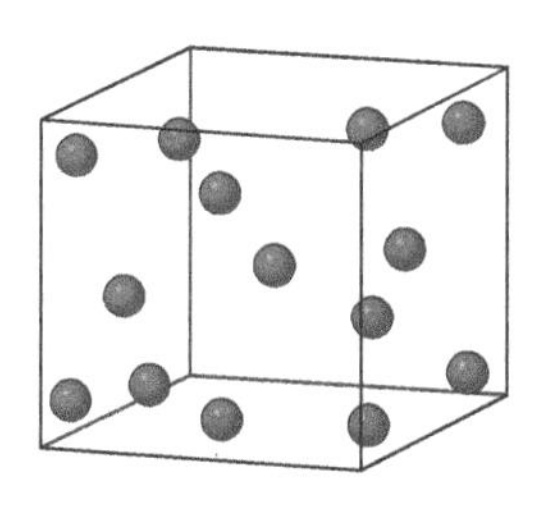

Figure 1.1.19 An expanded model view showing that there is the same number of particles in each box

1.1f Heating and changes of state

Learning objectives:

- describe how mass is conserved when the physical state changes
- describe how heating a system changes its internal energy
- explain that when a change of state occurs the internal energy changes but not the temperature.

KEY WORDS

boiling point
conserved
evaporation
melting point
sublimate

When a substance changes state, energy is transferred to change the arrangement of the particles. There is no change in temperature. To explain this we use the idea that there are forces between particles in the solid and liquid states.

Conservation of mass

Changes of state are physical changes. Unlike a chemical change, the change does not produce a new substance. If the change is reversed the substance recovers its original properties.

When substances change state, the mass is **conserved**. If you start with 1 kg of ice and melt it, you will have 1 kg of water. Nothing has been added or removed. The process is reversible. If you freeze the 1 kg of water, you will end up with 1 kg of ice again.

1. **What is sublimation?**
2. **What mass of iodine gas is produced by heating 45 g of solid iodine?**

DID YOU KNOW?

Dry ice is frozen carbon dioxide, which turns directly from a solid to a gas at a temperature of −78.5 °C. The fog you can see is a mixture of cold carbon dioxide gas and droplets of water vapour, condensed from the air as the dry ice **sublimates.**

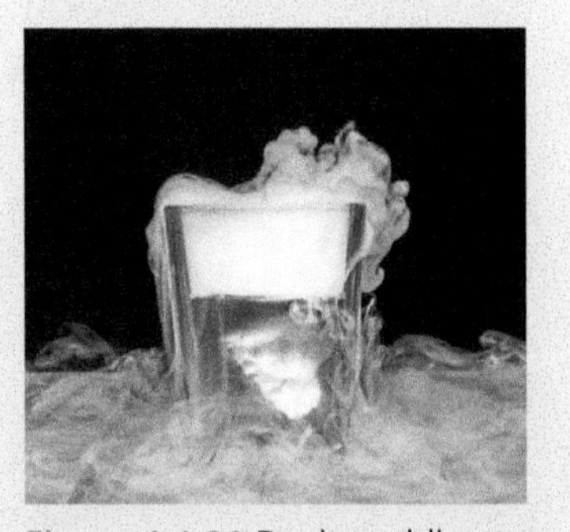

Figure 1.1.20 Dry ice sublimates

Explaining changes of state

Heating a solid at its **melting point** transfers energy to the particles which allows them to overcome the forces holding them in position. While a solid is melting, all the energy transferred is used to rearrange the particles, so the temperature does not rise.

Heating a liquid at its **boiling point** transfers energy to the particles so they can escape the forces of attraction and become a gas. The liquid stays at its boiling point until all the particles have broken free.

The amount of energy needed to change state from solid to liquid, or liquid to gas, is different for different substances. The stronger the forces between the particles, the more energy is needed, and the higher the melting point and boiling point.

MAKING LINKS

You will find out more about the forces involved in different types of bonding in topic 6.2 Structure and bonding.

3. **Olive oil melts at –6 °C and coconut oil at 25 °C. Explain which type of oil has stronger forces between its molecules.**
4. **Explain why water in a kettle does not all turn to gas when the water first reaches 100 °C.**

Internal energy

Energy is stored inside a system by the particles that are within it. This is called the internal energy.

Heating changes the internal energy of a system by increasing the energy of the particles that make it up. The hotter a material is, the faster its particles move and the more internal energy it has.

As a liquid warms up the average speed of the particles in it increases. But not all the particles of the liquid will be travelling at the same speed. The faster particles with more energy can escape, leaving behind the slower particles with less energy. This is **evaporation**.

When the faster particles escape, the total internal energy left behind decreases. This has a cooling effect on the material left behind.

KEY INFORMATION

Energy transfer by heating will *either* increase the temperature of a system *or* change its state.

5. **What happens to the internal energy of a system when thermal energy is supplied?**
6. **Suggest why being burned by steam is worse than being burned by hot water.**

Figure 1.1.21 The sand in this pot is wet. Evaporation takes energy away from the inner pot and keeps the food cool

1.1g Specific heat capacity

Learning objectives:

- define and explain specific heat capacity
- state the factors that are involved in increasing the temperature of a substance
- calculate specific heat capacity and energy changes when a material is heated.

KEY WORDS

joule
specific heat capacity

Different materials need different amounts of energy to raise their temperature by 1°C. The syrup on a sponge pudding absorbs more energy during cooking than the sponge, so stays warmer on the plate.

Figure 1.1.22 The syrup stays hotter than the sponge pudding

Heating up

When a liquid is heated, the particles move faster. They gain kinetic energy and the temperature rises. The particles are close together and attract each other strongly. Their motion opposes the forces of attraction and keeps them separated. As they are moving faster, they also separate a bit more and gain potential energy. So the liquid has more kinetic energy and more potential energy. Its internal energy has increased.

1. **What effect does heating have on a liquid?**
2. **What happens to particles of a gas when it is heated?**

Increasing temperature

When a liquid is heated below its boiling point, its temperature increases. The temperature rise depends on

- the mass of liquid heated
- the particular liquid being heated
- the energy input to the system.

3. **Dan is heating a large saucepan of water to boiling point. Explain why more energy is needed to do this than to heat a cup of water to the same temperature.**
4. **Milk does not need as much energy to raise its temperature by 10 °C as the same mass of water. Will hot milk give out less energy than the same amount of water at the same temperature when it cools down? Explain.**

Specific heat capacity

When an object is heated, energy is transferred to the object and its temperature rises. The amount of energy needed to change the temperature of an object depends on the material the object is made from. The property of a material that determines the energy required is called the **specific heat capacity.**

The specific heat capacity of a substance is the energy needed to raise the temperature of 1 kg of the substance by 1 °C.

It is given by the equation:

change in thermal energy = mass × specific heat capacity × change in temperature

$$\Delta E = m \times c \times \Delta\theta$$

where ΔE = change in thermal energy in J

m = mass in kg

c = specific heat capacity in J/kg °C

$\Delta\theta$ = temperature change in °C.

Example: A change in thermal energy of 18 kJ of energy was supplied to a 2 kg steel block and raised its temperature from 20 °C to 40 °C. Calculate the specific heat capacity of steel.

$\Delta E = mc\Delta\theta$

ΔE = 18 kJ = 18000 J

m = 2 kg

$\Delta\theta$ = (40 – 20) °C = 20 °C

Specific heat capacity, $c = \dfrac{\Delta E}{m\Delta\theta}$

$$= \frac{18000\text{ J}}{2\text{ kg} \times 20\text{ °C}} = 450\text{ J/kg °C}$$

Material	Specific heat capacity in J/kg °C
water	4200
ice	2100
aluminium	880
copper	380

Water has a very high specific heat capacity. This means that, for a given volume, it absorbs a lot of energy when it warms up. It also gives out a lot of energy when it cools down (Figure 1.1.23).

Figure 1.1.23 A domestic radiator contains water which is heated by a boiler

DID YOU KNOW?

Water needs lots of energy to heat it up and gives out lots of energy when it cools down. This is why a hot-water bottle is so effective in warming a bed. An equal volume of mercury would store only 1/30th as much energy, but would be 13 times heavier!

COMMON MISCONCEPTION

Do not confuse thermal energy transfer and temperature. Temperature is how hot an object is. A change in thermal energy does not always produce the same temperature change.

5 **Calculate how much energy is needed to heat 100 g of water from 10 °C to 40 °C.**

6 **Water has a very high specific heat capacity. Describe a practical use of this.**

7 **A 2 kW electric heater supplies energy to a 0.5 kg copper kettle containing 1 kg of water.**

a **Calculate the time taken to raise the temperature by 10 °C.**

b **What have you assumed in doing this calculation?**

8 **Suggest why some saucepans are made of copper.**

REQUIRED PRACTICAL

1.1h Investigating specific heat capacity

Learning objectives:

- use theories to develop a hypothesis
- evaluate a method and suggest improvements
- perform calculations to support conclusions.

KEY WORDS

energy store
energy transfer
specific heat capacity

Electric storage heaters store energy by heating up blocks during the night, when electricity is cheap. The energy is released the next day. Engineers can test materials to find those that can store and release the most energy.

These pages are designed to help you think about aspects of the investigation rather than to guide you through it step by step.

Using scientific ideas to plan an investigation

When a lump of brass is immersed in boiling water its temperature increases. It would end up at 100 °C (if we continued to heat the water). Thermal energy is transferred from the store in the water into the brass. If we then take the brass out of the boiling water and put it into another beaker of water at room temperature, thermal energy would transfer from the store in the brass to the water. The brass would cool down and the water would heat up until they were both at the same temperature.

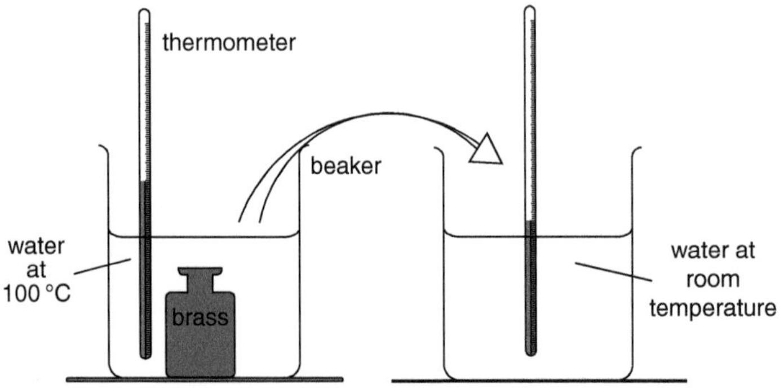

Figure 1.1.24 The lump of brass is transferred from the water at 100 °C to the water at room temperature

1. **How could we find out what the temperature of the water became when the brass was added to it?**
2. **How could we calculate the rise in temperature of the water?**
3. **How could we calculate the temperature drop of the brass?**
4. **What would happen to the temperature of the water and the brass in the second beaker if we left them for a long time (e.g. an hour)?**

SAFETY

It would be advisable to place the beaker in a tray/container that will hold the boiling water in case the brass lump is dropped into the beaker and it cracks, spilling the water. The thermometers should be clamped, and a stirring rod used to agitate the water and therefore obtain more accurate temperature readings.

Evaluating the method

This experiment is used to find the **specific heat capacity** of brass by assuming that all the thermal **energy transferred** from the hot water increases the temperature of the lump of brass.

We are equating the decrease in thermal **energy store** of the water to the increase in thermal energy store of the brass. If this assumption is not true, the method will not be valid.

5. **The lump of brass has to be moved from one beaker to the other. Consider how this step in the method could affect the accuracy of the results.**
6. **The energy transferred from the brass will cause the water in the second beaker to get hotter. Why will the energy transferred to the water not be stored there permanently?**
7. **What are the implications of your answers to questions 5 and 6 for the way the experiment is carried out?**
8. **Why is it important that the lump of brass is covered in water in the second beaker?**

Using the data to calculate a value for SHC

We find the specific heat capacity of brass by calculating the energy transferred into the water when its temperature increases and equating that to energy transferred out of the brass when its temperature decreases.

Decrease in thermal energy of brass = increase in thermal energy of water

$$m_{water} \times c_{water} \times \text{temperature increase}_{water}$$
$$= m_{brass} \times c_{brass} \times \text{temperature decrease}_{brass}$$

The final temperature of the water and brass is the same (they reach thermal equilibrium). As long as we know the values of the mass of water, mass of brass, specific heat capacity of water and initial temperatures of the water and brass, we can find the unknown value for the specific heat capacity of brass.

9. **There is 250 g of water (c_{water} = 4200 J/kg °C) in the second beaker and its temperature rises from 17 °C to 26 °C. Determine how much energy has been transferred into it.**
10. **How much energy can we assume has been transferred out of the block when it is put into the second beaker?**
11. **If the brass had been in boiling water by how much would its temperature have decreased?**
12. **The brass has a mass of 150 g. Calculate its specific heat capacity.**
13. **Explain why this method is likely to give a lower value for the specific heat capacity than its true value?**

DID YOU KNOW?

Stone Age man, Native American Indians and backwoodsmen all have used hot stones to boil water. Hot stones from a fire are dropped into a wooden bowl of cold water. Thermal energy stored in the stones is transferred to the water making it hot enough to boil. This is an example of a decrease in one energy store producing an increase in another. You can make use of this method to find the specific heat capacity of different materials.

KEY INFORMATION

When thinking about this experiment remember that energy tends to move from hotter areas to cooler ones.

1.1i Changes of state and specific latent heat

KEY WORDS

latent heat
specific latent heat
specific latent heat of fusion
specific latent heat of vaporisation

Learning objectives

- explain what is meant by latent heat and distinguish it from specific heat capacity
- perform calculations involving specific latent heat.

When a substance is heated, at certain temperatures the thermal energy transferred by heating goes into changing the state of the substance rather than continuing to increase the temperature. This can be shown on very characteristic graphs.

Latent heat

When you heat a solid, such as a lump of ice, its temperature rises until it starts to change to a liquid. At its melting point, the temperature stays the same until all the ice has melted (Figure 1.1.25).

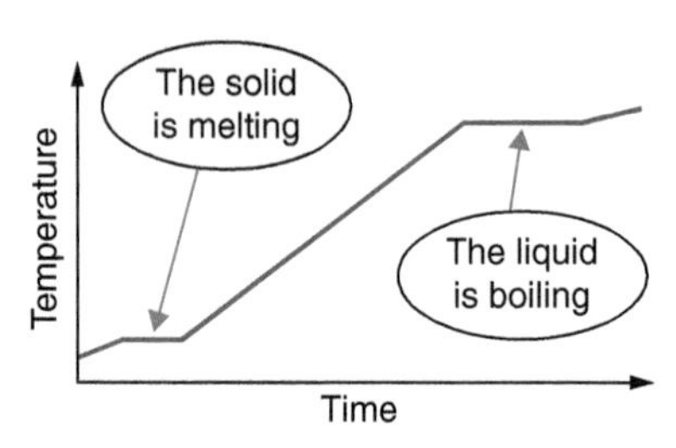

Figure 1.1.25 Temperature–time graph for heating a substance

The temperature of the liquid then rises until it starts to change into a gas. At its boiling point, the temperature stays the same until all the liquid has turned into a gas. If you continued to heat the gas, the temperature would start to rise again.

Latent heat is the energy needed to change the state of a substance without a change in temperature.

1. **Describe is happening when the graph flattens out.**
2. **Energy is needed to turn water into steam. Suggest how the particle model explains this.**

MAKING LINKS

Look back to topic 1.1d to remind yourself about particle theory.

Changes of state

The amount of energy needed to change the state of a sample of a substance, without a change in temperature, depends on the mass of the sample and the type of substance. All substances have a property called **specific latent heat.**

The specific latent heat is the amount of energy needed to change the state of 1 kg of a substance without a change in temperature. Its unit is J/kg.

It is given by the equation:

$$\boxed{E = mL}$$

where E = energy for a change of state in J

m = mass in kg

L = specific latent heat in J/kg

Specific latent heat of fusion refers to a change of state from solid to liquid. **Specific latent heat of vaporisation** refers to a change of state from liquid to vapour.

The specific latent heat of vaporisation is much greater than the specific latent heat of fusion. Most of this energy is used to separate the particles so they can form a gas, but some is required to push back the atmosphere as the gas forms.

Change of state	Specific latent heat in J/kg
ice to water	340 000
water to steam	2 260 000

3 **Explain why there is no change in temperature when a block of ice melts.**

4 **Why is the specific latent heat of vaporisation much greater than the specific latent heat of fusion?**

Latent heat calculations

Many calculations on specific and latent heat require you to find the energy needed to change the state as well as the temperature of an object. This involves using both specific heat capacity and latent heat equations.

$\Delta E = m\, c\, \Delta\theta \qquad E = m\, L$

5 **Calculate the energy transferred from a glass of water to just melt 100 g of ice cubes at 0 °C.**

6 **a** **Calculate the total energy transferred when 200 g of ice cubes at 0 °C are changed to steam at 100 °C.**

b **Sketch a temperature–time graph for this transfer.**

Figure 1.1.26 Machines in cafes use steam to heat milk quickly

KEY INFORMATION

In these calculations use the specific heat capacities given in the table in topic 1.1g.

KEY INFORMATION

Use the equation for specific heat capacity as well as the equation for latent heat.

DID YOU KNOW?

A jet of steam releases latent heat when it condenses (changes to a liquid). This can be used to heat drinks quickly (Figure 1.1.26).

KEY INFORMATION

Energy transfer does not always involve a change in temperature. When a change of state occurs, the energy supplied changes the energy stored (internal energy) but not the temperature.

MATHS SKILLS

1.1j Drawing and interpreting graphs

Learning objectives:

- draw a graph of temperature against time
- interpret a graph of temperature against time.

KEY WORDS

line of best fit
range
scale

A graph of temperature against time for a substance can show quite a lot about what is going on. If we know how to read a graph we can work out what that might be.

Drawing a graph

To set out the axes for a graph, the **range** of values has to be identified. The maximum and minimum values should be known.

The number of units per square on the graph paper should be chosen with care. It's essential to select a **scale** that is easy to interpret. One that uses five large squares for 100 units is much easier to interpret than a scale that has, say, three large squares for 100 units. Sometimes it is better to use a smaller scale (e.g. two large squares for 100 units) and have a scale that is easier to interpret than to try to fill the page.

A group of students conducted an experiment where they recorded the temperature of stearic acid every minute as it cooled down from 90 °C. The results are shown in the table below.

Time in minutes	Temperature in °C	Time in minutes	Temperature in °C
0:00	95.0	8:00	68.1
1:00	90.9	9:00	68.1
2:00	86.3	10:00	67.9
3:00	80.7	11:00	67.3
4:00	73.0	12:00	64.2
5:00	68.1	13:00	58.4
6:00	68.0	14:00	48.2
7:00	68.1	15:00	35.9
		16:00	25.0

1. **What are the maximum and minimum values that need to be plotted?**
2. **Draw and label axes for plotting this dataset.**
3. **Plot the points and draw a line of best fit.**
4. **Describe the features of the line of best fit.**

KEY INFORMATION

Suitable scales are 1, 2, 5 and 10 (or multiples of 10) units per square.

MATHS

Some graphs will have a line or curve of best fit. A line of best fit goes roughly through the centre of all the plotted points. It does not go through all the points. The **line of best fit** could be a curve. If all the points fall close to the line (or curve), it suggests that the variables are closely linked. Look at the points carefully and see what would fit them well.

Interpreting a graph

The graph in Figure 1.1.27 shows temperature against time for heating a lump of paraffin wax. The temperature was recorded using a datalogger. The rate at which energy was supplied to the wax stayed constant. The temperature of the solid wax increases until about 7 minutes and then stops rising. It then starts rising again at about 15 minutes.

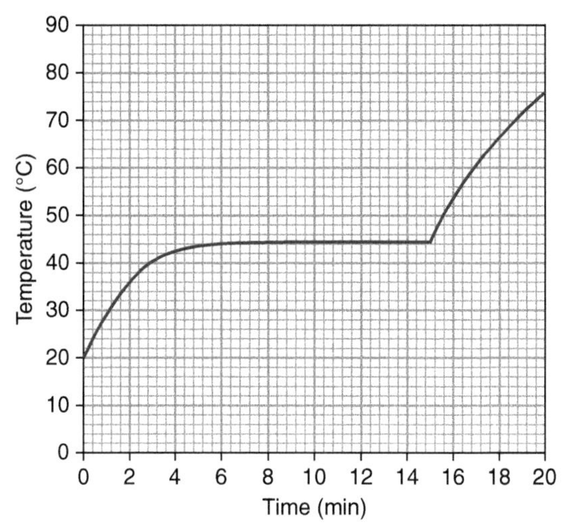

Figure 1.1.27 Graph of temperature against time for heating paraffin wax

5. **Why is there a curve and not a straight line when the energy is being supplied at a constant rate?**
6. **What is happening between 7 and 15 minutes?**
7. **What is the melting point of paraffin wax?**

Understanding the shape of the line

This graph shows a really important point. Whilst the wax is melting, energy is being supplied but the temperature isn't changing. The energy being supplied is being used to change the state of the wax and whilst this is happening, the temperature stays the same.

Now look at the graph you drew of temperature against time for stearic acid. This was cooling down but there are similarities.

8. **Suggest an explanation for the shape of the graph.**
9. **What is happening between about 5 and 9 minutes?**
10. **What is the melting point of stearic acid?**
11. **Describe the shape of the graph with reference to what is happening with regard to the internal energy of the stearic acid.**

DID YOU KNOW?

Steam at 100 °C can burn you more severely than the same mass of water at 100 °C. As well as the high temperature steam has latent heat – the energy it used to turn into a gas.

1.1k Meaning of purity

Learning objectives:

- explain what is meant by purity
- distinguish between the scientific and everyday use of the term 'pure'
- use melting and boiling point data to distinguish pure from impure substances.

KEY WORDS

pure
purity
unadulterated

Just because something is labelled as 'pure' does not mean that it really is. Chemists can investigate purity with some simple rules.

Meaning of purity

In everyday language, a 'pure' substance can mean a substance that has had nothing added to it, so it is in its natural state, or **unadulterated** – for example, pure milk. But this is not the chemical definition.

In chemistry, a **pure** substance is a single element or compound, not mixed with any other substance. A sample of gold with a **purity** of 100% only contains gold atoms – nothing else.

Figure 1.1.28 Separating 'pure' milk. This is not the chemistry definition of pure

A mixture consists of two or more elements or compounds not chemically combined. A mixture is not pure, even if the components are pure. The chemical properties of each substance in the mixture are unchanged.

DID YOU KNOW?

Pure and impure milk can both be separated into their components by centrifuging. This means spinning round at high speeds so that layers of substances can be separated. Blood can also be separated in this way.

1. **Identify which of these is a pure substance in the everyday sense of the word 'pure' and explain why: tea, salt, milk, cooking oil.**
2. **Draw out, using particle pictures, examples of:**
   - **a a pure single-atom element**
   - **b a mixture of two elements**
   - **c a pure compound, such as water (H_2O).**

The importance of purity

Pure substances are used in industry to make materials, food and medicines. Impurities in drugs might cause unwanted or dangerous side effects. Chemists who manufacture food, medicines or new materials such as plastics need to create products that are free from impurities that could be hazardous.

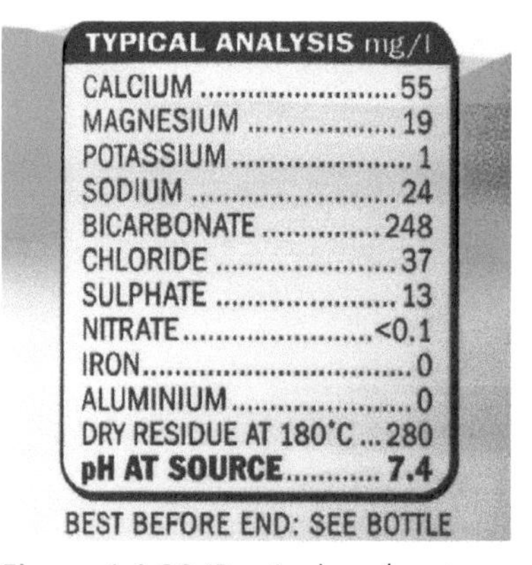

Figure 1.1.29 'Pure' mineral water contains many different substances

3 Look at the label in Figure 1.1.29. What proportion of the mineral water is actually other substances, and would be left behind if the water was evaporated off? (Hint: this is the dry residue value.) Give your answer as a percentage.

4 Assess the benefits and potential risks of buying a nutritional supplement advertised as pure and natural, but with no other labelling information.

Establishing purity

Pure elements and compounds melt and boil at specific temperatures. The purity of a compound can be established using data from its melting point or boiling point.

When there is a mixture of the pure compound with other substances, the impurities affect the melting and boiling points in three ways:

- reduce the melting point
- increase the boiling point
- increase the range over which melting and boiling takes place.

This is because the impurities are themselves substances that each melt or boil at different temperatures from each other. The greater the amount of an impurity, the bigger the differences from the true melting point and boiling point.

Because pure elements and compounds have definite melting points and boiling points, we can use official data to decide if a substance is indeed pure.

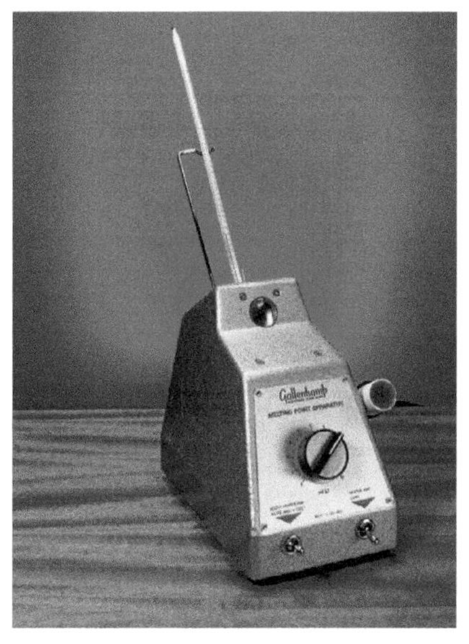

Figure 1.1.30 Using a melting point to establish purity. A small sample in a thin tube is inserted into a metal block in the apparatus alongside the thermometer. The metal block is slowly warmed up and the observer records the temperature at which the sample melts

5 Sam and Alex each made a sample of the same substance. They measured the melting point. They then crystallised their solid and measured the melting point again. They did this several times.

	Melting point in °C				
Sam	46.5	46.6	46.7	46.7	46.7
Alex	42.8	43.7	44.5	44.9	45.1

Identify who made the purest sample and explain your reasoning.

6 Akira and Ben tested their samples for their boiling points. The standard boiling point from data tables was 85.0 °C. Their results were: Akira 86.2 °C, Ben 87.1 °C. Identify who had the purer sample and explain your reasoning.

DID YOU KNOW?

Mixing salt with ice lowers the melting point, so the mixture melts at a temperature below 0 °C. This is why salt is spread onto icy roads in winter.

Check your progress

You should be able to:

☐ describe and explain the properties of solids, liquids and gases using the particle model. →	☐ use the model to explain why physical changes are reversible. →	☐ identify the strengths and limitations of the particle model.
☐ use density = mass/volume to calculate density. →	☐ use particle diagrams to communicate ideas about relative densities of different states. ☐ use the density equation to calculate mass and volume. ☐ manipulate apparatus to collect data to discover the density of an unknown substance. →	☐ link the particle model for solids, liquids and gases with density values in terms of the arrangements of the atoms or molecules.
☐ describe how, in the particle model the higher the temperature the faster the molecules move. →	☐ explain how a gas has a pressure. ☐ use the particle model to explain the effect on temperature of increasing the pressure of a gas at constant volume. →	☐ describe how the temperature of a gas is related to the average kinetic energy of the molecules. ☐ use the particle model to explain that increasing the volume of a gas, at constant temperature, can lead to a decrease in pressure.
☐ describe changes of state as physical changes. →	☐ describe how mass is conserved when substances change state. ☐ explain that changes of state are physical, not chemical, changes because the material recovers its original properties if the change is reversed. →	☐ explain using the particle model how changes of state conserve mass.
☐ describe how heating raises the temperature of a system. →	☐ recognise that heating raises the temperature or changes the state of a system but not at the same time. →	☐ explain that the internal energy of a system is stored by the particles that make up the system.
☐ describe the effect of an increase in temperature on the motion of particles. ☐ define and explain specific heat capacity. ☐ use appropriate apparatus safely to make and record a range of measurements accurately. →	☐ use the specific heat capacity equation to calculate the energy required to change the temperature of a certain mass of a substance. ☐ state the factors that are involved in increasing the temperature of a substance. ☐ perform calculations to find the specific heat capacity of a material. →	☐ use the specific heat capacity equation to calculate mass, specific heat capacity or temperature change. ☐ calculate energy transferred when a material of a certain mass is heated. ☐ evaluate the safety and procedures you have used in practically investigating specific heat capacity.
☐ state that when an object changes state there is no change in temperature. ☐ describe changes of state from a graph. →	☐ describe the latent heats of fusion and of vaporisation. ☐ use the equation $E = mL$. →	☐ use the particle model to explain why the latent heat of vaporisation is much larger than the latent heat of fusion.
☐ describe what is meant by a pure substance. →	☐ recognise that a pure substance has a definite melting and boiling point. →	☐ distinguish pure and impure substances using melting and boiling point data. ☐ discuss the risks of impurities in medicines and food.

Worked example

1 Kiran uses an electric immersion heater to supply 20 kJ of thermal energy to a 1.5 kg aluminium block. He recorded a temperature rise of 14.6 °C. Calculate the specific heat capacity of aluminium.

913

This is the correct numerical answer, but

a You have not given any units to the number. Always give the unit to the quantity.

b You have not shown any working. When working a calculation, write the equation, and show step by step how you do the calculation.

2 Explain how raising the temperature of a gas, keeping the volume constant, increases the pressure exerted by the gas.

The high temperature makes the molecules vibrate faster so there is a stronger force on the side of the container.

You are on the right lines. With a gas, the molecules are no longer vibrating but they are free to move around at speed. Heating gives them more kinetic energy. As the molecules move around they hit the side of the container, and it is the force of this that creates the pressure.

3 What does it mean to say that the changes of state are reversible?

It means that the changes go both ways.

This answer is correct, but you should give an example as well. Give an example of a change that goes both ways, e.g. water turns to steam and steam turns back to water.

4 Explain what heating does to the energy stores of a system.

Heating raises the total energy inside the system.

Yes it does, but again you should give more specific information about the system and mention the energy transfers that have taken place. A fuller answer would mention that the energy supplied by heating increases the speed of movement of the particles (their kinetic energy).

End of chapter questions

Specific heat capacity of water = 4200 J/kg °C

Specific latent heat of fusion of ice to water = 340 000 J/kg

Specific latent leat of vaporisation of water to steam = 2 260 000 J/kg

Getting started

1. Label these diagrams as solid, liquid and gas. **1 Mark**

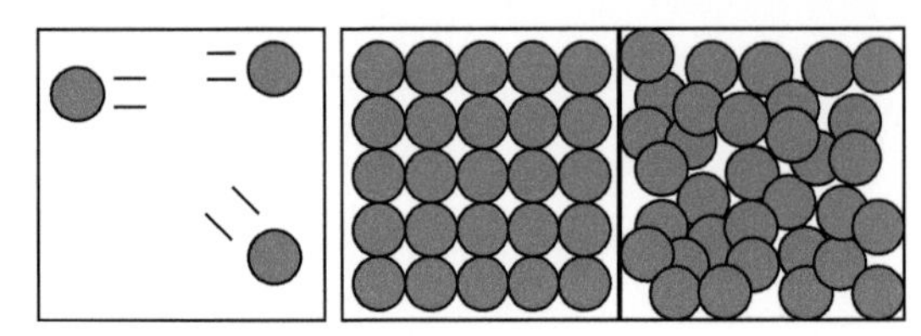

2. Describe how these models represent a solid, liquid and gas. **2 Marks**
3. Substance G has a melting point of –33 °C and a boiling point of 52 °C. Explain which state it is in at 20 °C. **1 Mark**
4. Two steel blocks, one with a mass of 100 g and the other with a mass of 200 g, are placed in boiling water for several minutes. Which of these statements is **not** true? **1 Mark**
   - a They are both at the same temperature.
   - b They are made of the same material.
   - c They have the same amount of energy.
   - d They remain solid.
5. What is the term given to the energy needed to change the state of a substance? **1 Mark**
6. Calculate the amount of energy needed to change the temperature of 2 kg of water by 10 °C. **2 Marks**
7. State the equation for density. **1 Mark**

Going further

8. Which of these does **not** affect the amount of energy needed to heat a material up? **1 Mark**
   - a Its colour.
   - b Its mass.
   - c Its specific heat capacity.
   - d The temperature rise.
9. An object has a mass of 100 g and a volume of 25 cm^3. Will it float in water? **2 Marks**

10. Calculate the energy needed to boil away 150 g water at 100 °C. **2 Marks**
11. Water in an ice cube tray is put into a freezer. Explain what happens to the energy stored inside the system. **3 Marks**
12. Using the particle model, explain how a gas exerts pressure on the container holding the gas. **2 Marks**

More challenging

13. Night storage heaters charge up during the night time and release energy during the day. Explain why they are made from a material with a high specific heat capacity. **2 Marks**
14. Calculate how much energy is needed to heat 100 g of water from 10 °C to 50 °C. **2 Marks**
15. 4.50 g of water was frozen to ice at 0 °C. How much energy had to be taken out of the water? **2 Marks**
16. Why is the specific latent heat of vaporisation of water to steam a much higher value than the specific latent heat of fusion of ice to water? [The values are given at the top of at the start of these questions.] **2 Marks**
17. Explain how raising the temperature of a gas, keeping the volume constant, increases the pressure exerted by the gas. **2 Marks**

Most demanding

18. Explain why changing the temperature of a material and changing its internal energy mean different things. **2 Marks**
19. A 2 kg block of copper is given 8.88 kJ of energy to raise its temperature by 10 °C. What is the specific heat capacity of copper? **2 Marks**
20. 50 g of steam condense to water at 100 °C. How much energy was given out? **2 Marks**
21. With reference to the particle model, explain how increasing the volume in which a gas is contained, at constant temperature, creates a decrease in pressure. **2 Marks**

Total: 37 Marks

ATOMIC STRUCTURE

IDEAS YOU HAVE MET BEFORE:

ELEMENTS, MIXTURES AND COMPOUNDS

- Elements cannot be broken down by chemical means.
- Compounds are made from elements chemically combined.
- A pure substance is a single element or compound not mixed with any other substance.

THE PARTICLE MODEL OF MATTER

- A simple model of matter uses solid spheres to represent atoms, molecules and ions.
- The particles are arranged in different ways in solids, liquids and gases.
- Different elements are made from different atoms.

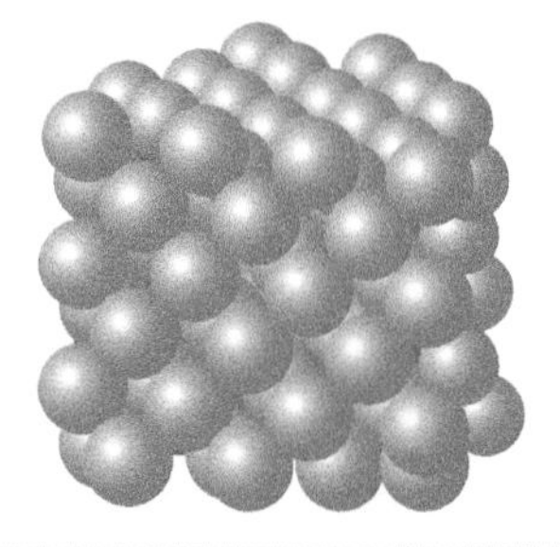

SIZE AND SCALE

- We use measurements expressed as numbers and units to compare properties of materials, such as length, mass, volume or density.
- SI units are an international system of units including the metre, kilogram and second.
- Atoms are so small that about 5 000 000 would fit side by side in a length of one millimetre.

1.2

IN THIS CHAPTER YOU WILL FIND OUT ABOUT:

ARE ALL ATOMS OF AN ELEMENT THE SAME?

- All atoms of the same element have the same atomic number.
- Atoms of the same element can have different numbers of neutrons; these atoms are called isotopes of the element.
- Atoms can be represented by symbols showing the mass number and atomic number.

$^{12}_{6}C$

HOW DID THE MODEL OF THE ATOM DEVELOP?

- Atoms used to be thought of as small unbreakable spheres.
- Experiments led to ideas of atoms with a nucleus and electrons.
- Electrons in shells and the discovery of the neutron came later.

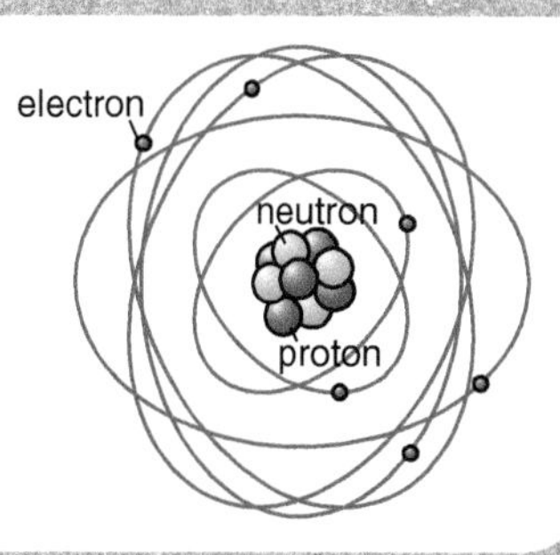

HOW DO WE WRITE VERY LARGE AND VERY SMALL NUMBERS?

- Sub-units use standard prefixes, such as 'milli' to mean one thousandth.
- Standard form uses powers of ten (such as 10^3 or 10^{-3}) to write very large or very small numbers.
- The radius of a typical atom is 0.1 nm (1×10^{-10} m).

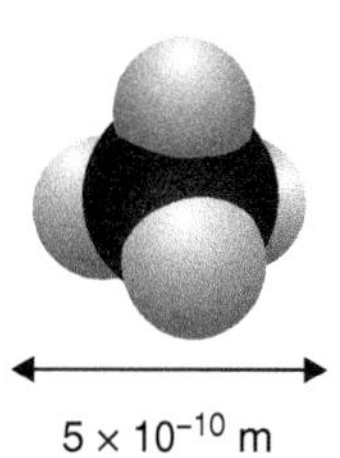

1.2a Scientific models of the atom

Learning objectives:

- describe how and why the model of the atom has changed over time
- explain how data support theories, and how new data lead to changes in theories.

KEY WORDS

atom
electron
neutron
nuclear model
plum pudding model
proton
sub-atomic particle

The early model of the atom was of a solid sphere that could not be split. The discovery of the negatively charged electron upset this simple model. Because atoms are neutral, scientists had to incorporate a positive part to the atom as well.

Developing the atomic theory

In 465 BC the Greek philosopher Democritus hypothesised that matter was made from **atoms**. These were solid, but invisible, and had different shapes and sizes. They could not be divided into smaller particles.

In 1804 John Dalton proposed his atomic theory. Dalton's atoms were tiny spherical particles which could not be broken up. He suggested that the particles of each element were special to that element, and were solid like tiny 'billiard balls'. His model helped to explain the properties of gases for the first time, and the formulae of compounds.

Figure 1.2.1 Dalton's idea of atoms: they were like tiny billiard balls

In 1897 J.J. Thomson discovered the negatively charged particle called the **electron**. The mass of the electron was about 1000 times smaller than the mass of the atom. This meant that there must be **sub-atomic particles** smaller than the atom.

Because atoms are neutral, Thomson proposed the **plum pudding** model, in which negative electrons are embedded in a ball of positive charge.

1. **Suggest why Dalton's atomic model did not include positive and negative charge.**
2. **Explain why the discovery of the electron changed the Dalton model of the atom.**

Changing theories

Dalton's theory lasted for nearly 100 years, but Thomson's model lasted for less than 20 years.

In 1909, a team of scientists working with Ernest Rutherford carried out an experiment where they aimed a beam of positive alpha particles (a form of radioactive decay) at a thin gold foil. Most alpha particles passed through but a few bounced back.

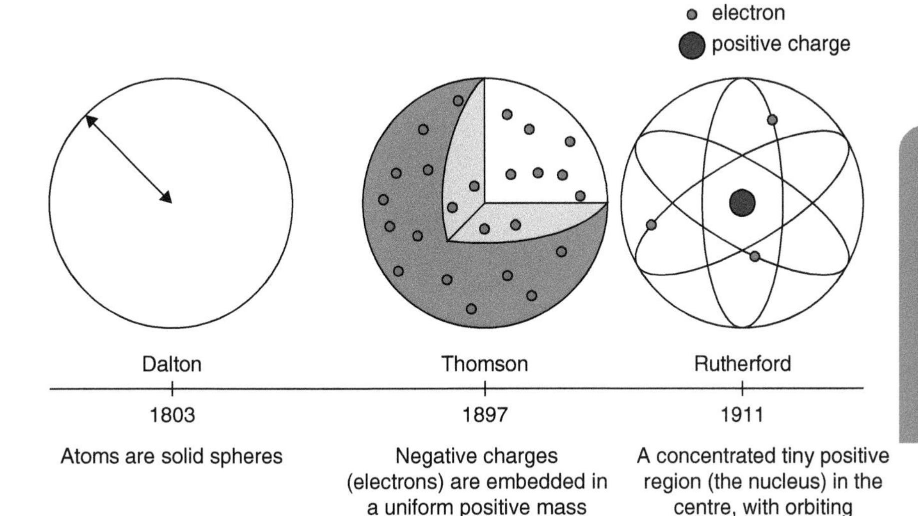

Figure 1.2.2 Timeline for the development of models of the atom

KEY INFORMATION

At each stage, the explanations of atomic theory were provisional until more convincing evidence was found to make the model better.

To explain this result, in 1911 Rutherford suggested the atom had a positively charged nucleus and much of the atom was empty space. This was the **nuclear model** of the atom.

Using the new model, scientists found that the tiny nucleus contained positively charged particles. These were first called **protons** in 1920. But further experiments showed the nucleus was too massive for the number of protons it contained.

In 1932 James Chadwick discovered the **neutron**, which had the same mass as the proton but no charge.

3. **Suggest how the discovery of the neutron changed ideas about the nucleus.**
4. **Explain why the model of the atom kept changing.**

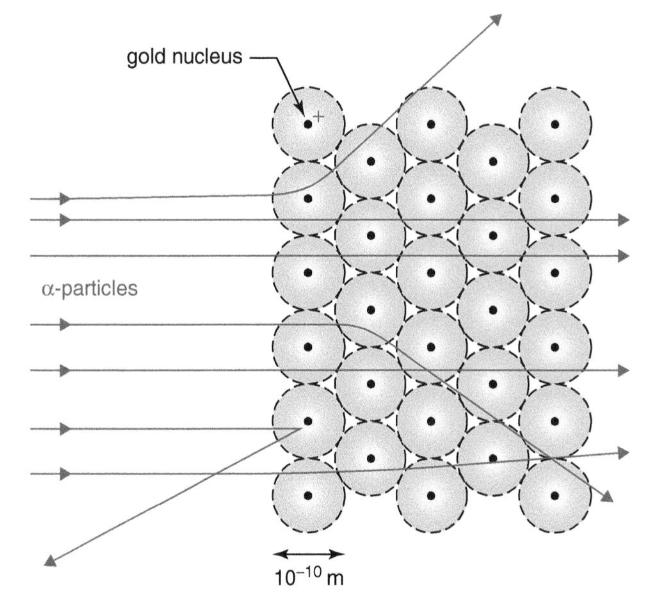

Figure 1.2.3 The positive nucleus repels the positive alpha particles

Challenging theories

The Rutherford alpha-particle experiment was very important – it changed the whole theory of the atom.

If Thomson's model was correct, the alpha particles should have passed through the foil, with nothing to attract or repel them. Yet some alpha particles bounced back. Rutherford's team had to think of a new theory to explain their unexpected observations.

5. **Explain why Rutherford's result did not support the Thomson model.**
6. **Suggest how a scientific theory becomes accepted.**

DID YOU KNOW?

The idea of atoms as small particles is not new. However, our ideas about the theory of atoms are still developing. Search on 'CERN LHC' to find out more.

1.2b The size of atoms

Learning objectives:

- recall the size and order of magnitude of atoms and small molecules
- recognise expressions in standard form
- estimate the size of atoms based on scale diagrams.

KEY WORDS

atomic radius
nanometre
order of magnitude

The size of the whole atom can be up to 100 000 times the size of the nucleus. So if you could imagine an atom to be the size of a football stadium, the nucleus would be a marble in the middle.

Sizes of atoms

Individual atoms are very small indeed. There are about five million of them in the full stop at the end of this sentence.

It depends on how it is measured, but the radius of an atom is about 0.000 000 000 1 m, which we can write as 1×10^{-10} m (this way of writing numbers is called 'standard form'). The **order of magnitude** is 10^{-10} m.

To describe such a small size, we use a unit called the **nanometre**, abbreviated to nm. 1 nm = 0.000 000 001 m, or 10^{-9} m. So the typical **atomic radius** is about 0.1 nm. The very smallest atoms have a radius of just 0.031 nm.

The nucleus is in the centre of the atom and the electrons surround it, but most of the atom is empty space.

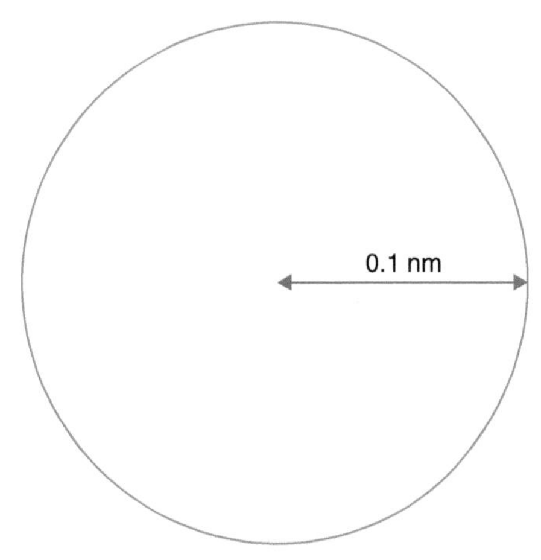

Figure 1.2.4 Typical atomic radius

DID YOU KNOW?

The radius of the atom, where the electrons are orbiting, is much larger than the radius of the nucleus in the centre of the atom. The nucleus has a radius of about 1×10^{-14} m. So there are about 4 orders of magnitude difference between the size of the atom and the size of the nucleus.

1. **What is between the nucleus and the outer part of the atom?**
2. **What is the order of magnitude of the radius of the smallest atoms?**

REMEMBER!

Always look carefully to see whether the question is talking about the radius or the diameter when it gives the measurements of an atom!

Sizes of molecules

Although a typical atomic radius is about 0.1 nm (1×10^{-10} m), atoms are not all the same size. The outer boundary of the atom is not fixed, so its position can only be measured approximately.

The radius of a small molecule such as methane is about 0.5 nm, or 5×10^{-10} m. This is in the same order of magnitude as the radius of an atom.

You can see from Figure 1.2.5 that the methane molecule as a whole is not much more than the size of the five individual atoms.

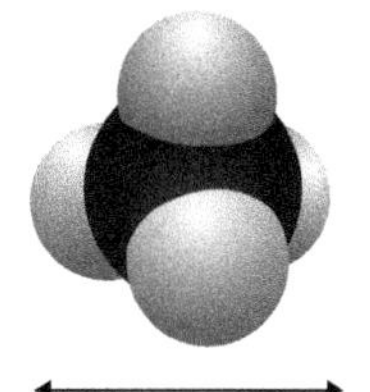

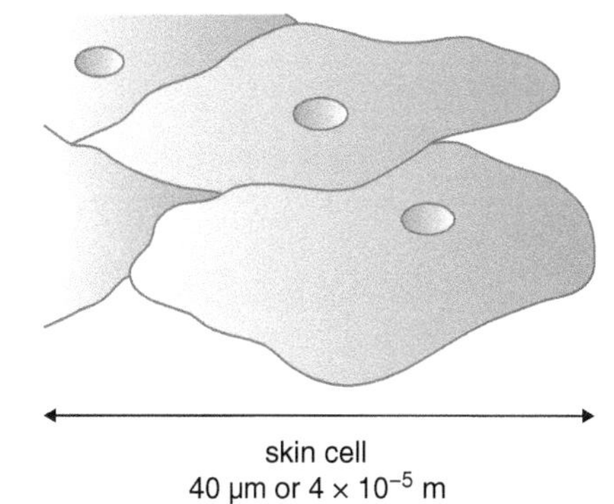

Figure 1.2.5 The size of a molecule of methane compared to a skin cell

3. **What is the size of the skin cell in nanometres?**
4. **What is the order of magnitude difference in size between the methane molecule and the skin cell?**

Estimating atomic sizes

A useful skill is to estimate the sizes of objects using scale drawings.

A scale bar is usually provided for you to measure against.

5. **Look at Figure 1.2.6 below. If the radius of the sodium atoms is 0.18 nm, what is their *diameter*?**
6. **What is the *radius* of the chlorine atoms?**

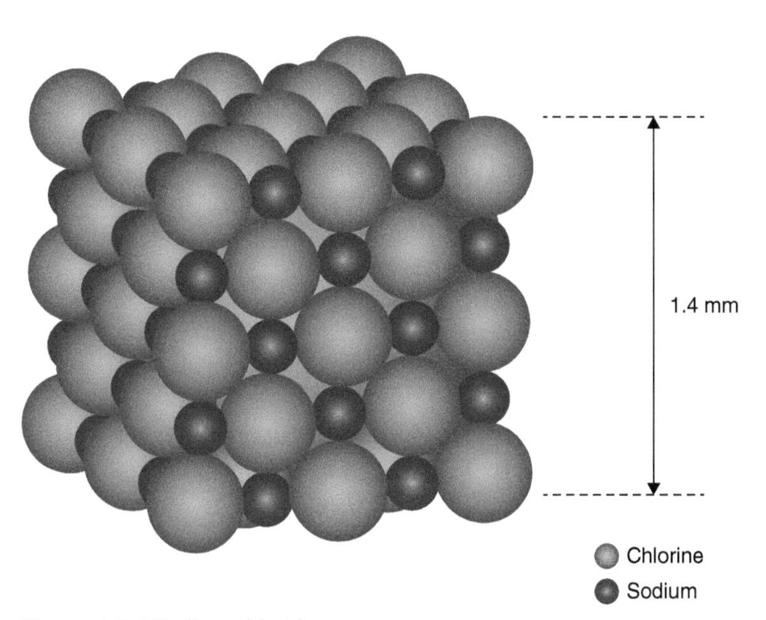

Figure 1.2.6 Sodium chloride

MAKING LINKS

See topic 5.1g to learn how sodium atoms and chlorine atoms react to form sodium chloride.

MATHS SKILLS

1.2c Standard form and making estimates

Learning objectives:

- recognise the format of standard form
- convert decimals to standard form and vice versa
- make estimates without calculators so the answer in standard form seems reasonable.

KEY WORDS

decimal point
standard form

When we talked earlier about an atom we used a model to describe it. We imagined it as a sphere with a radius of about 0.0000000001 m. We also saw that the radius of the nucleus of the atom is about 0.00000000000001 m. It is very awkward to keep writing so many zeros – it is easy to lose count and it is not so easy to see the comparison between one number and the other. Another way of writing these numbers uses **standard form**.

KEY INFORMATION

Standard form is used to represent very large or very small numbers.
A number in standard form is written in the form $A \times 10^n$, where $1 \leq A < 10$ and n is an integer. For numbers less than 1, n is negative.

Positive powers of ten for very large numbers

We write 1, 10 and 100 knowing what we mean. We can also write them as 1, 1 × 10 and 1 × 10 × 10. We also know that 10 × 10 is 10^2. So 100 is 10^2. We can write the numbers 1, 1 × 10 and 1×10^2.

Standard form	M	HTh	TTh	Th	H	T	U	.	t
							1	.	0
1 × 10						1	0	.	0
1×10^2					1	0	0	.	0
1×10^3				1	0	0	0	.	0
1×10^4			1	0	0	0	0	.	0
1×10^5		1	0	0	0	0	0	.	0
1×10^6	1	0	0	0	0	0	0	.	0

1. **Write 1000000000 in standard form.**
2. **Write out the number 1×10^8.**

10^6 is NOT 10 multiplied by itself 6 times. It is 10 multiplied by itself 5 times.

What about writing bigger numbers in standard form?

The **decimal point** is fixed and the position, or place value of the most significant digit, shows how big a number is.

To write 1 000 000 in standard form take the first number on the left, which is 1. Looking at the table, how many places do we have to move the 1 to the right to reach the decimal point?

REMEMBER!

1000 can be written as 1 × 10 × 100, which is the same as 1 × 10 × 10 × 10. How many tens? Three. So 1000 is written 1×10^3 (one times ten to the power of three). The number 3 tells you how many tens are in the multiplication.

We have to move the 1 six places to the right. So in standard form 1 000 000 is 1×10^6. The number 6 tells you how many tens there are when you write the number as a multiplication of 10 ($10 \times 10 \times 10 \times 10 \times 10 \times 10$).

Negative powers of ten for very small numbers

It is also possible to write numbers smaller than 1 in this form. If 1 is divided by 10 it is 0.1. The number 1 has moved one place to the right of the decimal point. This is written as 1×10^{-1} in standard form. What is 1 divided by 100? The number 1 moves two places to the right of the decimal point to be 0.01. In standard form this is 1×10^{-2}.

standard form	U	.	t	h	th	Tth	Hth	milliionth
1×10^{-1}	0	.	1					
1×10^{-2}	0	.	0	1				
1×10^{-3}	0	.	0	0	1			
1×10^{-4}	0	.	0	0	0	1		
1×10^{-5}	0	.	0	0	0	0	1	
1×10^{-6}	0	.	0	0	0	0	0	1

3 **Write 0.000 000 000 000 000 001 in standard form.**

4 **Write out the number 1×10^{-9}.**

Working with numbers to standard form

Standard form can also be used to represent numbers where the most significant digit is not one. For example, the ordinary number 6000 can be written as 6×1000, or 6×10^3, in standard form.

Remember that standard form always has exactly one digit bigger than or equal to one but less than 10. 0.3×10^4 is not in standard form. It is 3×10^3 in standard form.

The table gives some big and small numbers.

Distance from Earth to the Sun (m)	Speed of light (m/s)	Atomic radius (m)	Nuclear radius (m)	Mass of a gold atom (g)
1.5×10^{11}	3×10^8	1×10^{-10}	1×10^{-14}	3.3×10^{-22}

When you calculate with big and small numbers using a calculator it is essential that you first estimate what your answer should look like. Making an estimate of the result of the calculation can save you from making mistakes with your calculator. The best way to estimate the answer without a calculator is to round the numbers sensibly and then carry out the calculation in your head.

5 **Calculate:**

a) $6 \times 10^9 \times 3 \times 10^3$

b) $6 \times 10^9 \times 4 \times 10^{-2}$

c) $\dfrac{6 \times 10^8}{2 \times 10^2}$.

6 **Convert the distance from Earth to the Sun from metres to kilometres.**

7 **Calculate the mass of 3.0×10^{26} gold atoms using the mass of a single gold atom given in the data table.**

KEY INFORMATION

To multiply two numbers in standard form you simply add the indices or powers of the tens. For example, $2 \times 10^{15} \times 3 \times 10^9$ is 2×3 with 10^{15+9}, which is 6×10^{24}. With smaller numbers $2 \times 10^{-15} \times 3 \times 10^{-9}$ is 6×10^{-24}.

1.2d Sub-atomic particles

Learning objectives:

- interpret and draw diagrams of the structure of atoms
- recall that the radius of a nucleus is less than 1/10 000 that of the atom (about 1×10^{-14} m)
- recall the relative charges and masses of protons, neutrons and electrons
- calculate the number of protons, neutrons and electrons in atoms.

KEY WORDS

atomic number
electron
neutral
neutron
proton
relative mass
relative charge

Using expressions in standard form helps us to understand the very small size of atoms. When we compare atoms and sub-atomic particles, it can be easier to use relative masses and charges, rather than the actual values.

KEY INFORMATION

The word 'particle' is used in science to describe a tiny unit of matter. Sometimes this can mean atoms, as in the particle model (see topic 1.1a), and sometimes units that are even smaller, such as sub-atomic particles.

Structure of atoms

We saw how ideas about the structure of atoms developed in topic 1.2a. Currently, scientists believe that an atom is made up of a tiny nucleus that is surrounded by **electrons** occupying a larger space.

- The nucleus carries a positive charge.
- The electrons that surround the nucleus each carry a negative charge.

The nucleus of an atom is made up of protons and neutrons.

- **Protons** have a positive charge.
- **Neutrons** have no charge.

An atom always has the same number of protons (+) as electrons (–) so atoms are always **neutral**.

The **atomic number** is the number of protons in an atom. The atomic number for helium is 2 because it has two protons. Sometimes the atomic number is called the proton number.

All the atoms of a particular element have the same number of protons. Atoms of different elements have different numbers of protons.

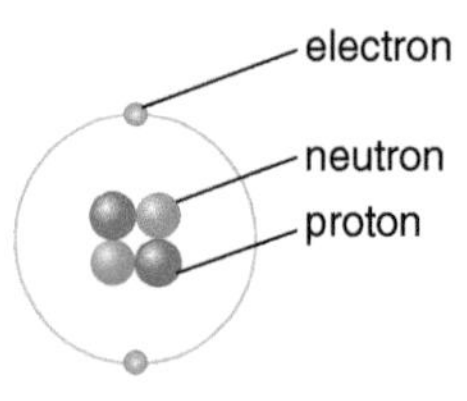

Figure 1.2.7 The structure of a helium atom. There are the same number of protons and electrons. Please note that this diagram is not drawn to scale

1. **What type of charge does an electron carry?**
2. **Why is the charge on an atom neutral?**
3. **An atom of calcium has 20 protons. How many electrons are in a calcium atom?**

Comparing sub-atomic particles

The nucleus of an atom is made up of particles (protons and neutrons) that are very much heavier than electrons. The nucleus contains almost all the mass of the atom and the electrons contribute very little.

The **relative masses** and **relative charges** of electrons, protons and neutrons are shown in the table. These are the masses and charges compared to the mass and charge of the proton.

Name of particle	Relative mass	Relative charge
proton	1	+1
neutron	1	0
electron	very small	–1

The radius of a nucleus is less than one ten-thousandth of the radius of the atom (about 1×10^{-14} m).

4 **A fluorine atom has 9 positive charges, 9 negative charges and a mass of 19. Describe the structure of its atom.**

5 **A chlorine atom has 17 electrons and a mass of 35. Describe the structure of its atom.**

6 **Explain why the radius of the nucleus is much smaller than the radius of the whole atom.**

DID YOU KNOW?

Electrons have such a small relative mass that it is usually treated as zero.

Atomic number and mass number

The nucleus of an atom is made up of protons and neutrons.

- The atomic number is the number of protons in an atom.
- The mass number of an atom is the total number of protons and neutrons in an atom.

If an atom has an atomic number of 11, a mass number of 23 and a neutral charge, it must have:

- 11 protons, because it has an atomic number of 11
- 11 electrons, because there are 11 protons and the atom is neutral
- 12 neutrons, because the mass number is 23 and there are already 11 protons (23 – 11 = 12).

Here are some more examples.

	Atomic number	Mass number	Number of protons	Number of electrons	Number of neutrons
carbon	6	12	6	6	6
fluorine	9	19	9	9	10
sodium	11	23	11	11	12
aluminium					

7 **Copy the table and complete the row for an atom of aluminium, Al.**

8 **How many protons, electrons and neutrons are there in an atom with an atomic number of 15 and a mass number of 31?**

MATHS SKILLS

1.2e Sizes of particles and orders of magnitude

Learning objectives:

- identify the scale of measurements of length
- explain the conversion of small lengths to metres
- explain the relative sizes of nuclei and atoms
- make order of magnitude calculations.

KEY WORDS

nanometre
order of magnitude
scale

All sub-atomic particles are so small that we use numbers in standard form to describe them. We can also compare them to other objects using orders of magnitude.

Scale of objects

Placing a tennis ball, golf ball, basketball and table tennis ball in order of size is easy.

unit	basketball	tennis ball	golf ball	table tennis ball
cm	25.0	6.8	4.1	4.0
m	0.25	0.068	0.041	0.04

Figure 1.2.8 It is easy to put these in order of diameter

We can measure objects smaller than these in millimetres.

1 m = 1000 mm   1 mm = 0.001 m   or   1 mm = 10^{-3} m

We can even see objects in the next set of smaller units, the *micrometre*. We measure the width of a human hair in this unit.

1 m = 1000000 μm   1 μm = 0.000001 m   or   1 μm = 10^{-6} m

After that we need instruments to help us see and measure lengths. Atoms and small molecules are on the 'nano-scale'. The unit is the **nanometre**.

1 m = 1 000 000 000 nm   1 nm = 0.000000001 m or
1 nm = 10^{-9} m

1 **Calculate the number of basketballs that would fit in a kilometre.**

2 **A carbon nanotube has a length of 2 × 10^{-9} m. Calculate the number of nanotubes that would fit in 1 mm.**

MATHS

1 cm = 0.01 m is the long way to write the conversion.
1 cm = 10^{-2} m is the conversion into *standard form.*

Atoms and sub-atomic particles

Going one step further down into the atomic scale:

- the radius of an atom is measured in *picometres* (pm), 10^{-12} m
- the radius of a nucleus is measured in *femtometres* (fm), 10^{-15} m

Why is the radius of a nucleus so much smaller than the radius of an atom?

Between the nucleus and the electrons of the atom there is mostly empty space, so neutrons and protons have radii measured in femtometres (fm).

3 **The hydrogen atom has a radius of 2.5×10^{-11} m and its nucleus a radius of 1.75×10^{-15} m. Calculate how many times larger the atom is compared to the nucleus.**

4 **What is a radius of 0.000000000001 m in standard form?**

Order of magnitude

Two numbers that have the same **order of magnitude** are about the same size, and are on the same **scale**. If a number is one order of magnitude larger than another, it is 10 times bigger. If two numbers differ by two orders of magnitude, then one is 100 times larger than the other.

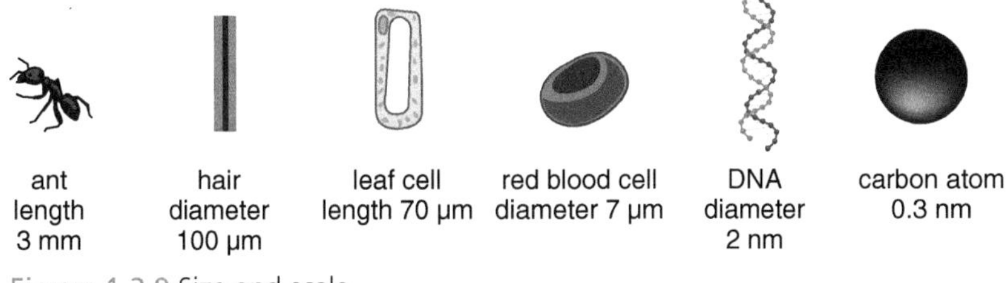

Figure 1.2.9 Size and scale

When comparing sizes, scientists often refer to differences in order of magnitude. That's the difference calculated in powers of 10.

So, to find the difference in order of magnitude between an ant and a carbon atom:

The ant is 3 mm in length = 3×10^{-3} m.

The carbon atom has a diameter of 0.3 nm = 3×10^{-10} m.

$$\frac{3 \times 10^{-3}\text{nm}}{3 \times 10^{-10}\text{nm}} = 1 \times 10^{7}$$

The difference in order of magnitude is 10^7, expressed as 7 orders of magnitude.

5 **What is the order of magnitude difference between 0.5 m and 500 000 m?**

6 **A white blood cell measures 1.2×10^{-5} m. An egg cell measures 1.2×10^{-4} m. Calculate the difference in order of magnitude.**

KEY INFORMATION

These measurements are within the 'human' scale from:

- the diameter of a human hair to the diameter of a water droplet 10^{-6} m to 10^{-3} m
- the diameter of a pin-head to the diameter of a basketball 10^{-3} m to 10^{-1} m
- the length of a car to the height of the Shard building, London 1 m to 10^3 m
- the height of Ben Nevis to the length of the Great Wall of China 10^3 m to 10^6 m.

After that we measure on an 'astronomical' scale.

Ratio between values	Order of magnitude difference
0.0001	10^{-4}
0.001	10^{-3}
0.01	10^{-2}
0.1	10^{-1}
1	10^{0}
10	10^{1}
100	10^{2}
1000	10^{3}
10000	10^{4}

1.2f Isotopes

KEY WORDS

isotope
symbol

Learning objectives:

- recognise that atoms of the same element can have different masses because they have different numbers of neutrons
- calculate the number of protons, neutrons and electrons in isotopes
- interpret symbols representing the mass number and atomic number of an atom.

All atoms of an element have the same number of protons and the same number of electrons, but they do not all have the same number of neutrons. Chlorine has some atoms with mass number 35 and some with mass number 37, but they all have the same chemical properties.

Explaining isotopes

Most elements can exist in more than one form. These have the same atomic number, otherwise you'd have a different element, but they have different mass numbers. These different forms are called **isotopes**.

Carbon is a very common element that has different isotopes. *All* carbon atoms have 6 protons, so carbon's atomic number is 6. The left-hand carbon atom in Figure 1.2.10 has 6 neutrons, so it has a mass number of 6 + 6 = 12.

Another form of carbon (on the right in Figure 1.2.10) has an atomic number of 6 (6 protons) and a mass number of 14. It must therefore have 8 neutrons (14 – 6 = 8).

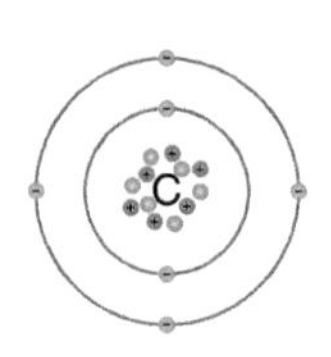

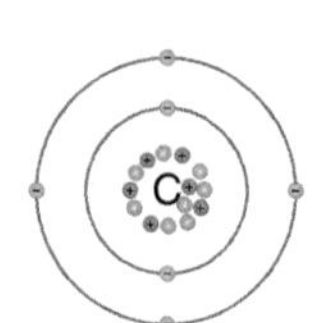

Figure 1.2.10 Two different isotopes of carbon. The letter in the centre of this kind of diagram represents the element

1. **What is an isotope?**
2. **Lithium has an atomic number of 3. There are two isotopes of lithium, one with a mass number of 6 and one with a mass number of 7.**

   Describe the difference in number of protons and number of neutrons between the two isotopes.

Using symbols to represent atoms

Atoms can be represented by the chemical **symbol** for the element together with the atomic number and mass number. The atomic number is the number of protons in the atom.

Using this notation, lithium-3 is:

mass number $^{7}_{3}\text{Li}$
atomic number

For a general atom we can use the symbol $^{A}_{Z}X$, where:

- *A* is the mass number
- *Z* is the atomic number (or proton number)
- X is the chemical symbol for the element.

Z is the number of protons in the nucleus, so the number of neutrons is (*A* – *Z*).

$^{14}_{6}C$ has a mass number of 14. So the sum of the neutrons and protons in its nucleus is 14. The atomic number is 6, so it has 6 protons. The number of neutrons in the nucleus is 14 – 6 = 8 neutrons.

3 How many protons and neutrons are in the nucleus of each of the isotopes of nitrogen, $^{14}_{7}N$ and $^{15}_{7}N$?

4 How many protons and neutrons are in the nucleus of each of the isotopes of uranium $^{235}_{92}U$ and $^{238}_{92}U$?

Identifying patterns

Most elements have two or more isotopes. The oxygen you breathe is mostly $^{16}_{8}O$, but there is also $^{17}_{8}O$, and $^{18}_{8}O$. All do the same chemical job in your body!

The table below shows the three common isotopes of hydrogen.

Isotope	Electrons	Protons	Neutrons	Mass number
$^{1}_{1}H$	1	1	0	1
$^{2}_{1}H$	1	1	1	2
$^{3}_{1}H$	1	1	2	3

DID YOU KNOW?

A radioactive isotope of oxygen is used as a medical tracer in PET scans. The patient breathes in oxygen-15. Oxygen is used in respiration in the brain, so by recording where this isotope is present in the brain, the PET scan shows which areas are respiring the most.

5 Explain the similarities and differences between the three isotopes of hydrogen.

6 What pattern, if any, is noted between:

a the atomic number, the number of protons and the number of electrons in any element?

b the atomic number, the number of protons and the number of neutrons in any element?

1.2g Electrons in atoms

Learning objectives:

- recall that in atoms with more than one electron, the electrons are arranged at different distances from the nucleus
- recognise that the energy associated with an electron shell increases with distance from the nucleus
- explain how electrons occupy shells in an order.

KEY WORDS

electronic structure
electron shell

Isotopes of the same element are different because they have different numbers of neutrons. Neutrons have no charge – they are neutral. The number of protons is the same for all isotopes of an element, so the number of electrons is also constant. The number of electrons balances the number of protons.

Electron shells

The protons and neutrons are fixed in the nucleus, and do not move. Overall, the nucleus has a positive charge. Orbiting the nucleus are electrons, which have a negative charge. The number of electrons equals the number of protons in any stable atom.

Electrons move around the nucleus in 'shells' at different distances from the nucleus. The space between the nucleus and the **electron shells** is empty space.

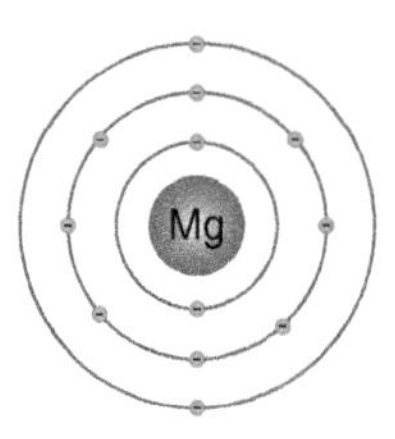

Figure 1.2.11 Electrons orbit in shells around the nucleus. In this type of diagram, the symbol for the element is given in the nucleus at the centre

1. **What is between the nucleus and the orbiting electrons?**
2. **Explain why the charge on the atom is neutral.**

Electronic structure

The shell closest to the nucleus has the lowest energy. This innermost shell can only hold two electrons. Electrons in an atom occupy the lowest available energy levels first.

Next, the second shell fills with electrons, which holds up to 8 electrons. When this is filled, electrons go into the third shell. The fourth shell starts to fill once there are 8 electrons in the third shell.

Fluorine has 9 electrons. The first two go into the first energy level. As the first shell is now full, the next 7 go into the second shell. This is written using numbers as the **electronic structure** 2,7. Information about the electrons can be written as numbers or drawn in a diagram showing the energy levels.

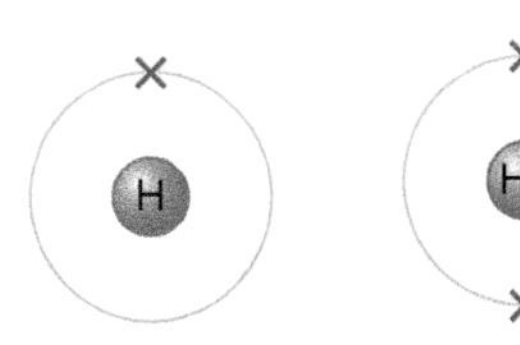

Figure 1.2.12 The structure of the first energy level. In this energy level diagram the electrons are shown by crosses

3. **Figure 1.2.13 shows the electronic structure of phosphorus (P). How many electrons does an atom of phosphorus have?**
4. **Use numbers to write the electronic structure of phosphorus.**

Developing patterns

The table shows the number of electrons in each shell for the first 20 elements in the Periodic Table. The number of electrons goes up by one each time and corresponds to a different element.

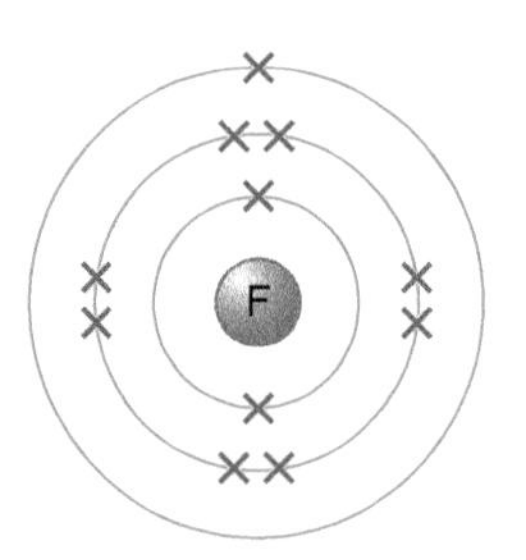

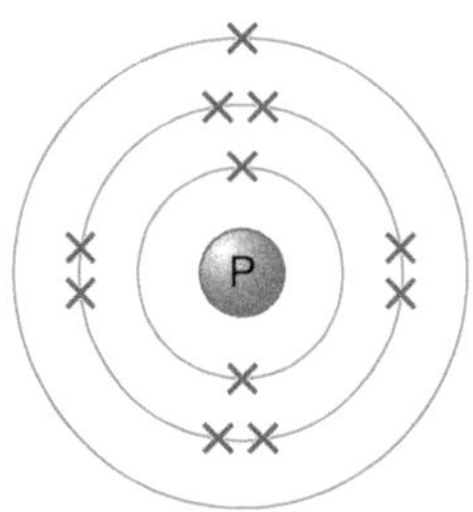

Figure 1.2.13 Electrons in the second and third energy levels. The energy increases as the distance from the nucleus increases

Element	Symbol	Number of electrons				
		Shell				Total
		1	2	3	4	
hydrogen	H	1	–	–	–	1
helium	He	2	–	–	–	2
lithium	Li	2	1	–	–	3
beryllium	Be	2	2	–	–	4
boron	B	2	3	–	–	5
carbon	C	2	4	–	–	6
nitrogen	N	2	5	–	–	7
oxygen	O	2	6	–	–	8
fluorine	F	2	7	–	–	9
neon	Ne	2	8	–	–	10
sodium	Na	2	8	1	–	11
magnesium	Mg	2	8	2	–	12
aluminium	Al	2	8	3	–	13
silicon	Si	2	8	4	–	14
phosphorus	P	2	8	5	–	15
sulfur	S	2	8	6	–	16
chlorine	Cl	2	8	7	–	17
argon	Ar	2	8	8	–	18
potassium	K	2	8	8	1	19
calcium	Ca	2	8	8	2	20

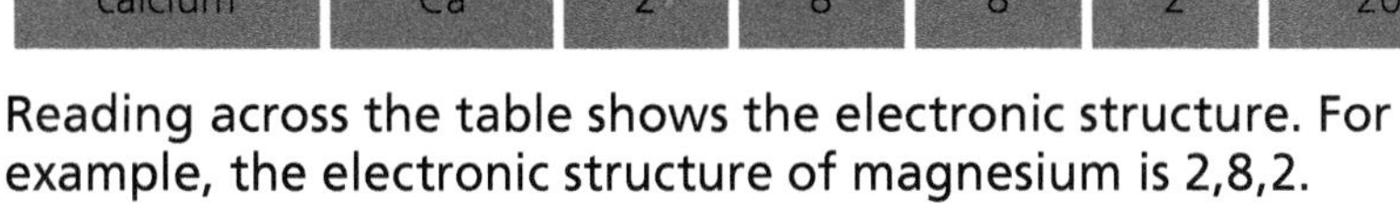

Reading across the table shows the electronic structure. For example, the electronic structure of magnesium is 2,8,2.

DID YOU KNOW?

It is the electrons of an atom that take part in chemical reactions. The different number of electrons in each shell makes a difference to the reactions that can take place.

5 **Use numbers to write the electronic structure of argon.**

6 **Draw the electron shell diagrams for magnesium and argon, using crosses for electrons.**

Check your progress

You should be able to:

☐ explain that early models of the atom did not have shells with electrons. ☐ Describe different models of the atom.	→	☐ explain that early models of atoms developed as new evidence became available.	→	☐ describe what the evidence was that led to the atomic model changing over time.
☐ recall the size and order of magnitude of atoms and small molecules.	→	☐ recognise and interpret expressions in standard form.	→	☐ estimate the size of atoms based on diagrams.
☐ use SI units and the prefix nano.	→	☐ convert decimals to standard form and vice versa.	→	☐ make estimates without calculators so that the answer in standard form seems reasonable.
☐ describe the structure of atoms.	→	☐ recall that the radius of a nucleus is less than 1/10000 that of the atom (about 1×10^{-14} m).	→	☐ explain why atoms are neutral.
☐ draw a diagram of a small nucleus containing protons and neutrons with orbiting electrons at a distance.	→	☐ recall the relative charges and masses of protons, neutrons and electrons.	→	☐ represent atoms symbolically showing their atomic information.
☐ know that elements can differ in mass number due to having different numbers of neutrons.	→	☐ calculate the mass number from the numbers of protons and neutrons in the nucleus.	→	☐ complete data tables showing the atomic numbers, mass numbers and numbers of sub-atomic particles from symbols.
☐ recall that in each atom the electrons are arranged at different distances from the nucleus.	→	☐ understand that there is an amount of energy associated with each level or shell.	→	☐ explain how electrons occupy shells in an order.

Worked example

1 a What is the atomic number of an element?

The number of protons.

This is correct, but a fuller answer would include that the protons are in the nucleus.

b What is the mass number of an element?

Protons plus electrons.

This is incorrect. Electrons have very little mass. The mass number is the number of protons plus the number of neutrons, all in the nucleus.

2 a What is an isotope?

Different numbers of neutrons.

This answer could be clearer. Isotopes are atoms of the same element with different numbers of neutrons, but the same number of protons.

b Which of these atoms are isotopes of the same element?

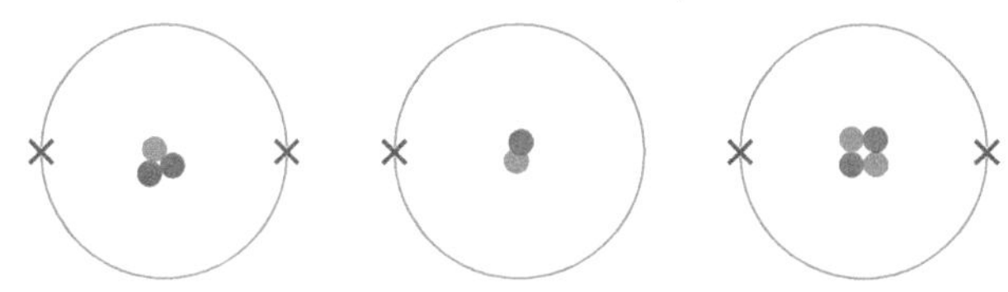

c Explain your reasoning.

The first and third atoms. Same number of protons, different neutrons.

This is correct. The red dots must be the protons, as they balance the number of electrons. The green dots are therefore neutrons, and the first and third diagrams have different numbers of neutrons in the nucleus, but the same number of protons.

3 a After the discovery of the electron in the early 20th century, the 'plum pudding' model was used to explain the structure of the atom.

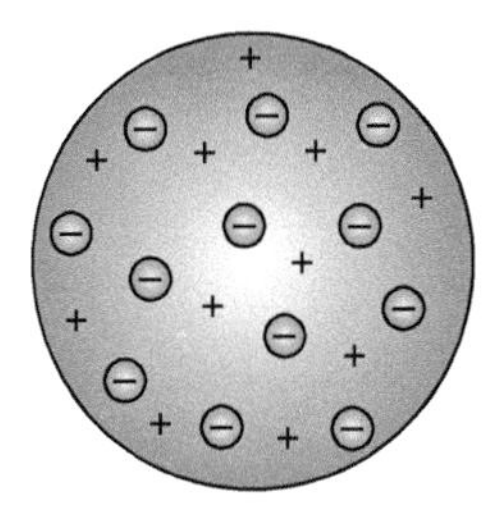

What did scientists think that the 'pudding' part of the atom was?

The positive bit.

It would be more scientific to say that it was a uniform positively charged mass.

b What causes scientists to change a scientific model?

They find new data.

Through experimental enquiry and/or calculations, they find new data, which the previous model does not explain, so they have to adapt the model, or create a new one that does.

End of chapter questions

Getting started

1. Describe the structure of an atom using a diagram. 1 Mark
2. An atom has 3 protons, 4 neutrons and 3 electrons. What is its atomic mass? 1 Mark
   - a 3
   - b 6
   - c 7
   - d 10
3. What is the relative charge on: 3 Marks
   - a an electron
   - b a neutron
   - c a proton
4. Determine the electron arrangement in an atom with 10 electrons. 1 Mark
5. Write out 1 000 000 in standard form. 1 Mark
6. What is the typical size of the radius of the atom in metres? 1 Mark
7. Describe the Dalton model of the atom. 1 Mark

Going further

8. Define the atomic number of an element. 1 Mark
9. Sodium can be represented by the notation $^{23}_{11}Na$. 1 Mark

   What are the numbers 23 and 11 and what do they stand for?
10. What is an isotope? 1 Mark
11. Write out 0.000 000 000 1 in standard form. 1 Mark
12. What discovery led to the Dalton model of the atom being changed? 1 Mark

More challenging

13. Identify the number of electrons in an atom of $^{31}_{15}P$. 1 Mark
14. There are two atoms $^{28}_{14}Si$ and $^{30}_{14}Si$.

    Work out how many neutrons each atom contains. 1 Mark
15. Calculate: 2 Marks
    - a $4 \times 10^3 \times 8 \times 10^6$
    - b $5 \times 10^8 \times 3 \times 10^{-2}$

Most demanding

16. **Explain how the evidence from Geiger and Marsden's scattering experiment led to the development of the nuclear model of the atom.** 4 Marks
17. **Suggest how a scientific theory becomes accepted.** 3 Marks
18. **Deduce the number of sub-atomic particles that make up the atom $^{31}_{15}P$.** 2 Marks

Total: 28 Marks

CELLS IN ANIMALS AND PLANTS

IDEAS YOU HAVE MET BEFORE:

ALL LIVING ORGANISMS ARE MADE OF CELLS.

- Cells are the building blocks of life.
- Cells contain specialised structures.
- Organisms such as bacteria are unicellular.
- Most plants and animals are multicellular.

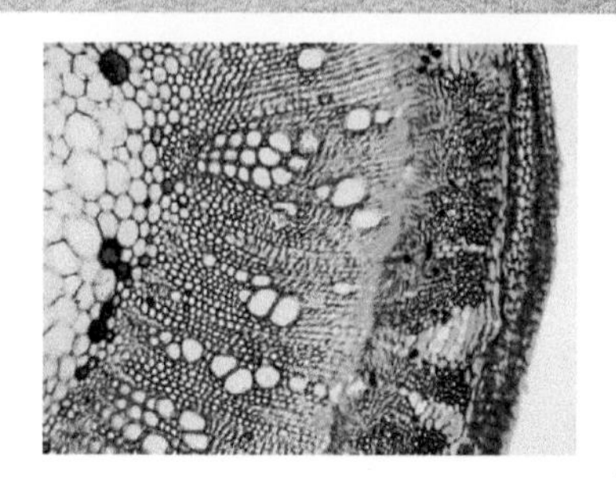

MOLECULES MOVE BY DIFFUSION.

- Diffusion is the net movement of molecules from a higher concentration to a lower concentration until they are equally distributed.
- Different factors can affect the rate of diffusion.
- The steepness of a concentration gradient affects the rate of diffusion.

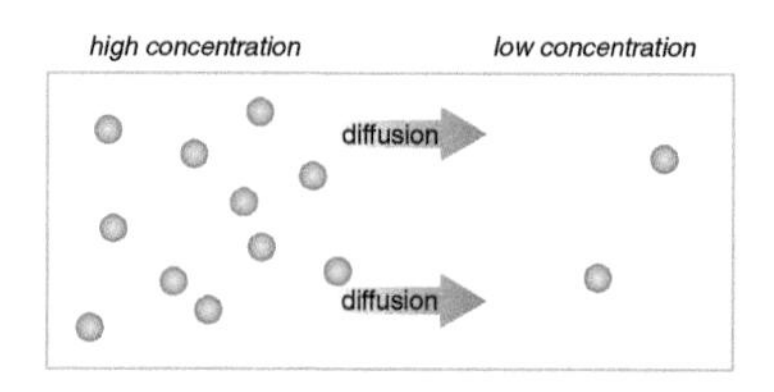

IN MULTICELLULAR ORGANISMS, CELLS HAVE TO DIVIDE.

- Cells divide as we're growing, and to replace cells that are injured, worn out or have died.
- This type of cell division is called mitosis.
- When a cell divides by mitosis, two daughter cells are produced, each with an identical number of chromosomes and identical DNA.

IN MULTICELLULAR ORGANISMS CELLS BECOME SPECIALISED.

- Specialised cells have a particular job to do.
- Specialised cells are organised into tissues, tissues into organs, and organs into body systems.

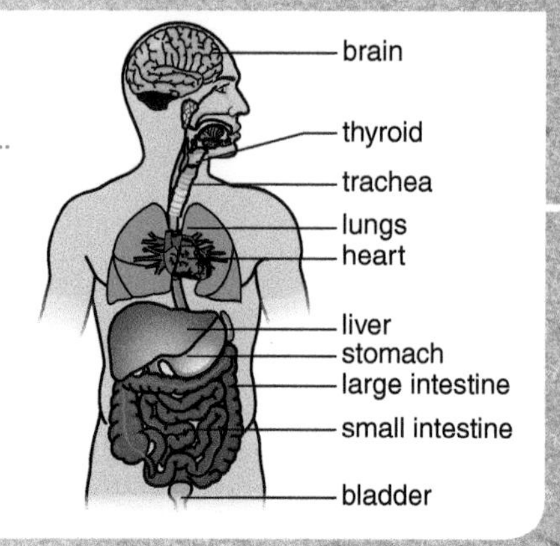

1.3

IN THIS CHAPTER YOU WILL FIND OUT ABOUT:

HOW HAVE SCIENTISTS DEVELOPED THEIR UNDERSTANDING OF CELL STRUCTURE AND FUNCTION?

- The structures inside cells and the jobs they do within the cell.
- The study of cells using different types of microscopes.
- The cells of bacteria, and how they are different to the cells of plants and animals.

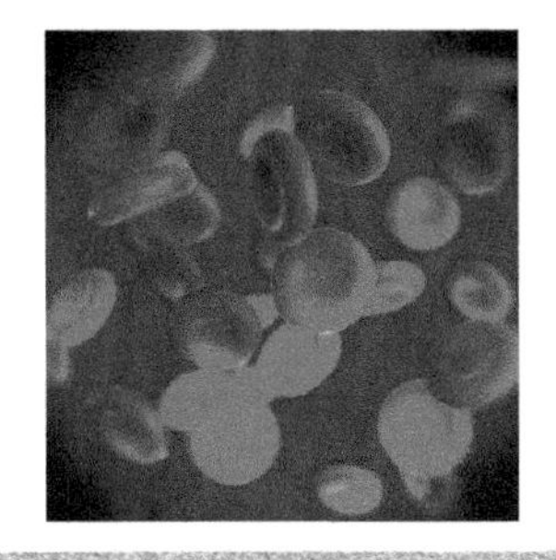

DO ALL MATERIALS MOVE BY DIFFUSION?

- In living tissues, water moves by osmosis from a high concentration of water to a lower concentration of water, across partially permeable membranes.
- The movement of water can affect the turgidity of living cells.
- Some substances that living cells need can be moved against a concentration gradient, by active transport.

HOW DOES CELL DIVISION PRODUCE CELLS FOR REPRODUCTION?

- In asexual reproduction, only one parent is involved. No sex cells are produced and cells divide by mitosis.
- During sexual reproduction, a cell divides by meiosis to produce four gametes, each with half the number of chromosomes.
- Meiosis ensures that the chromosome number stays constant – 46, or 23 pairs in humans – in each generation.
- Meiosis also produces gametes that are genetically unique, leading to variation between individuals.

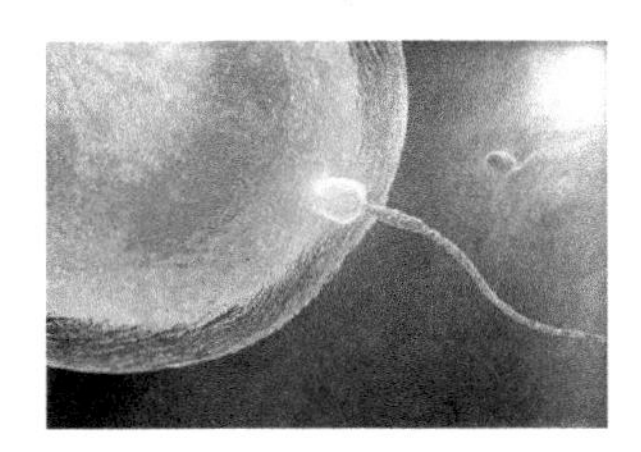

HOW DO WE DEVELOP INTO A COMPLEX ORGANISM FROM JUST A FERTILISED EGG CELL?

- How and when the body's cells divide, and that the newly formed cells are identical to the existing cells.
- How cells differentiate to become specialised, and how specialised cells are organised.
- Cells that are unspecialised in the embryo, and cells that remain unspecialised in adults, are called stem cells.

1.3a Electron microscopy

Learning objectives:

- identify the differences in the magnification and resolving power of light and electron microscopes
- explain how electron microscopy has increased our understanding of sub-cellular structures
- carry out calculations involving magnification, real size and image size
- use estimations and make order of magnitude calculations
- use prefixes centi, milli, micro and nano and interconvert units.

KEY WORDS

magnification
micrograph
resolving power
scanning electron microscope (SEM)
transmission electron microscope (TEM)

An electron microscope uses an electron beam to create a highly magnified image of a specimen. Because an electron beam has a much shorter wavelength than visible light, the electron microscope can produce images of much higher resolution than a light microscope.

The light microscope

Microscopes magnify the specimen you are looking at, making it look bigger than it is.

In light microscopes the magnified image is produced by light passing through lenses. Very high magnifications are not possible because the power of the lenses and the amount of light that can enter the microscope are limited. The highest magnification is around ×1500.

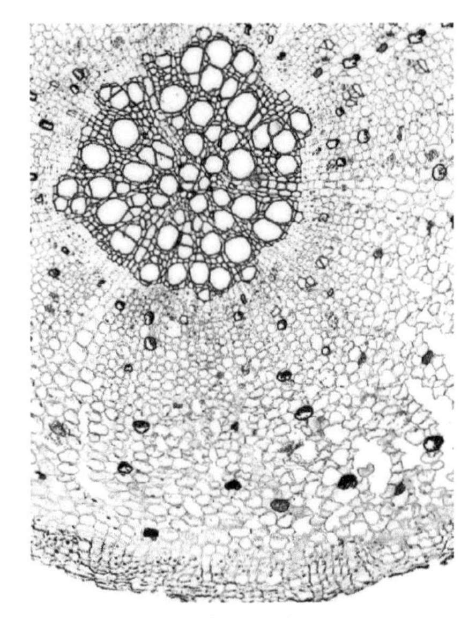

Figure 1.3.1 An image from a light microscope is called a light micrograph. In this light micrograph of a cross-section of a plant root, the magnification is ×100 and coloured dye has been added to show up the details

Using higher **magnification** does not always mean that you can see greater detail. To see more detail in an image, you need higher **resolving power**, or resolution. At a higher resolution two dots close together can be seen as separate points, when with a lower resolution they would look like a single blob.

The maximum resolving power of a light microscope is around 0.2 mm, or 200 nm. This means that, in a light **micrograph**, you cannot separately pick out two points closer than 200 nm apart.

1. **What is the maximum resolving power of the light microscope?**
2. **What is the maximum magnification possible with a light microscope?**

The electron microscope

An **electron microscope** has a much higher resolving power than a light microscope. This means that it can be used to study cells in much finer detail.

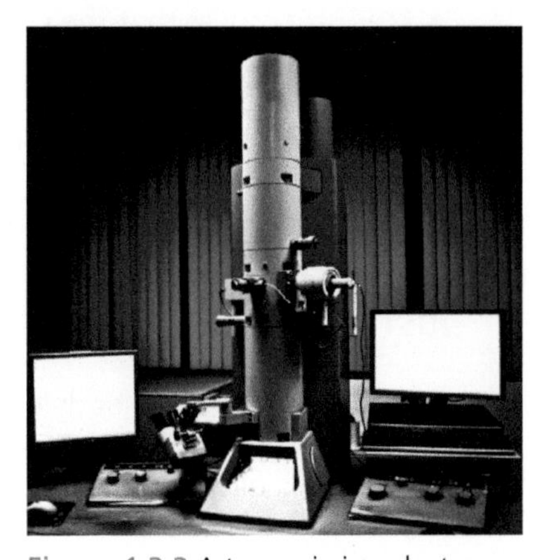

Figure 1.3.2 A transmission electron microscope. The electrons are displayed as an image on a fluorescent screen

A **transmission electron microscope** (TEM) shoots a beam of electrons through an extremely thin section (slice) of cells. The image is formed by the electrons that pass through the specimen. The highest possible magnification from a TEM is around ×1 000 000. The maximum resolving power is less than 1 nm.

The **scanning electron microscope** (SEM) bounces electrons off the surface of a specimen coated in an ultra-thin layer of a heavy metal – usually gold. A narrow electron beam scans the specimen. Images are formed by the scattered electrons.

SEMs can reveal the surface shape of structures such as small organisms and cells. The resolving power is lower than for a TEM, and magnifications are also often lower.

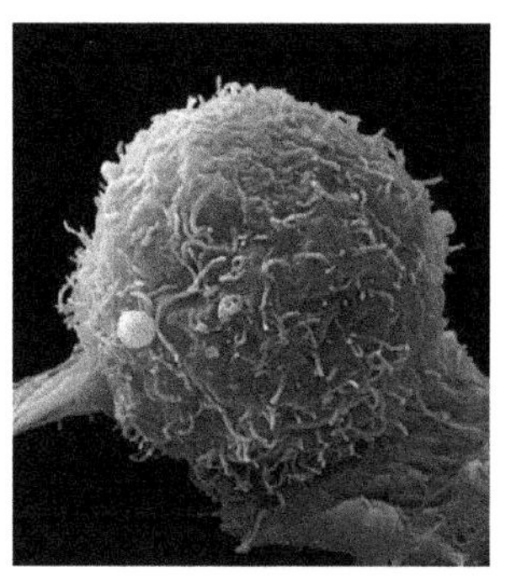

Figure 1.3.3 A scanning electron micrograph of a cancer cell (x4500). Electrons do not have a colour spectrum like the visible light used in a light microscope, so they can only be 'viewed' in black and white. Here, false colours have been added

3. **Estimate how many times greater a TEM's maximum resolving power is compared to that of a light microscope. What order of magnitude is this?**
4. **What types of samples would a TEM and an SEM be used to view?**

HIGHER TIER ONLY

Magnification of images

The magnification of an image is the number of times bigger it is than the object being viewed. Micrograph images must show the magnification to be meaningful.

$$\text{magnification of image} = \frac{\text{size of image}}{\text{size of real object}}$$

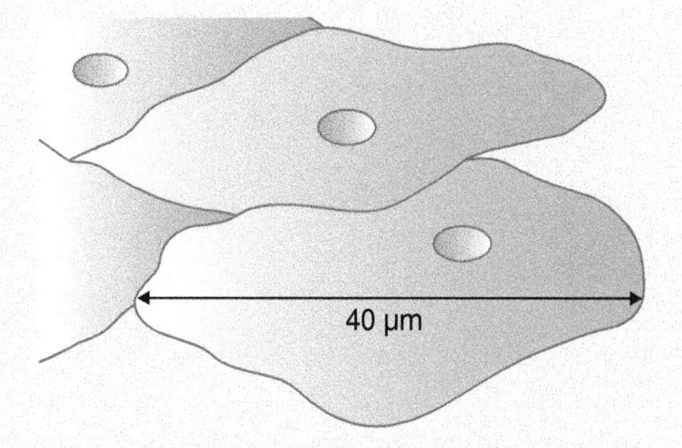

Figure 1.3.4 A drawing of a micrograph of a cell

The cell drawn in Figure 1.3.4 is 50 mm across on the page. In real life it measures 40 μm. To calculate the magnification, the values must all be in the same units. First convert 50 mm into μm (or convert 40 μm to mm).

50 mm = 50 000 μm

The cell measures 40 μm. Therefore:

$$\text{magnification of image} = \frac{50000\ \mu\text{m}}{40\ \mu\text{m}} = \times 1250$$

5. **In an SEM image of a leaf surface the length of a pore (stoma) was measured as 19.5 mm. If the image magnification was ×1500, calculate the actual length of the pore, in nanometres (nm). Write your answer in standard form (see topic 1.2c).**
6. **A TEM image of a plant cell in a book is 6.0 cm long. The plant cell measures 1.2×10^2 mm long. Calculate the magnification.**
7. **How do you think electron microscopy could improve our understanding of cells?**

COMMON MISCONCEPTIONS

Do not confuse magnification, which is how much bigger we can make something appear, with resolving power, which is the level of detail we can see.
Think about a digital photo. You can make it as big as you like, but at a certain point you will not be able to see any more detail.

1.3b Cell structures

Learning objectives:

- describe the structure of eukaryotic and prokaryotic cells and explain how the sub-cellular structures are related to their functions
- carry out calculations involving magnification, real size and image size including numbers written in standard form
- use estimations and make order of magnitude calculations
- use prefixes centi, milli, micro and nano and interconvert units.

KEY WORDS

cell membrane, cellulose, cell wall, chlorophyll, chloroplast, chromosome, cytoplasm, DNA, eukaryotic, mitochondrion, nucleus, plasmids, prokaryotic, ribosome, vacuole

Unlike light microscopy, electron microscopy reveals fine detail in tiny sub-cellular structures, because of its high resolving power. This has helped us understand how cell structures function and interact.

Eukaryotic cells

Almost all organisms are made up of cells. Plant and animal cells have a basic structure. This type of cell, containing a true nucleus in the cytoplasm, is called a **eukaryotic** cell.

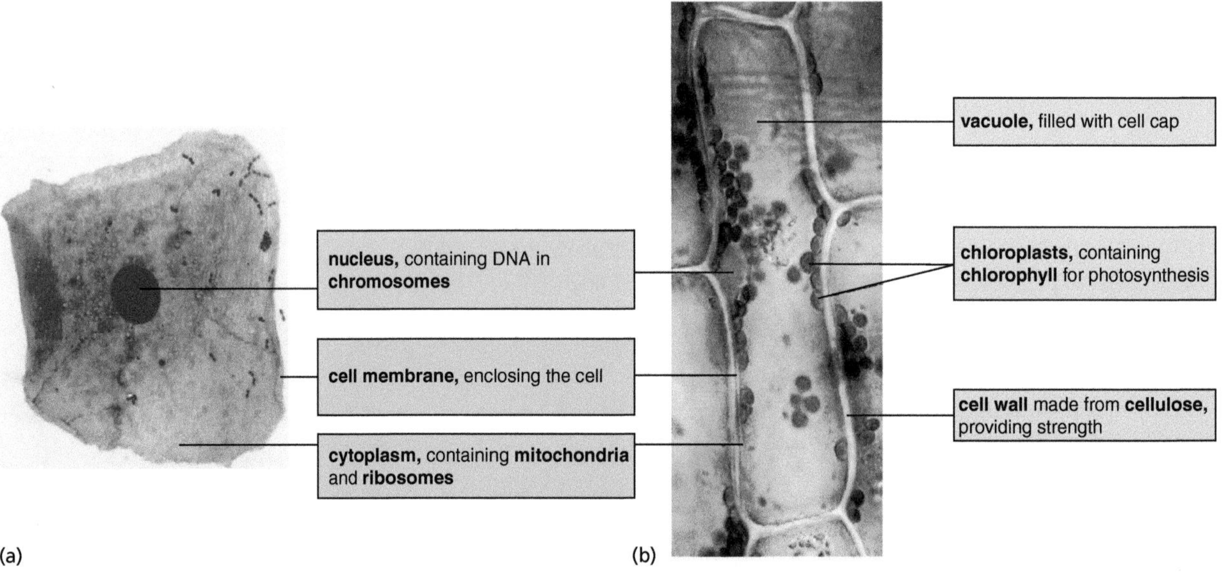

Figure 1.3.5 Light micrographs of (a) a simple animal cell and (b) a plant leaf cell, with colour added to the images

1. **List the sub-cellular structures found in both plant and animal cells.**
2. **Which sub-cellular structures are found only in plant cells?**
3. **What is the function of:**
   - **the nucleus?**
   - **the cell membrane?**
4. **What structure gives strength to a plant cell?**

Prokaryotic cells

Bacteria are among the simplest of living things. Along with bacteria-like organisms called Archaeans, they belong to a group called the Prokaryota. These are single-cell organisms with a **prokaryotic** cell structure.

Prokaryotic cells are much smaller than eukaryotic cells, around 0.1–5.0 µm in diameter. Their **DNA** is not enclosed in a nucleus. It is found as a single molecule in a loop. They may also have one or more small additional rings of DNA called **plasmids**.

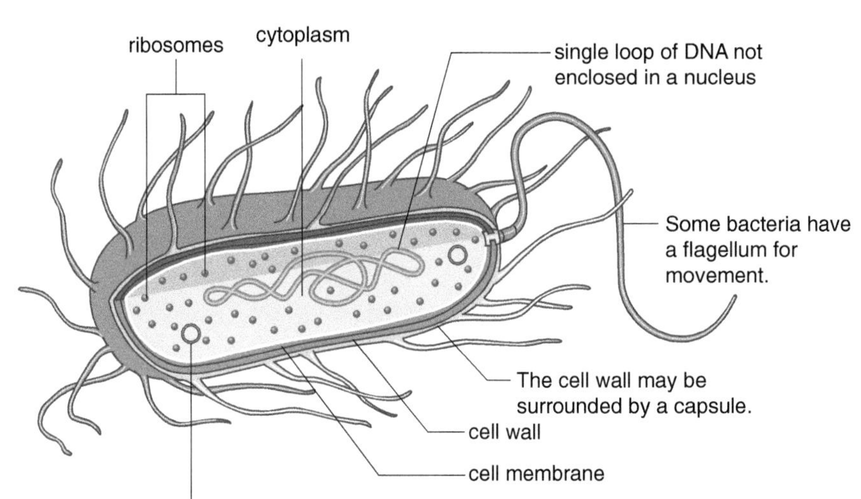

Figure 1.3.6 A diagram to show the structure of a prokaryotic cell

Cell ultrastructure

We can just about see some sub-cellular structures such as mitochondria and chloroplasts with the light microscope, but the electron microscope reveals their internal structure in more detail. Other structures, such as ribosomes, can only be seen using electron microscopes.

7 **The animal cell in Figure 1.3.5 is about 60 mm across, while the plant cell is about 30 mm wide.**

a **Calculate the magnification of each micrograph in Figure 1.3.5.**

b **How many times larger than a typical prokaryotic cell is the animal cell in Figure 1.3.5? What order of magnitude is this?**

HIGHER TIER ONLY

8 **Use the magnifications given in Figure 1.3.7 to estimate the actual lengths of the mitochondrion and the chloroplast shown. Give your answers in micrometres (mm), and also in nanometres (nm) using standard form (see topic 1.2c).**

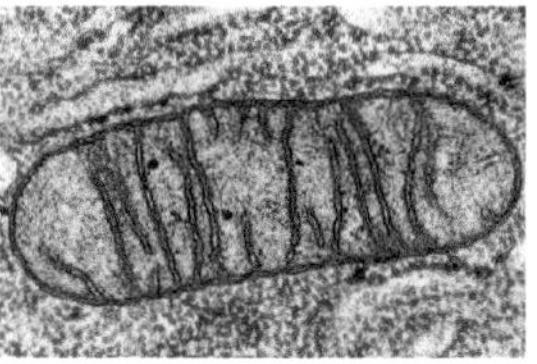

(a)
Mitochondria are where aerobic respiration takes place in the cell. A mitochondrion has a double membrane. The internal membrane is folded.

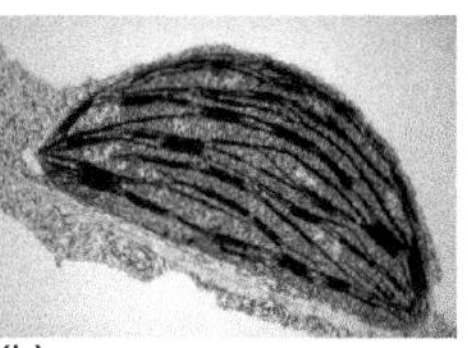

(b)
Chloroplasts are the structures in the plant cell where photosynthesis takes place. Like mitochondria, they also have a complex internal membrane structure.

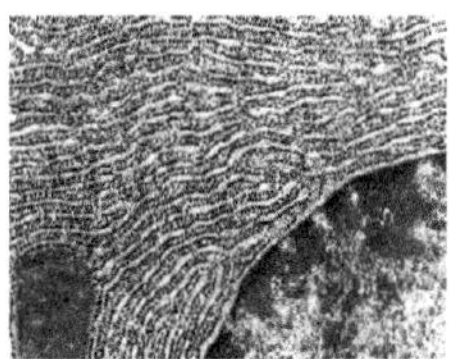

(c)
Ribosomes are tiny structures where protein synthesis takes place. You can see them as dots in the micrograph. They can either lie free in the cytoplasm or may be attached to an internal network of channels within the cytoplasm.

Figure 1.3.7 Viewing (a) a mitochondrion, (b) a chloroplast and (c) ribosomes by transmission electron microscopy, with false colour added to show up the details

5 **List the differences between prokaryotic and eukaryotic cells.**

6 **Where is DNA found in prokaryotic cells?**

DID YOU KNOW?

Scientists sometimes investigate the ratio of the area of the cytoplasm to that of the nucleus in micrographs. A high ratio of cytoplasmic : nuclear volume can indicate that the cell is about to divide. A low ratio can be characteristic of a cancer cell.

REQUIRED PRACTICAL

1.3c Observing cells under a light microscope

Learning objectives:

- use appropriate apparatus to record length and area
- use a microscope to make observations of biological specimens and produce labelled scientific drawings
- use estimations to judge the relative size or area of sub-cellular structures
- carry out calculations involving magnification, real size and image size.

KEY WORDS

field of view
scale

These pages are designed to help you think about aspects of the investigation rather than to guide you through it step by step.

Many scientists use electron microscopes to observe fine detail in cells. But much of the microscope work carried out – including in hospital and forensic science labs – is done with the light microscope.

Preparing cells for microscopy

Live cells can be mounted in a drop of water or saline on a microscope slide.

Most cells are colourless. We must stain them to add colour and contrast. In the school laboratory, you may have used methylene blue to stain animal cells or iodine solution to stain plant cells.

Figure 1.3.8 A glass coverslip is carefully lowered onto the cells or tissue, taking care to avoid trapping air bubbles. The coverslip keeps the specimen flat, and retains the liquid under it

1. **Write an equipment list for looking at cheek cells with a microscope. State why each piece of equipment is used.**
2. **Suggest why it's better to mount the cells in saline than in water.**
3. **The micrograph of the frog's blood in Figure 1.3.9 shows red blood cells and two types of white blood cell.**
   - a **Label the different types of cell and the cell structures that are visible. Hint: use a photocopy or printout of the page.**
   - b **How is the structure of the frog's red blood cells different from that of human red blood cells?**

SAFETY

A risk assessment is required for this practical work.

Some local authorities do not allow students to make cheek cell slides. Where allowed, suitable disinfectant must be provided for the used cotton buds and slides/cover slips.

High and low power

The slide is first viewed with low power. This is because:

- the field of view with high power is small. It would be difficult to locate cells if starting with the high power objective.
- it enables you to see the layout of cells within the tissue.
- it's useful when estimating the numbers of different types of cell on the slide or in a tissue (though here, high power may be needed).

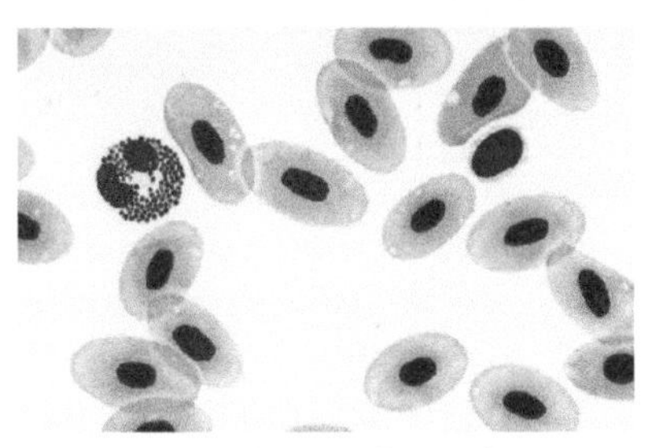

Figure 1.3.9 Blood cells in a frog, stained with false colour

A low power digital image (or drawing) can be used to show the arrangement of cells in a tissue. This includes regions of the tissue but not individual cells.

If required, the cells or tissue can then be viewed with high power to produce a detailed image of a part of the slide.

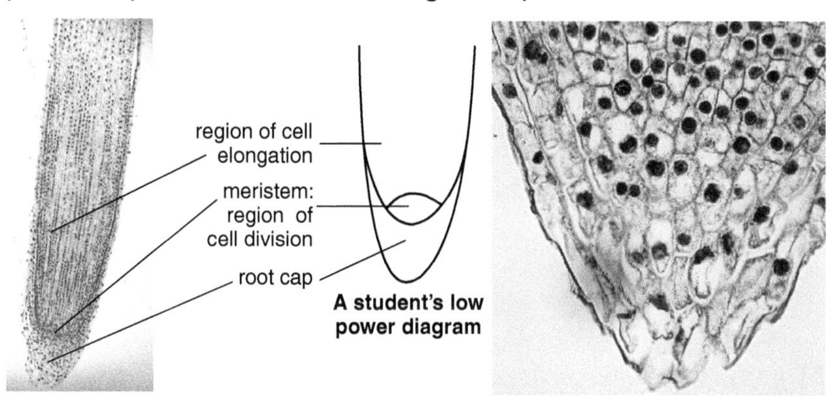

Figure 1.3.10 Low and high power micrographs, with false-colour staining, and a student diagram, of a plant root

DID YOU KNOW?

These slides are temporary. If a permanent slide of cells is required, the cells or tissue must be dehydrated, embedded in wax and cut into thin slices called sections before staining.

4. **Why is a slide viewed with low power first?**
5. **On a printout of the low power plan of the root, label the root cap, meristem (the region of cell division) and the region of cell elongation.**

Recording images

As you have seen in topic 1.3a, a microscope drawing or micrograph is of little value if it gives no indication of size.

It's usual to add a magnification to the image. We can then envisage, or work out, the true size of a specimen.

Alternatively, we can use a **scale bar**. Any scale bar must be:

- drawn for an appropriate dimension
- a sensible size in relation to the image.

Look at Figure 1.3.11. For the top micrograph, the magnification is *x1000*, which means that a 10 *millimetre* scale bar can be drawn to represent 10 *micrometres*.

You will find out how scientists measure, or sometimes estimate, the size of cells in section 1.3d.

6. **Complete the scale bar for the bottom micrograph in Figure 1.3.11.**
7. **Calculate the length of the *Paramecium* in the bottom micrograph.**

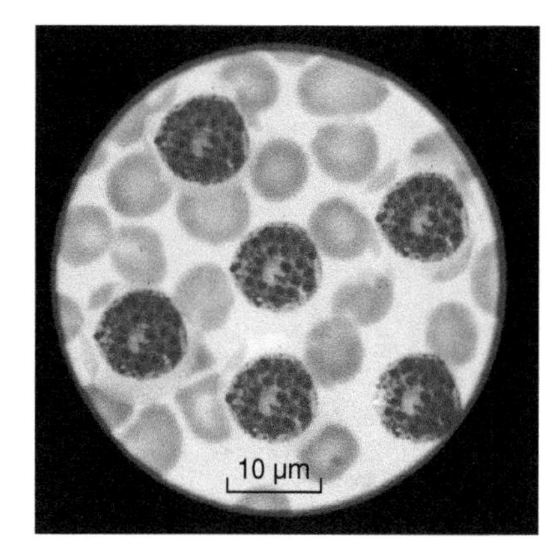

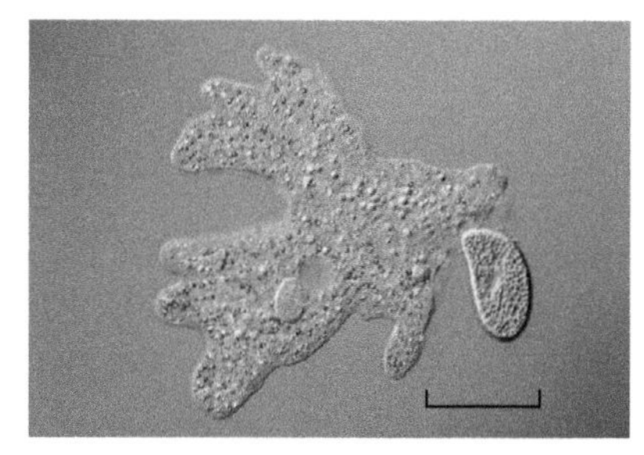

Figure 1.3.11 Light microscopy is also used to examine small organisms such as protists. The top image shows six blood cells infected with the malarial parasite. The bottom image shows two protists found in pond water – *Amoeba* on the left; *Paramecium* on the right (at x200 magnification).

MATHS SKILLS

1.3d Size and number

KEY WORDS

calibrate
graticule
standard form

Learning objectives:

- use appropriate apparatus to record length and area
- recognise and use expressions in decimal and standard form
- use ratios, fractions and percentages
- make estimates for simple calculations.

The sizes of structures are important in biology, from whole organisms to molecules, because their functions depend on their relative sizes. Using microscope images, scientists can estimate the sizes of tiny structures, or measure them accurately.

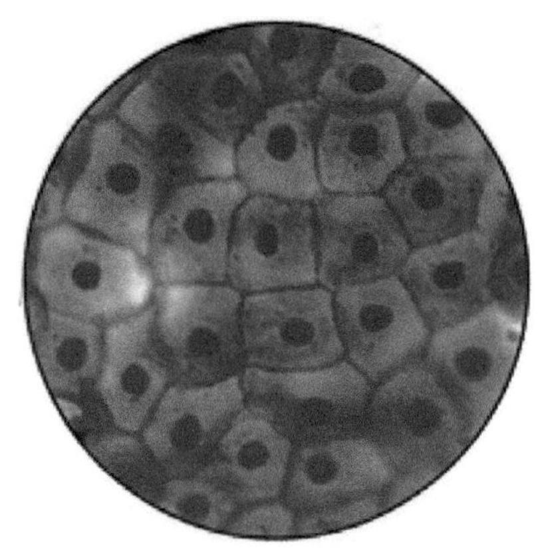

Figure 1.3.12 In this image, *approximately* five cells fit across the field of view. We round numbers up or down to make calculations straightforward

Estimating cell size

Accurate measurements are often essential. But estimating cell size or number is sometimes sufficient and may be quicker.

To estimate cell size, we can count the number of cells that fit across a microscope's field of view.

$$\text{Size of one cell} = \frac{\text{diameter of field of view}}{\text{number of cells that cross this diameter}}$$

If the field of view of this microscope, at this magnification, is 0.3 mm, or 300 μm, we can do a quick calculation without a calculator.

Each cell must be roughly (300 ÷ 5) μm, or 60 μm across. This is an approximation, but could be important.

1. **Suggest how to estimate the field of view of a microscope.**
2. **State one advantage of estimating cell size over exact measurement.**

DID YOU KNOW?

Scientists *estimate* cell or organism numbers when it is impossible or unnecessary to count them all.

Measuring cell size

To make accurate measurements of cell size a scientist **calibrates** their microscope. A **graticule** – piece of glass or plastic onto which a scale has been drawn – is placed into the eyepiece of the microscope.

A stage micrometer is placed on the microscope stage. This is simply a microscope slide onto which an accurate scale has been etched.

In Figure 1.3.13, 36 divisions on the eyepiece graticule are equivalent to 100 μm on the stage micrometer: 1 division is equivalent to $\frac{1}{36} \times 100$ μm = 2.8 μm

REMEMBER!

The decimal point remains fixed. It is the digits that move as a number is multiplied or divided by powers of 10. So, as a number gets larger, the digits move to the left (and vice versa).

The cell highlighted in the right-hand diagram is 20 eyepiece divisions across: the width of the cell = (20 × 2.8) µm = 56 µm

❸ **What would be the diameter of a cell that was 65 divisions on this graticule?**

❹ **How many graticule divisions would a cell that was 35 µm across take up?**

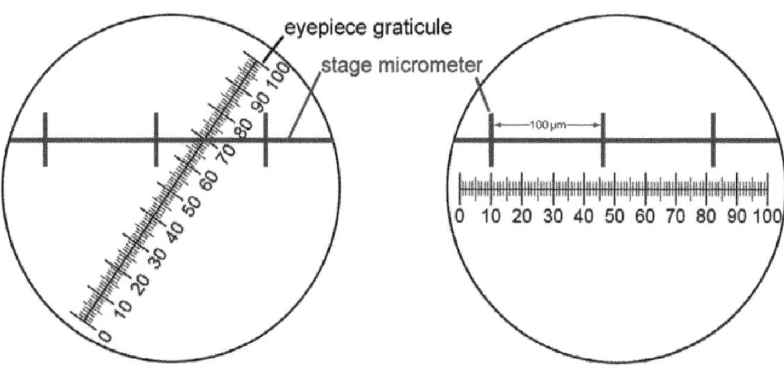

The graticule is enclosed within the eyepiece.
A stage micrometer is placed on the stage of the microscope.

The divisions on the eyepiece graticule and stage micrometer are lined up.

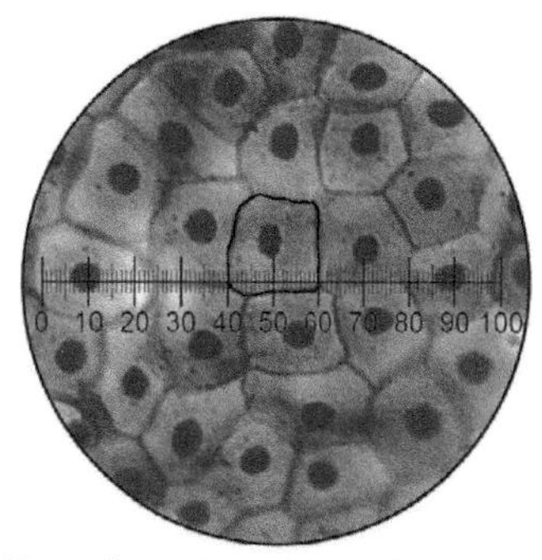

The calibrated eyepiece graticule can be used to make measurements of any cells or other structures viewed with that microscope.

Figure 1.3.13 Calibrating, then using an eyepiece graticule

Working with numbers in standard form

We can convert measurements in units of mm or µm to sizes in metres expressed in standard form. This makes it easier to compare sizes.

The table gives the sizes of some types of white blood cell in metres, written in standard form. A large capillary (a type of blood vessel) is about 40 µm across. Estimate how many small lymphocytes could fit side by side across a large capillary.

1 µm = 0.000 001 m = 1×10^{-6} m

40 µm = $40 \times (1 \times 10^{-6})$ m = 4.0×10^{-5} m

So the capillary is 4.0×10^{-5} m wide.

Dividing the capillary width by the width of the lymphocyte:

number of lymphocytes = 4.0×10^{-5} m / 7.5×10^{-6} m

= $(4.4 \div 7.5) \times 10^{-5} - (-6)$

= 0.53×101

= 5.3

So about 5 lymphocytes could fit across the capillary

Blood cell type	Width of an average cell (m)
Lymphocyte (small)	7.5×10^{-6}
Macrophage	5.0×10^{-5}
Megakaryocyte	1.5×10^{-4}
Neutrophil	1.2×10^{-5}

REMEMBER!

When you are working with numbers in standard form:
- multiply – multiply numbers and add powers
- divide – divide numbers and subtract powers.

❺ **Look at the table of cell sizes. Arrange the cell types in descending order of size. (You do not need to memorise the names of the cells shown here.)**

❻ **How many times larger is a megakaryocyte than a lymphocyte?**

1.3e Diffusion into and out of cells

Learning objectives:

- explain how substances are transported into and out of cells by diffusion
- identify the factors that affect rate of diffusion
- explain what the term 'partially permeable membrane' means.

KEY WORDS

concentration gradient
diffusion
partially permeable membrane

Living cells constantly need to take in important substances and get rid of other substances that may be harmful. Some of these substances move across the cell membrane by diffusion.

Diffusion in living systems

Diffusion is a spreading out and mixing process. It is sometimes called **passive transport**. This is because it happens due to the **random** motion of particles. No extra energy is required.

All that is needed for diffusion to happen is a **concentration gradient**. In diffusion, there is a net movement of particles from a region where they are in higher concentration to a region where their concentration is lower, due to the random movement of particles. Diffusion continues until the concentration is the same in both regions.

Cells are made largely of water containing many dissolved substances. These substances and water enter and leave the cells through cell membranes. Cell membranes allow some particles through, but block others. They are called **partially permeable membranes** (or selectively permeable membranes). Diffusion can occur across a partially permeable membrane, but it can also happen without any membrane separating the regions.

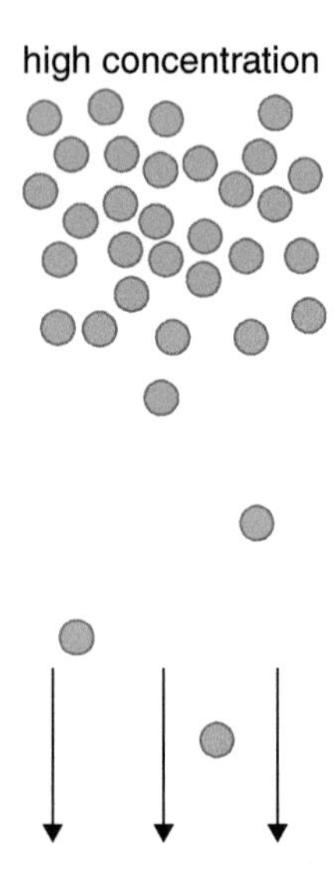

Figure 1.3.14 This diagram illustrates a concentration gradient, showing the net movement of particles from a higher concentration to a lower concentration

1. **Why is diffusion into and out of cells also called passive transport?**
2. **Why does diffusion stop when the concentration in both regions is the same?**

Factors affecting diffusion

Factors that affect the rate of diffusion across a membrane are:

- the **concentration gradient** – the difference in the concentrations of a substance on each side of the membrane
- the **temperature** – the higher the temperature, the higher the rate of diffusion, because the particles have more energy and their random movements are therefore greater

- the **surface area of the membrane** – the greater the membrane's surface area, the higher the rate of diffusion.

3 What effect do you think increasing the concentration gradient will have on the speed of diffusion?

4 Why do you think a larger surface area of membrane means a higher rate of diffusion?

Diffusion and cells

Cells need continual supplies of dissolved substances such as oxygen and glucose for cellular activities. Waste products such as carbon dioxide need to be removed. To enter or leave a cell by diffusion, dissolved substances have to be small enough to pass through the partially permeable cell membrane.

Cell membranes are a bit like football nets. Large footballs cannot pass through the netting, but small golf balls easily pass through. In the same way, large molecules, for example starch, cannot pass through the cell membrane. However, small glucose molecules can easily pass through cell membranes. Because cell membranes are very thin, some substances can easily diffuse through them.

5 Explain how glucose molecules pass in and out of cells. Use a diagram to help you.

6 Will diffusion ever stop completely? Explain your answer.

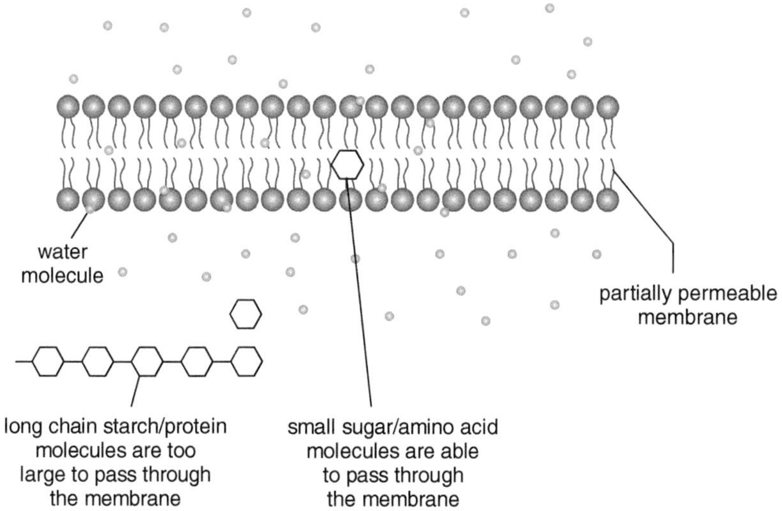

Figure 1.3.15 A cell membrane model. What is happening to the particles?

1.3f Osmosis

Learning objectives:

- describe how water moves by osmosis in living tissues
- identify factors that affect the rate of osmosis.

KEY WORDS

flaccid
osmosis
plasmolysis
solute
solvent
turgid

Water moves across cell membranes by a special kind of diffusion called osmosis. Cell membranes are partially permeable – they allow small molecules such as water through but not larger molecules.

The diffusion of water

A solution consists of a **solvent,** such as water, and a dissolved substance called a **solute,** such as sugar. A dilute solution has a high water concentration, and a low solute concentration. A more concentrated solution has a lower water concentration, and a higher solute concentration.

Osmosis is the diffusion of water molecules from a dilute solution to a more concentrated solution across a partially permeable membrane. Most solute particles are too large to pass through the tiny holes in a partially permeable membrane, but water molecules can pass through.

1 Cell membranes are partially permeable. How does water move in and out of living cells?

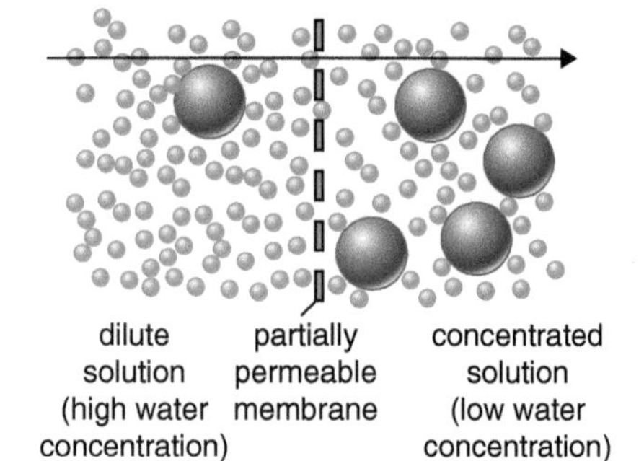

Figure 1.3.16 Osmosis is the net movement of water molecules from a dilute solution to a more concentrated one, across a partially permeable membrane

Osmosis and cells

Look at Figure 1.3.17, which shows model 'cells'. The partially permeable membrane bags represent cell membranes.

In (a) the cell has a concentrated solution. Water molecules enter by osmosis from the surrounding dilute solution.

In (b) the cell has a dilute solution. Water molecules move out to the surrounding concentrated solution.

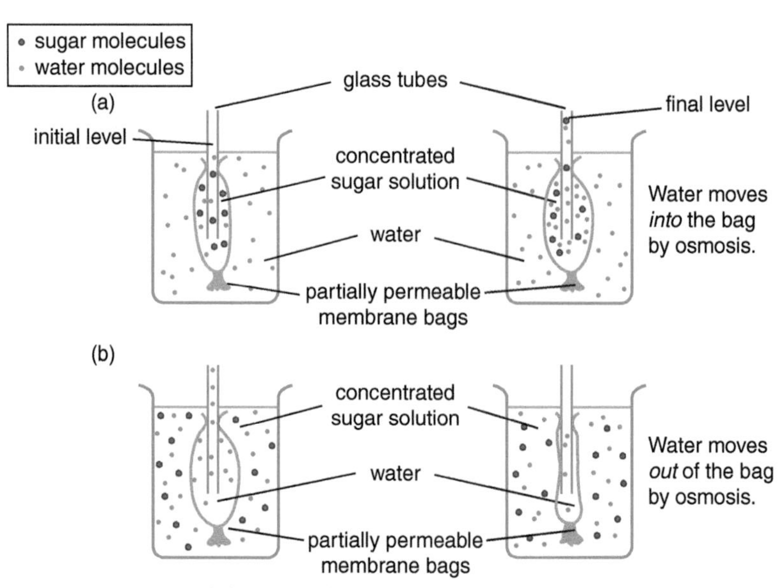

Figure 1.3.17 Model cells can be used to investigate osmosis

Living cells must balance their water content to work efficiently. Chemical reactions in cells use water. If the cytoplasm becomes concentrated, water enters by osmosis. If the cytoplasm becomes too dilute, water leaves the cell by osmosis.

Problems occur in animal cells when the external solution is more dilute than that inside the cell. Water enters; the cells swell and may burst.

When the external solution is more concentrated than that inside the cell, water moves out by osmosis. The cell shrinks and shrivels.

Plant cells have inelastic cell walls. Water enters the cell by osmosis and fills the vacuole. This pushes against the cell wall, making the cell swollen, or **turgid.**

If water moves out the cell by osmosis, the vacuole shrinks and the cell becomes floppy, or **flaccid.**

If too much water leaves the cell, the cytoplasm shrinks and moves away from the cell wall.

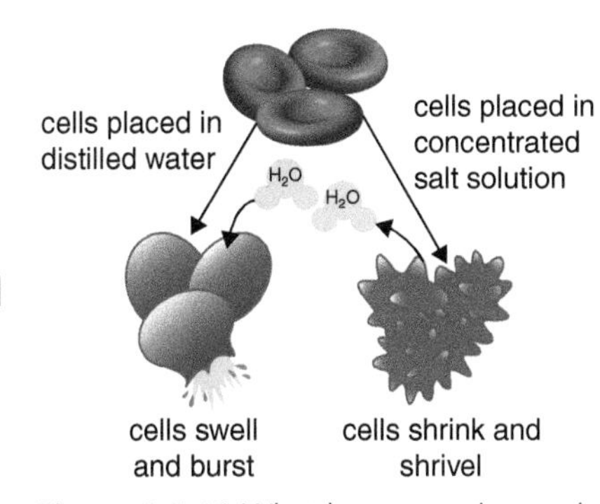

Figure 1.3.18 What happens when red blood cells are placed in a very dilute or a very concentrated solution.

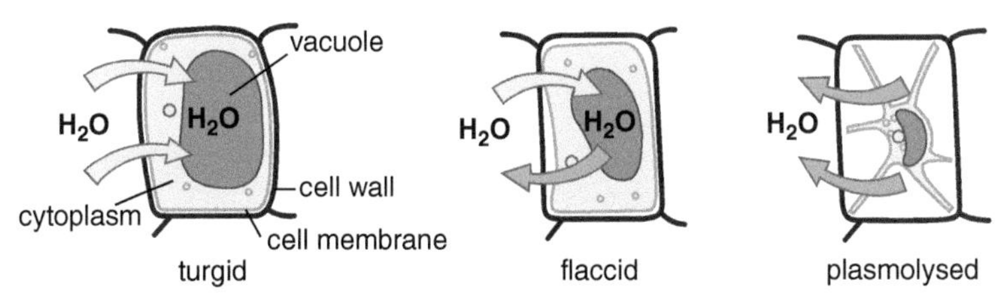

Figure 1.3.19 Movement of water in plant cells placed in a dilute solution, a more concentrated solution and a very concentrated solution

REMEMBER!

Osmosis is only the movement of water across a partially permeable membrane. No other molecules move by osmosis.

2. **How are osmosis and diffusion similar and different?**
3. **Describe how osmosis can affect plant cells.**
4. **Describe how osmosis affects animal cells.**

Explaining osmosis

Water molecules and sugar molecules in a solution move around randomly. When a sugar molecule hits the membrane, it bounces away. When a water molecule hits the membrane, it can pass through a hole to the other side.

In Figure 1.3.20, there are more water molecules on the left, so more water molecules can pass through the membrane to the right-hand side than can pass in the opposite direction. The water molecules move both ways, but the *net movement* is from left to right.

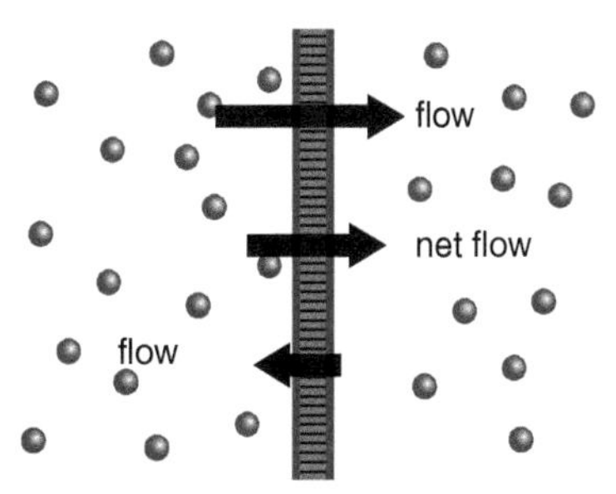

Figure 1.3.20 Water particles move both ways across the membrane. Note that for simplicity the sugar molecules are not shown in this diagram. The concentration of sugar is higher on the right of the membrane than on the left

5. **Explain why there are differences in the effects of water on plant and animal cells.**
6. **Explain osmosis.**

REQUIRED PRACTICAL

1.3g Investigating osmosis

Learning objectives:

- use scientific ideas to develop a hypothesis
- plan experiments to test a hypothesis
- draw conclusions from data and compare these with hypotheses made.

KEY WORDS

osmosis
partially permeable membrane
plasmolysis

A concentration gradient between the solutions inside and outside a cell means that water moves by osmosis from the more dilute solution to the more concentrated solution. As a result, the cell either gains or loses mass. In plant tissue, where cell walls prevent the cells bursting or collapsing, we can measure the mass changes quite easily and investigate how they depend on the concentration of the outside solution.

These pages are designed to help you think about aspects of the investigation rather than to guide you through it step by step.

Developing a hypothesis

Gill and Aidan are going to investigate the effect of putting some onion cells into water and some into salt solution. They will examine the cells through a microscope. If the cell is short of water the cytoplasm comes away from the cell wall. It is said to be **plasmolysed**.

Before they do the investigation they are going to produce a hypothesis. Hypotheses need to be developed using previous knowledge or observations. They know that:

- the concentration gradient between the solutions inside and outside a cell causes water to move by **osmosis**
- water moves towards a higher solute concentration through a **partially permeable membrane**
- the salt solution is more concentrated than the cytoplasm in the onion cells.

1. **When the onion cells are put in water, how will the concentration inside the cell compare with that outside?**
2. **When the cells are put into salt solution, how will the concentration inside the cell compare with that outside?**
3. **When the cells are put into salt solution, in which direction will the water move?**
4. **Suggest a hypothesis for the experiment that Gill and Aidan are about to do.**

SAFETY

A risk assessment is required for this practical work.

Planning an investigation

Lily and Ahmed investigated the effect of a range of salt solutions on pieces of potato. They weighed the potato pieces, placed them in salt solutions of various concentrations for 15 minutes and then reweighed them. They used five pieces in each solution. Before they reweighed the potato pieces, they carefully dried them using a paper towel.

SAFETY

A risk assessment is required for this practical work.

5. **Make a list of apparatus required for this investigation.**
6. **What were the independent and dependent variables in the investigation?**
7. **Why did Lily and Ahmed dry the potato pieces?**
8. **Lily and Ahmed collected these results:**

Concentration of NaCl solution (g/dm³)	Mean starting mass (g)	Mean final mass (g)	Mean change in mass (g)	Mean percentage change in mass (%)
0.00	15.9	17.0		
9	19.2	20.1		
18	24.1	23.3		
26	20.7	19.2		
35	24.1	22.0		
44	14.9	13.5		

Copy and complete the table by calculating the mean change in mass and mean percentage change in mass of the pieces.

9. **The table shows the mean results for five potato pieces in each solution.**
   - a **Explain how to calculate the mean of five values.**
   - b **Why is it better to use five pieces in each solution and find the mean, rather than just one piece?**
10. **Why do Lily and Ahmed use percentage change in mass?**
11. **Plot a graph of mean percentage change in mass of the potato pieces against concentration of salt solution.**
    - a **Are any of the results anomalous?**
    - b **How should anomalous results be dealt with?**
    - c **Suggest what conclusion Lily and Ahmed can make from this data.**

Evaluating the experiment

Lily and Ahmed had written a hypothesis before they did their experiment. They thought that 'the more concentrated the salt solution was that the potato was put in, the greater the loss of mass would be from the potato'.

12. **Was their hypothesis supported or disproved?**
13. **Explain why they got the result that they did.**
14. **What possible changes would you suggest that Lily and Ahmed make to their experiment, to improve repeatability and accuracy of results and to consider aspects of health and safety?**
15. **Describe in detail how similar apparatus could be used to investigate the rate of water uptake by osmosis in plant tissue. Develop a hypothesis to test, and decide on the dependent, independent and controlled variables. Show your plan to your teacher, and then carry out your investigation. Plot a graph of your results and evaluate your experiment.**

MATHS SKILLS

1.3h The spread of scientific data

Learning objectives:

- be able to calculate means and ranges of data
- be able to use range bars on graphs
- understand how to estimate uncertainty from a set of measurements.

KEY WORDS

estimate
mean
precision
range
range bar
uncertainty

When results are collected, how they are spread out is important. It helps us to make judgements about the quality of the data we have collected. This is important when attempting to identify trends in data.

The spread of data

A person's blood glucose level was measured. The measurement was repeated three times on the same blood sample. The following values were obtained:

6.2 mmol/dm^3 6.1 mmol/dm^3 6.0 mmol/dm^3

If you carry out *any* experiment and then do it again, you often get a slightly different result. This may not be because you've used the equipment wrongly. In any measurement there are always random errors that cause measurements to be vary in unpredictable ways. This is why it's best to repeat measurements and find the mean.

A **mean** reduces the effect of random errors and gives you the best **estimate** of the true value.

The **mean** value of a set of measurements is the sum of the values divided by the number of values:

mean =

$$\frac{6.2\text{ mmol glucose}/\text{dm}^3 + 6.1\text{ mmol glucose}/\text{dm}^3 + 6.0\text{ mmol glucose}/\text{dm}^3}{3}$$

= 6.1 mmol glucose/dm^3

The **range** is a measure of spread. It is calculated as the difference between the largest and smallest values

6.0 – 6.2 mmol/dm^3, or 0.2 mmol/dm^3.

1 **Another set of readings, using a blood sample from a different person, were:**

9.6 mmol/dm^3
9.5 mmol/dm^3
9.8 mmol/dm^3

What is the mean of these values?

2 **What is the range of this set of values?**

The spread of data on graphs

Data that are consistent are said to be **precise**; the narrower the range of a set of data, the higher the degree of **precision**. We can be more confident of conclusions we draw from data with a high degree of precision.

Estimating uncertainty in data

The **best estimate** of the true value of a quantity is the mean of repeated measurements. When calculating a mean, include all the values for data you have collected, unless you have any anomalous results. Anomalous results are measurements that do not fit into the pattern of the other results. You could check if a result was an anomaly by repeating it. The more repeated readings you take, the better estimate you'll get of the true value. Three to five repeats are often suggested.

The table shows the data collected on the effect of the hormone thyroxine on heart muscle tissue:

Time in minutes	Oxygen uptake, in cm^3 oxygen/g of heart muscle tissue					
	Experiment 1	Experiment 2	Experiment 3	Experiment 4	Experiment 5	Mean
10	3.3	3.9	3.6	3.8	3.9	3.7
20	7.8	8.3	8.0	7.8	8.1	8.0
30	11.7	11.2	11.5	11.6	11.5	11.5
40	14.6	14.8	15.1	14.9	14.6	14.8
50	17.7	17.2	17.5	17.6	17.5	17.5
60	26.7	26.0	26.5	27.4	26.9	

The effect of thyroxine on heart muscle was measured by the oxygen uptake by the cell.

For the data collected after ten minutes:

The mean is 3.7 cm^3 oxygen/g of heart muscle.

The range *about* the *mean* gives an estimate of the level of **uncertainty** in the data collected.

For this set of data, the upper limit is 3.9 cm^3/g; the lower limit 3.3 cm^3/g.

The range is 0.6 cm^3/g. So, according to our data, the true value could be up to 0.3 cm^3/g above the mean, or 0.3 cm^3/g below the mean.

Uncertainty is therefore calculated by:

$$\text{uncertainty} = \frac{\text{upper limit of range} - \text{lower limit of range}}{2}$$

So, as we have seen above, uncertainty $= \frac{3.9 - 3.3}{2} = \frac{0.6}{2} = 0.3$

Uncertainty is written next to the mean. In this instance, it is:

3.7 $cm^3/g \pm 0.3$ cm^3/g

Note that the units are written *both* after the mean and value of uncertainty. The value of uncertainty has the same units and number of decimal places as the measurements.

3. **Calculate the mean for the set of data collected at 60 minutes.**
4. **Calculate the uncertainty in these measurements.**

1.3i Active transport

Learning objectives:

- describe active transport
- explain how active transport is different from diffusion and osmosis
- explain why active transport is important.

KEY WORDS

active transport

Diffusion and osmosis explain how gases and water move down a concentration gradient to enter (and sometimes exit) living things. Minerals are taken up into the root hair cells of a plant by another method.

Active transport

Cells can absorb substances that are at low concentration in their surroundings by **active transport**. These substances move *against* the concentration gradient, for example when plants absorb nitrate ions through their root hairs from the soil water. The concentration of nitrate ions in soil water is usually less than the concentration of nitrate ions inside the root hair cells.

Nitrate ions naturally diffuse down their concentration gradient, out of the cell and into the soil, but plants transport the nitrates *into* their cells using active transport.

Active transport moves substances from a more dilute solution to a more concentrated solution (against a concentration gradient). This requires energy from respiration.

KEY INFORMATION

Ions are charged atoms or molecules; see topic 5.1.

1. **Describe how minerals are absorbed by plants.**
2. **How is active transport different to diffusion and osmosis?**

Active transport and respiration

Investigations have shown that a plant can absorb different minerals in different amounts. The plant can select which minerals it needs. Look at the graph. Algae absorb a lot of chloride, but only a little calcium in comparison.

In 1938, scientists discovered that an increase in mineral uptake by a plant happened at the same time as an increase in its respiration rate. This gives evidence that the process needs energy – because the minerals are absorbed against the concentration gradient.

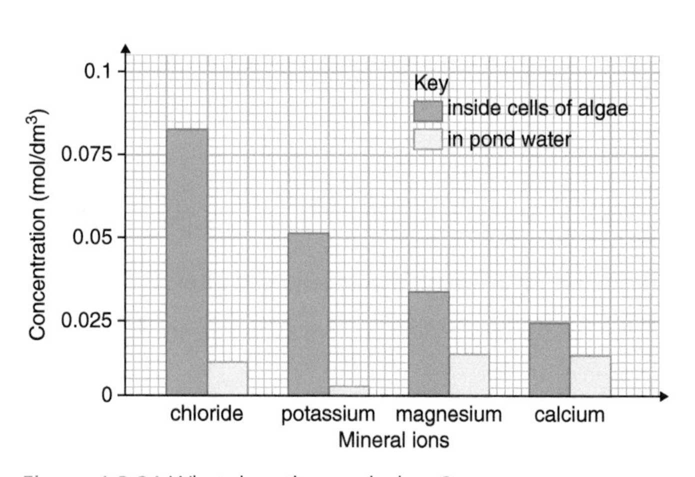

Figure 1.3.21 What does the graph show?

The greater the rate of cellular respiration, the more energy is available for active transport to happen.

Different cells use active transport for different purposes:

- The villi in the small intestine absorb glucose and amino acids into the blood, where they may already be in high concentration – after a meal, for example.
- Marine fish have cells in their gills that can pump salt back into the salty sea water.
- Cells in the thyroid gland take in iodine to use in the production of hormones.
- Cells in the kidney reabsorb sodium ions from urine.
- Crocodiles have salt glands in their tongues that remove excess salt from their bodies.

3 Suggest why cells that use active transport need to carry out this process.

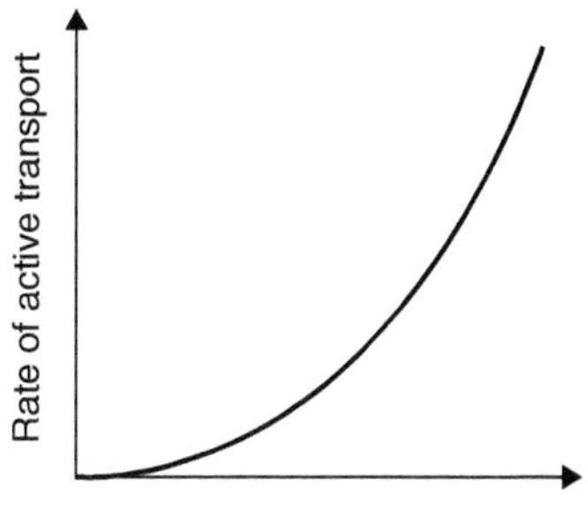

Rate of active transport

Rate of respiration

Figure 1.3.22 Explain the relationship between active transport and respiration

KEY INFORMATION

Active transport needs energy from respiration to move substances from a low concentration to a high concentration.

How does active transport work?

Look at the diagram. Special carrier molecules take mineral ions and other substances across the cell membrane. Different carriers take different substances. A carrier that moves glucose will not move calcium ions.

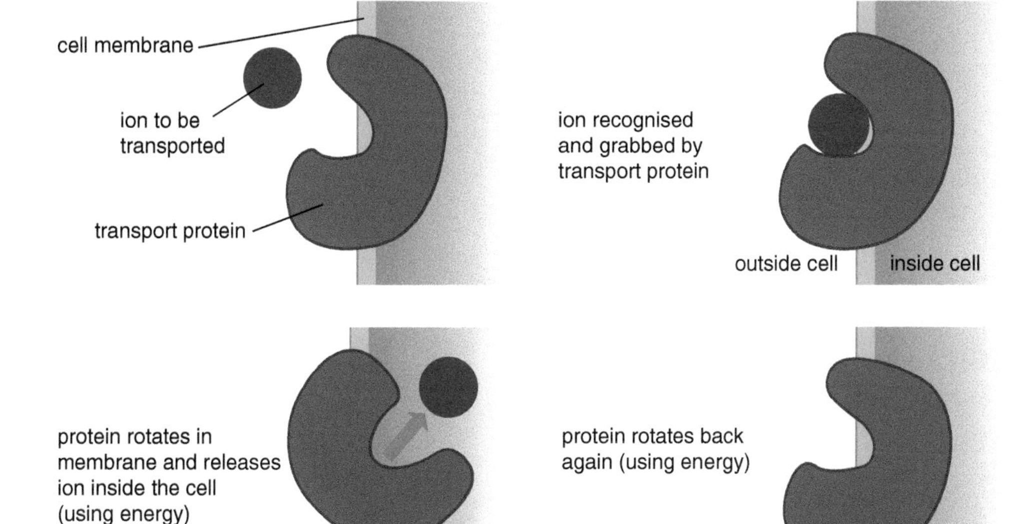

Figure 1.3.23 Each carrier molecule is specific to a certain mineral ion

4 Describe how ions are transported across a membrane in active transport.

5 Explain the similarities and differences between diffusion, osmosis and active transport.

DID YOU KNOW?

Many animals that live in the sea have salt glands near their eyes. These excrete a salt solution that is six times stronger than urine.

1.3j Mitosis and the cell cycle

Learning objectives:

- describe mitosis as part of the cell cycle
- describe the role of mitosis in growth and tissue repair
- describe how the process of mitosis produces cells that are genetically identical to the parent cell.

KEY WORDS

cell cycle
daughter cell
mitosis

Cells grow to a certain size, and then divide to form two new daughter cells. This process of cell division allows single fertilised egg cells to grow into multicellular organisms, and also produces new cells to repair damaged tissue.

Chromosomes

As we grow, the cells produced by cell division must all contain the same genetic information.

The genetic information of all organisms is contained in chromosomes, made of a molecule called DNA. The DNA in resting cells is found in the nucleus as long, thin strands, which can't be seen with a light microscope. For cell division, these strands form condensed chromosomes. Condensed chromosomes look thicker and can be seen with a light microscope.

Human body cells have 46 chromosomes, in 23 pairs. A chromosome carries a large number of genes. Each chromosome in a pair has the same pattern of genes along its length.

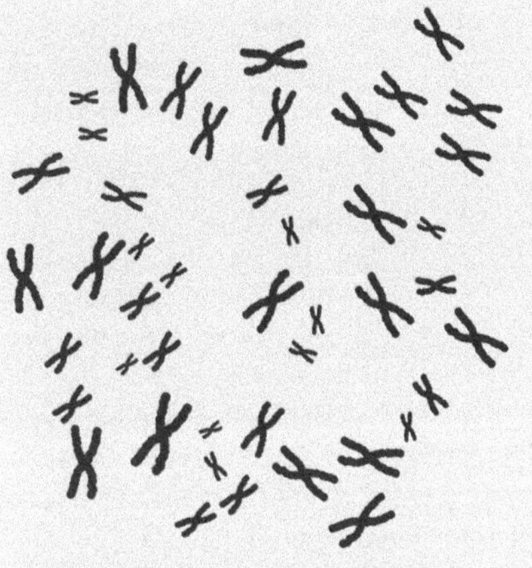

If stained, chromosomes condensed become visible during cell division

The pairs are arranged so that Pair 1 has the longest chromosomes, and Pair 22 has the shortest. The sex chromosomes are Pair 23.

Figure 1.3.24 A profile of a set of chromosomes, called a karyotype

1. **How many chromosomes are found in human body cells?**
2. **How are the chromosomes arranged in a karyotype?**

Mitosis

New cells have to be produced for growth and development, and to replace worn out and damaged body cells in injured tissues, for example.

When these new cells are produced, they must be the same as the parent cell. A parent cell divides to produce two new **daughter cells,** identical to each other and to the parent. This type of cell division is called **mitosis.**

3. **Why are new cells produced?**
4. **In this type of cell division:**
   - **how many chromosomes do daughter cells have?**
   - **how many daughter cells are produced?**

MAKING CONNECTIONS

To come up with a figure for how many cells there are in the human body, scientists must *estimate* by adding up cell counts from different organs.

Before a cell can divide it must grow, and make copies of all its organelles such as mitochondria and ribosomes. The 46 chromosomes in the nucleus are also replicated, so each chromosome consists of two identical molecules of DNA.

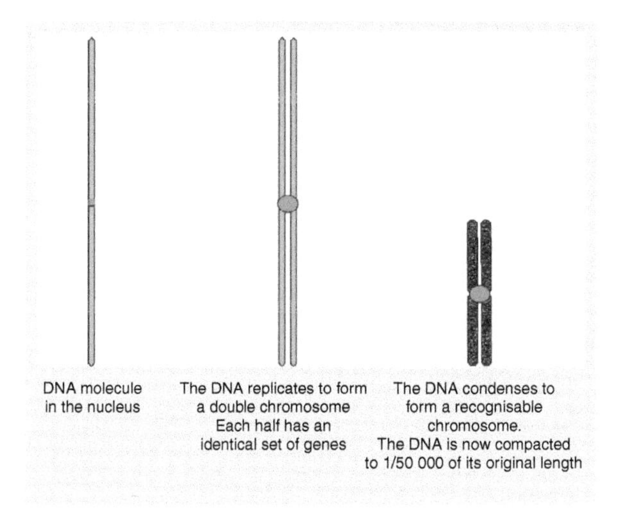

Figure 1.3.25 'Double chromosomes', consisting of two copies of DNA joined near the middle, often look 'X-shaped' in micrographs of cells

DID YOU KNOW?

Using radioactive carbon (^{14}C) dating of a cell's DNA, researchers in Sweden have been able to estimate the lifespan of different types of cells.

During mitosis, the double chromosomes are pulled apart as each new set of 46 chromosomes moves to opposite ends of the cell (Figure 1.3.26). Two nuclei then form. The cytoplasm and cell membrane then divides and two cells are produced.

5 Why do chromosomes appear double, or X-shaped, in micrographs?

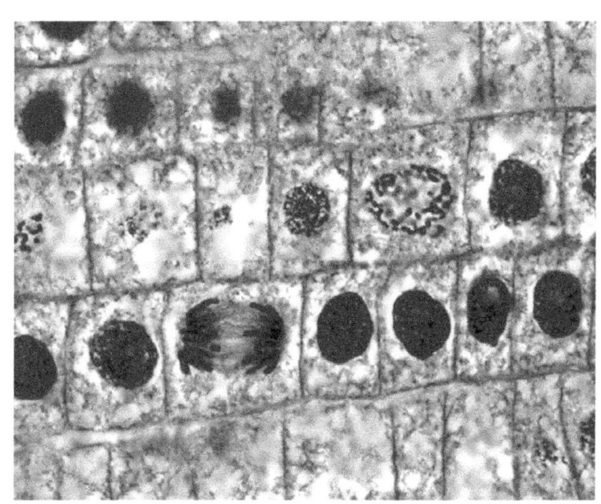

Figure 1.3.26 This light micrograph shows stained cells at different stages of the cell cycle in an onion root (x510). In one cell you can see the condensed chromosome copies being pulled to each end of the cell during mitosis

6 In the micrograph in Figure 1.3.26, you can't see chromosomes in all the cells. Explain why this is.

7 Why is it essential that each new cell produced in mitosis contains the same DNA as the parent cell?

8 Mitosis occurs rapidly in a newly formed fertilised egg. Suggest another situation in the body where you might expect cells to be actively dividing by mitosis.

The cell cycle

A cell that is actively dividing goes through a series of stages called the **cell cycle**. Mitosis is the part of the cell cycle in which the cell divides into two. Before that can happen, the cell grows, the DNA replicates and new organelles are produced. All these processes together form the cell cycle.

1.3k Meiosis

Learning objectives:

- explain how meiosis halves the number of chromosomes for gamete production
- explain how a new cell with the normal number of chromosomes is made at fertilisation
- understand that the four gametes produced by meiosis are genetically different.

KEY WORDS

gamete
genetic variation
meiosis

In sexual reproduction, two sex cells (one from each parent) join to make a new cell, which grows into a new organism. The new cell must have the normal number of chromosomes – so the sex cells must each have half the normal number. A special kind of cell division called meiosis creates sex cells with half the chromosomes of the parent cell.

Meiosis is a reduction division

Mitosis is the type of cell division used during growth, or when old or damaged cells need replacing.

Meiosis is another form of cell division, which take place when sex cells, or **gametes**, are produced in the ovaries and testes in animals, and in the carpels and stamens of plants.

During **meiosis**:

- four **gametes** are produced from one parent cell
- each gamete has half the number of chromosomes of the parent cell (in humans, that's 23 chromosomes instead of 23 pairs).

1. **How many gametes are produced from one cell during meiosis?**
2. **How many chromosomes does a gamete have?**

What happens during meiosis?

In meiosis, the DNA of each chromosome is copied, just as in mitosis. But it then divides *twice*, so the chromosome number is *halved*.

So, in mitosis, there is one replication of DNA and one division, but in meiosis, there is one replication of DNA and two divisions.

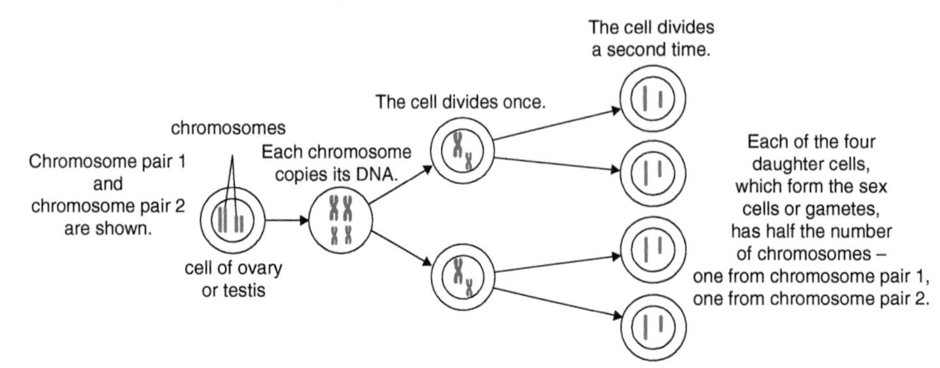

Figure 1.3.27 An overview of meiosis. Only two chromosomes are shown for clarity, but this process occurs for all chromosomes

When the gametes join at fertilisation, the new cell that is formed has the normal number of chromosomes. The diagram shows what happens at fertilisation in humans, showing only two pairs of chromosomes for simplicity. The new cell then divides by mitosis to grow into a new organism.

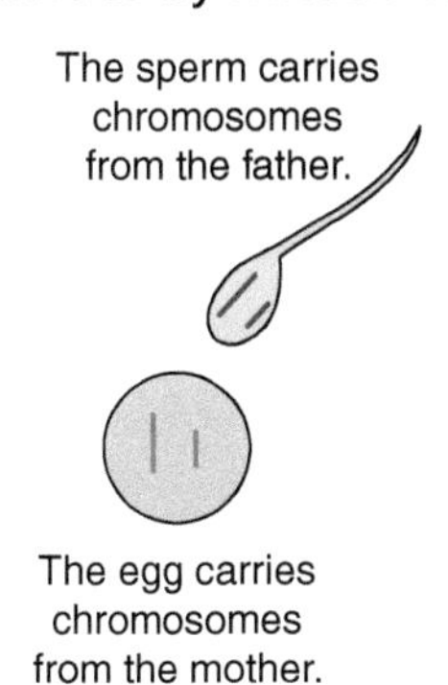

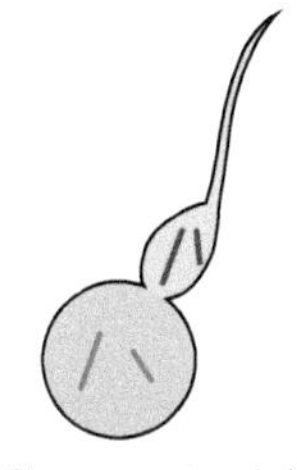

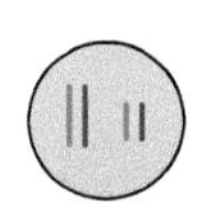

Figure 1.3.28 At fertilisation, the chromosomes from the gametes add together to make the normal number of chromosomes in the new cell

REMEMBER!

It is important to be able to recognise or describe differences and similarities between meiosis and mitosis.

3 How many replications of DNA occur in meiosis?

4 How many divisions occur in meiosis?

All gametes are genetically different

Of each pair of chromosomes in your cells, one chromosome was inherited from your mother and the other from your father. But when meiosis happens, which chromosome from each pair goes into each gamete is completely random. This means that the four gametes are genetically different.

Another thing that makes each gamete different is that there's some exchange of genetic material between the chromosomes in a pair when they line up together during meiosis. So when a gamete is produced, the chromosomes it receives may be different from those of the parent cell – they may be altered because parts of the DNA molecules have been swapped between two chromosomes in a pair. This also contributes to **genetic variation** because it means that every single gamete is unique.

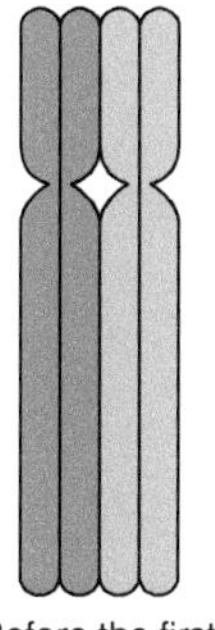

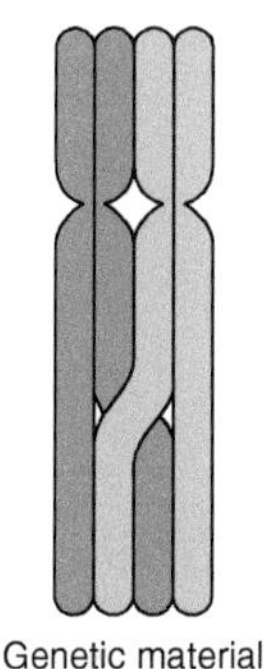

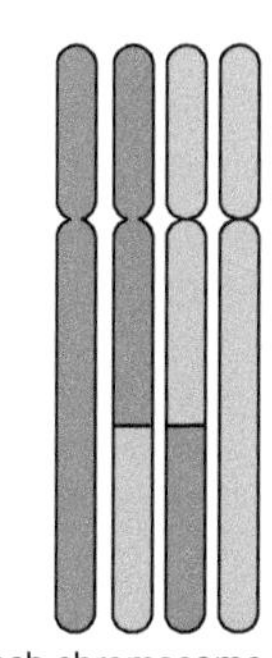

Figure 1.3.29 The set of chromosomes in a gamete is unique

5 Explain how meiosis leads to genetic variation.

6 In mitosis, the pairs of chromosomes don't line up next to each other as they do in meiosis. Suggest why this means the two daughter cells are genetically identical in mitosis.

1.3l Cell differentiation

Learning objectives:

- explain the importance of cell differentiation
- describe the function of stem cells in embryonic and adult animals.

KEY WORDS

differentiation
specialised
stem cells

For the first four or five days of our lives, the cells produced as the fertilised egg divides by mitosis are identical. Then, some of our cells start to become specialised to do a particular job.

Cell adaptations

As an embryo develops from a single fertilised egg into a multicellular organism, the cells need to take on different roles. Different types of cell are needed to ensure that the organism functions properly and as a whole.

As cells divide, new cells acquire certain features that make them **specialised** for a specific function. This is **differentiation.** A cell's size, shape and internal structure are adapted for its role. Most animal cells differentiate at an early stage of embryonic development.

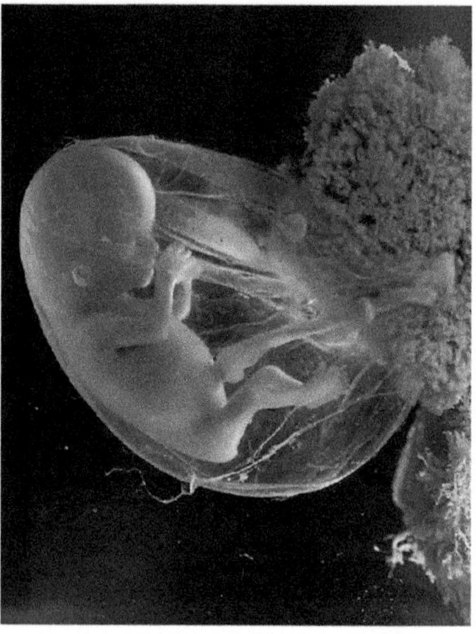

Figure 1.3.30 By around 13 weeks of embryonic development, a fetus has developed many of the 200 different cell types in the human body. It is around 7.5 cm long

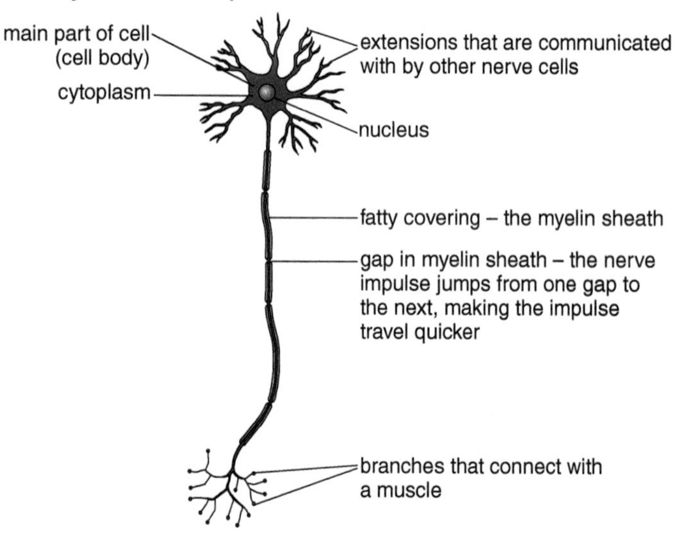

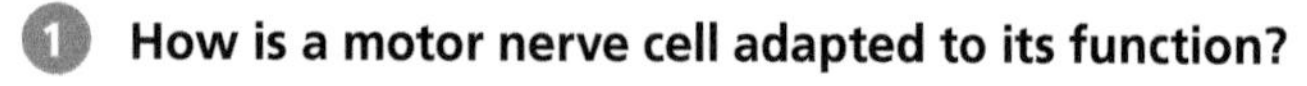

Figure 1.3.31 Nerve cells carry messages, or electrical impulses, from one part of the body to another. This diagram shows a motor nerve cell, which brings about movement of the skeleton by stimulating muscle cells to contract. Motor nerve cells can be 1 m or more in length, and 1–20 μm in diameter)

1. **How is a motor nerve cell adapted to its function?**
2. **Red blood cells are specialised to carry oxygen around the body. They are small and flexible, and so packed with haemoglobin that they don't have room for a nucleus. Suggest how these features adapt red blood cells to their function.**

Stem cells

At first, the cells in an embryo can grow and divide to form any type of cell. They are called **stem cells.**

Stem cells are unspecialised cells that can produce many different types of cell. Stem cells are found in the developing embryo, and some remain, at certain locations in our bodies, as adults.

3 Describe the function of embryonic stem cells.

Figure 1.3.32 The UK has a shortage of blood donors. In the summer of 2015 the NHS announced that it planned to start giving people blood transfusions using artificial blood by 2017. The red blood cells in the artificial blood will be produced using stem cells

Adult stem cells

Cells that have become specialised cannot later differentiate into different kinds of cells. However, there are some stem cells in most adult tissues that are able to start dividing to replace old cells or to repair damage in the tissues where they are found.

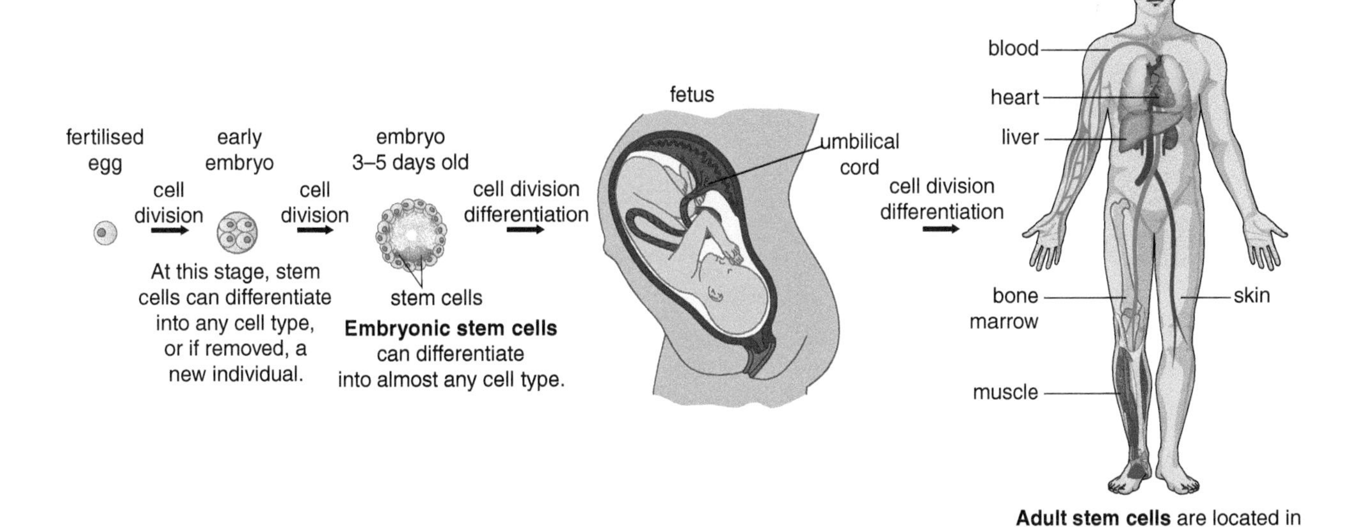

Figure 1.3.33 Stem cells in the human body

4 What is the function of adult stem cells?

5 Which can differentiate into more cell types – embryonic stem cells or adult stem cells?

6 Suggest why artificial blood transfusions produced using stem cells might not be acceptable to some people.

Check your progress

You should be able to:

calculate the magnification of a light or electron micrograph.	describe the differences in the magnification and resolving power of light and electron microscopes.	explain limitations of light microscopy and advantages of electron microscopy.
describe the functions of sub-cellular structures found in eukaryotic cells. describe the structure of a prokaryotic cell.	understand the size and scale of cells and be able to use and convert units. describe the differences between eukaryotic and prokaryotic cells.	carry out order of magnitude calculations when comparing cell size. calculate with numbers in standard form.
know the definition of diffusion.	explain diffusion using the idea of particles.	explain how substances pass in and out of cells.
recall that osmosis describes water movement in and out of cells.	explain osmosis as the movement of water through a partially permeable membrane, from a high water concentration to a lower water concentration.	predict water movement during osmosis. explain the words flaccid and turgid.
recall the theory of osmosis to create hypotheses on plant tissue.	plan and use appropriate apparatus and techniques to carry out an experiment to measure osmosis and test a hypothesis on plant tissue.	plot, draw and interpret appropriate graphs of results from an experiment to measure osmosis, evaluate the method and suggest possible improvements and further investigations.
recall that cells must divide for growth and replacement of cells.	describe how chromosomes double their DNA and are pulled to opposite ends of the cell, before the cytoplasm divides, during mitosis.	describe broadly the events of the cell cycle and explain the synthesis of new sub-cellular components and DNA.
identify meiosis as the cell division used to produce gametes.	explain the need for meiosis in producing gametes.	explain that the gametes produced by meiosis are genetically unique.
recall where stem cells are found. recall that organism development is based on cell division and cell specialisation.	describe how cells are specialised for their functions.	explain the importance of differentiation.

Worked example

1 **a** **What is a stem cell?**

A cell in a developing embryo that can divide to form any type of cell. After cells have become specialised they can't turn into other sorts of cell.

This describes embryonic stem cells well, but doesn't mention that there are also stem cells in adult tissues that can divide to replace old cells or repair damage in the tissue.

2 **The diagram represents the movement of water molecules across a cell membrane. The solution to the right of the membrane has a higher solute concentration than the solution to the left.**

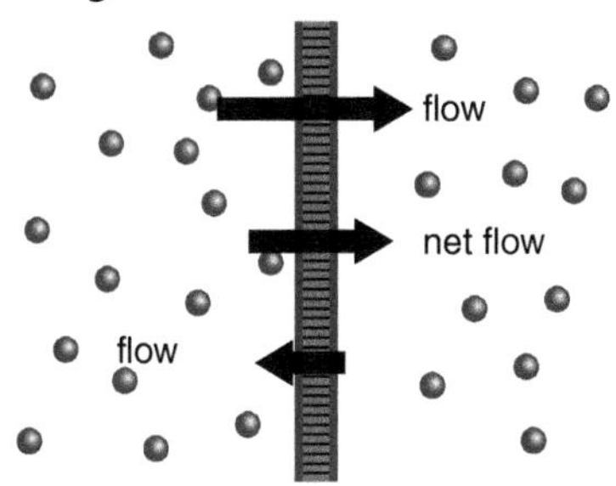

a **Explain what 'net flow' means here.**

Water is moving in the direction of the net flow.

Not a full answer. Water molecules are moving in both directions, but the rate of movement is higher in one direction than in the other. The net flow is the overall, combined result of the movements.

b **Why is there a net flow from left to right in this diagram?**

The concentration is higher on the left so more water flows to the right.

Be careful here. The *solute concentration* on the left is *lower* (more dilute), which means there is a higher water concentration. The net movement of water molecules is from a higher water concentration to a lower water concentration – from left to right in this case. Be sure to be clear in your answer.

c **Write a definition for osmosis.**

The diffusion of water.

Not a full answer. Osmosis is the diffusion of water across a partially permeable membrane.

d **Explain why there are differences in the effects of water on plant and animal cells.**

When water enters animal cells by osmosis the cells swell and may burst. When water moves out the cells shrivel up.

When water enters a plant cell it cannot burst because of the cell wall. It becomes turgid. If water moves out the cell becomes flaccid but doesn't shrivel up completely.

A clear, full answer.

3 **a** **Explain how dissolved minerals such as nitrate ions are absorbed by plants.**

By active transport through root hair cells.

A good answer, but you should explain that active transport is necessary to move mineral ions into the cell against their concentration gradient.

b **Explain the similarities and differences between diffusion, osmosis and active transport.**

Diffusion	Osmosis	Active transport
passive	passive	needs energy
down concentration gradient	down concentration gradient	up concentration gradient
no membrane needed	membrane	membrane

Table is a good idea. Don't forget also that osmosis is the movement of water molecules only – no other substance.

End of chapter questions

Getting started

1 The diagrams below show an animal cell and a bacterial cell.

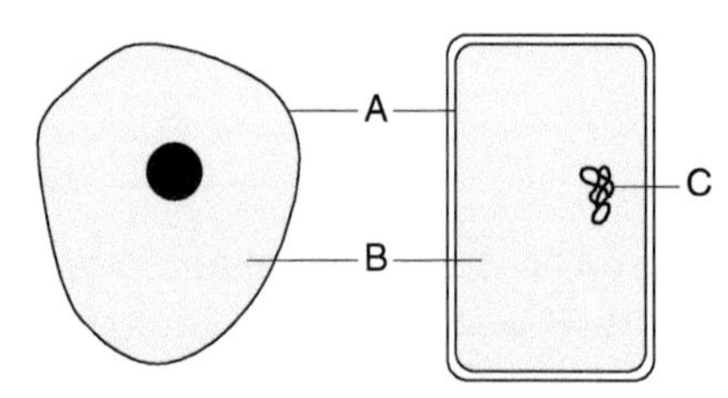

a The parts of the cell labelled A and B are found in both animal cells and bacterial cells.
Name cell parts A and B. — 2 Marks

(i) cell membrane
(ii) cell wall
(iii) cytoplasm
(iv) nucleus
(v) vacuole

b What is the name of chemical C? — 1 Mark

(i) cellulose
(ii) chlorophyll
(iii) DNA
(iv) protein

2 Explain how you know that the cell shown is a plant cell. — 2 Marks

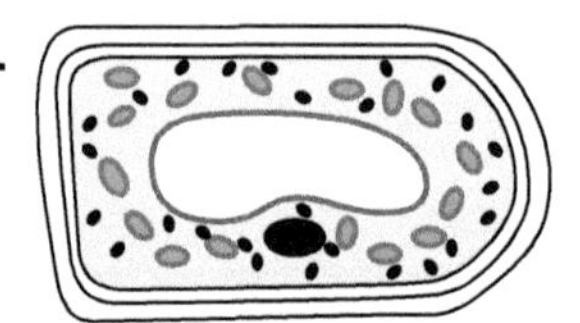

3 The diagrams below show some different cells. The length of each is included on the diagram. Arrange them in order of length. — 2 Marks

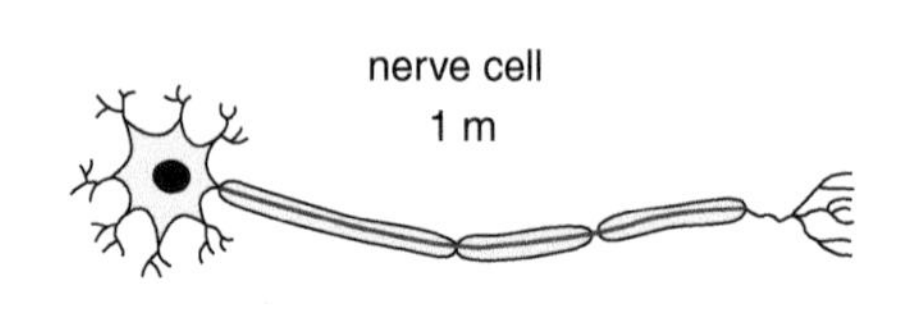

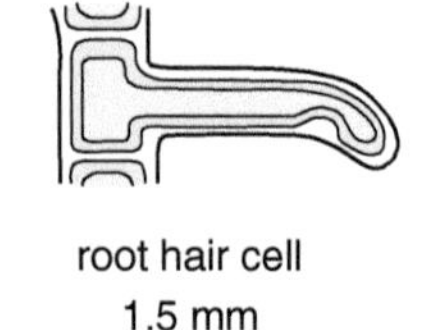

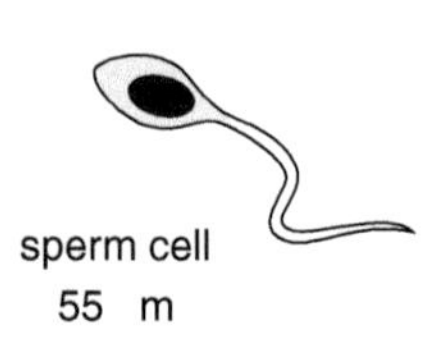

4 A bacterium divides into two every 30 minutes.

Starting with one bacterium, how many cells will there be after $1\frac{1}{2}$ hours? — 1 Mark

5 Gemma is using a light microscope to observe some dividing cells. A drawing of one of the cells is shown below. Describe what is happening in the cell. — 2 Marks

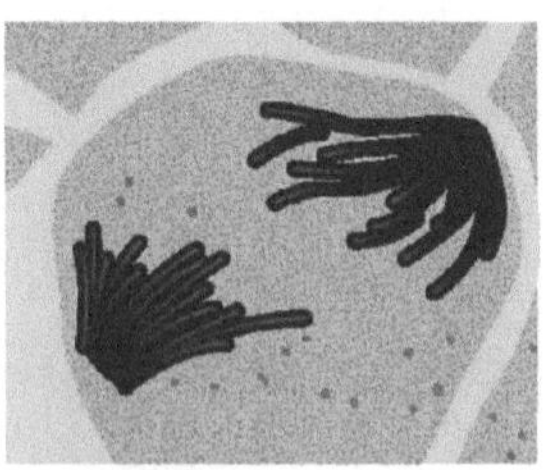

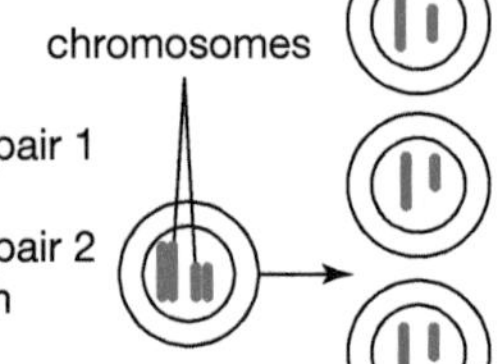

daughter cells

6 The diagram shows a process that is taking place when pollen grains are being produced. — 2 Marks

What conclusion can you draw? Explain your answer.

Going further

7. **Draw lines to match the two cell structures with their function.** 2 Marks

Cell structure	Function
	Controls what enters and leaves cells
Mitochondrion	Respiration
Ribosome	Protein synthesis

8. **Draw two lines to match the types of cell division with where they occur.** 2 Marks

Cell division	Where it occurs
	Asexual reproduction in bacteria
Meiosis	**Growth of a human embryo**
Mitosis	**Sperm production**

9. **Amoeba is a single-celled organism that lives in fresh water. It has a vacuole that fills with water, moves to the outside of the cell and bursts. A new vacuole starts to form. Use osmosis to explain why amoeba needs a vacuole.** 4 Marks

More challenging

10. **Explain why mitosis is important in plants and animals.** 2 Marks

11. **Explain how plant cells become turgid.** 3 Marks

12. **A teenage girl had her heartbeat measured at 74 beats per minute. Each beat pumped 70 cm^3 of blood. Calculate how much blood will be pumped in 10 minutes. Give your answer in litres.** 3 Marks

Most demanding

13. **Explain how meiosis leads to the production of gametes that are different genetically.** 6 Marks

14. **Name the processes by which a plant absorbs carbon dioxide for photosynthesis, essential minerals, and water, and explain the differences between them.** 6 Marks

Total: 40 Marks

WAVES

IDEAS YOU HAVE MET BEFORE:

DESCRIBING WAVES

- Frequencies of waves are measured in hertz (Hz).
- Waves travel at different speeds in different materials.
- Sound waves are longitudinal.
- Sound waves are produced by vibrations.
- Sound needs a medium to travel.
- Water waves and light waves are transverse.
- Light waves can travel through a vacuum.

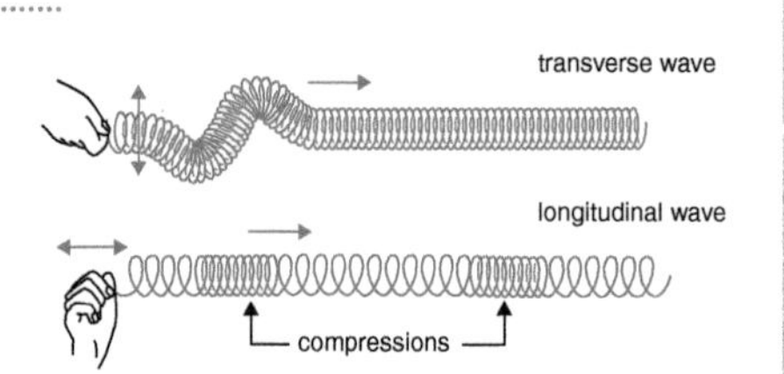

VISIBLE LIGHT

- White light is made up of a mixture of many different colours.
- Each colour of light has a different frequency.
- Different colours are absorbed or reflected by different surfaces.

LIGHT WAVES AND WATER WAVES HAVE SOME THINGS IN COMMON

- All waves carry energy from one place to another.
- When waves hit an object they may be absorbed by it, transmitted or reflected back.
- Waves may change direction (refract) at the point where two different materials meet.

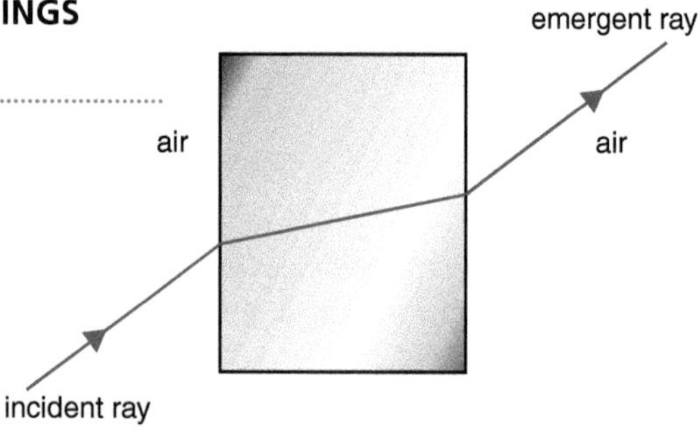

1.4

IN THIS CHAPTER YOU WILL FIND OUT ABOUT:

WHAT CHARACTERISTICS OF WAVES CAN BE MEASURED?

- We can measure the speed, wavelength and frequency of waves.
- We can calculate any of these three properties using the other two.
- The amplitude of a wave is its maximum displacement from its rest position.
- Echo sounding uses reflections of sound waves to detect objects hidden from view.

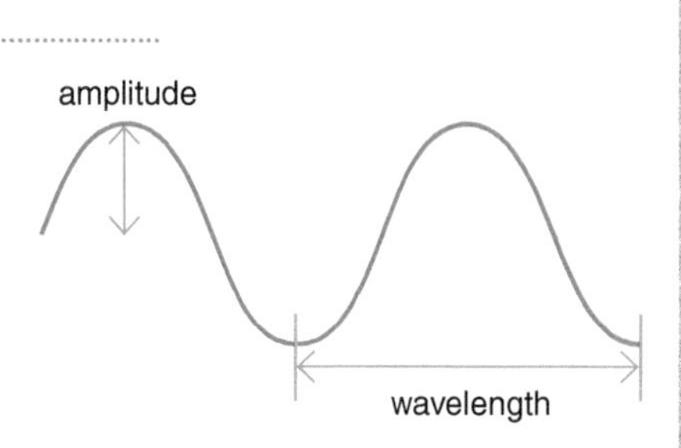

ARE THERE ANY WAVES BEYOND THE VISIBLE SPECTRUM?

- The visible spectrum is only a small part of a much wider spectrum called the electromagnetic spectrum.
- The invisible waves beyond red are called infrared and those beyond violet are called ultraviolet.
- Gamma rays, X-rays and ultraviolet rays have higher frequencies (smaller wavelengths) and transfer more energy.

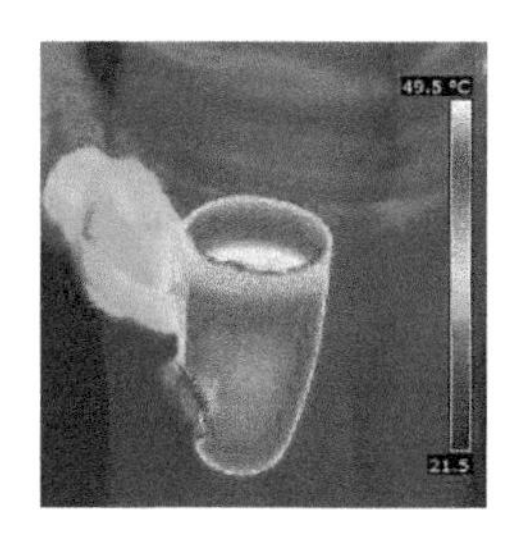

IN WHAT WAYS DO OTHER ELECTROMAGNETIC WAVES BEHAVE LIKE LIGHT?

- All electromagnetic waves are transverse waves that transfer energy.
- All electromagnetic waves can be absorbed, reflected and refracted at a boundary between two different media.
- All electromagnetic waves can be refracted when they go from one medium into another medium of a different density.

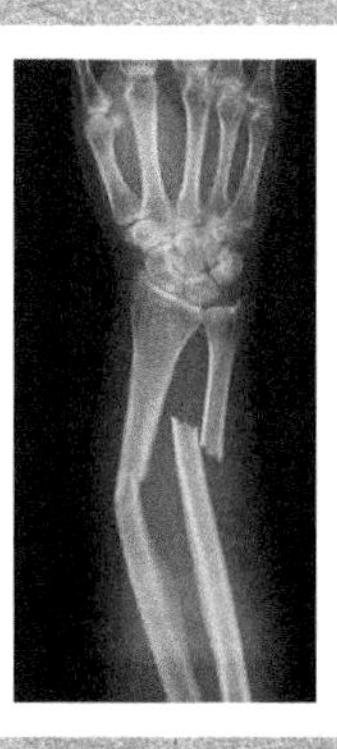

KEY CONCEPT

1.4a Transferring energy and information by waves

KEY WORD

amplitude

Learning objectives:

- to understand that all waves have common properties
- to understand how waves can be used to carry information
- to understand various applications of energy transfer by different types of electromagnetic waves.

A wave is a regular vibration that carries energy. Some waves carry energy in a form that can be decoded to extract information – an example of this is the picture and sound information carried by radio waves from a transmitter to your television set at home. Ripples on the surface of a pond, sound in air, ultrasound, visible light, X-rays and infrared rays are all types of wave. In water, the surface just moves up and down, but the energy is carried outwards from the source. A Mexican wave in a stadium is caused by spectators just standing up and sitting down but the wave travels all round the stadium (Figure 1.4.1).

Why not look on YouTube at the world record of a Mexican wave?

Figure 1.4.1 A Mexican wave

Common properties of waves

Waves all transfer energy. When we watch a firework display, sound waves travel slowly compared with light waves. We see the flash of the explosion first, when light waves transfer energy to sensors in our eyes. Then we hear the bang, when sound waves transfer energy from the explosion to our eardrums.

Amplitude measures the height of the wave above or below its rest point. The larger the amplitude of a water wave, the more energy it can transfer. An underwater earthquake can create a tsunami with waves 30 m high. This huge amplitude can transfer enough energy to power the whole of the UK for a year.

1. **Why do you hear the sound of thunder after you see the flash of lightning?**
2. **Draw a diagram to show two transverse waves with different amplitudes.**
3. **What would you notice if the amplitude of a sound wave increased?**

 Link: 1.4a → 1.4b

Using waves to transmit information

Because waves can carry energy, we use them to transmit information by varying the amount of energy carried by the wave. This can be done by simply switching the wave source on and off to create a pulsed code, as in Morse code using flashes of light, or by varying the frequency or amplitude of the wave.

In the past humans could only communicate using sound and light, but now we can also use transmitters and receivers to send information using a whole range of wavelengths that we cannot see or hear.

Figure 1.4.2 The beam of a lighthouse transmits information out to sea, warning sailors of dangerous rocks. Each lighthouse has its own distinct series of flashes

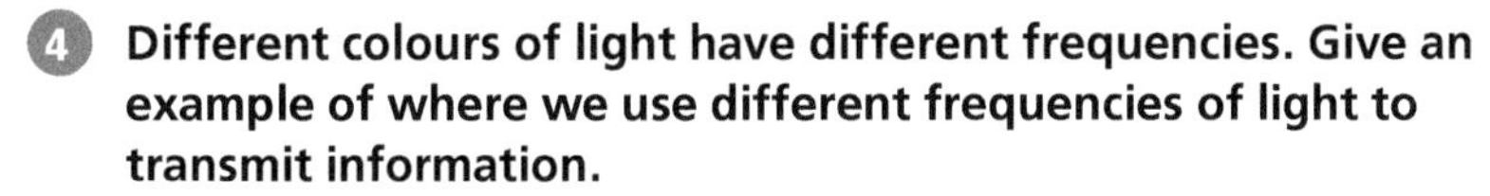

4 **Different colours of light have different frequencies. Give an example of where we use different frequencies of light to transmit information.**

5 **Give an example where information is transmitted by changing the amplitude of a wave, for example making a sound louder or a light brighter.**

Electromagnetic waves

Visible light is only a small part of a set of waves known as electromagnetic waves. The full range of electromagnetic waves is called the electromagnetic spectrum. These waves all travel at the same speed and can travel through empty space.

Each different part of the electromagnetic spectrum is used to transfer energy.

- Microwaves can transfer energy to cook food.
- An electric fire transfers energy to our bodies by infrared waves warming us up.
- Some energy from the Sun is transferred by ultraviolet rays.
- Not all energy from an X-ray machine is transferred.
- Some is absorbed by the body when an X-ray image is produced.
- Energy from radioactive sources can be transferred by gamma rays, and can be used to sterilise medical equipment or kill cancer cells.

Figure 1.4.3 Ultraviolet light can damage your eyes, and in extreme cases can even cause snow blindness. Skiers need to wear sunglasses that block out the ultraviolet, which can burn the surface of the eye

6 **For each of the examples of electromagnetic waves in the list above, suggest one piece of evidence that shows the energy transferred by the wave can be either absorbed or reflected.**

1.4b Transverse and longitudinal waves

KEY WORDS

compression
longitudinal wave
rarefaction
transverse wave

Learning objectives:

- compare transverse and longitudinal waves
- describe water waves as transverse waves and sound waves as longitudinal waves
- describe evidence that the wave travels along, but not the medium itself
- describe how to measure the speed of water waves.

All waves transfer energy, but not all in the same way. Ripples on water transfer energy as water particles move up and down. In sound waves, the particles move closer together and further apart, rather than up and down.

Transverse waves

Transverse and **longitudinal** waves can be produced on a Slinky spring.

In a transverse wave on a spring the vibrations of the particles are at right angles to the direction of the energy transfer (Figure 1.4.4). If the particles move up and down vertically, the energy carried in the wave is transferred horizontally, away from the energy source creating the wave. The wave moves but the spring oscillates about a fixed position.

Ripples on water are also transverse waves (Figure 1.4.5). The wave is moving outwards but the water particles move up and down.

A cork floating in water bobs up and down on a water wave. It only has vertical motion and it is not carried along by the wave.

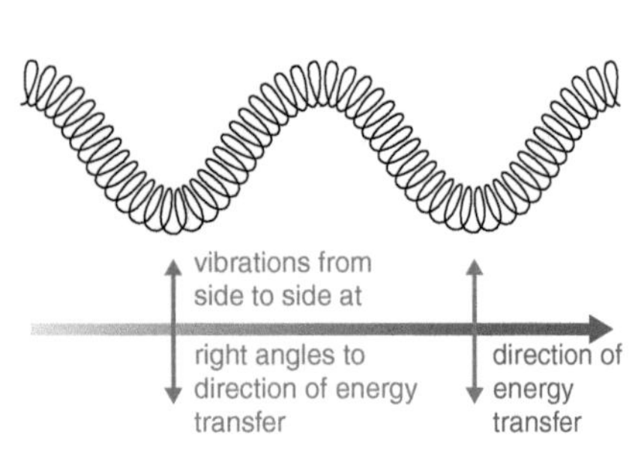

Figure 1.4.4 Transverse waves on a Slinky spring shaken from side to side on the floor

Figure 1.4.5 Ripples caused by a stone being dropped into water If the water moved outwards, it would leave a hole in the centre

1. **What evidence suggests water waves are transverse waves?**
2. **Alex sees a stick floating on a still pond. A boat creates waves that spread across the surface toward him. Alex expects the stick to move toward him too, but it does not. Suggest why this is.**

Longitudinal waves

In a longitudinal wave the vibrations of the particles are parallel to the direction of energy transfer (Figure 1.4.6). Longitudinal waves show areas of **compression** and rarefaction. A compression is when the waves bunch up. A **rarefaction** is

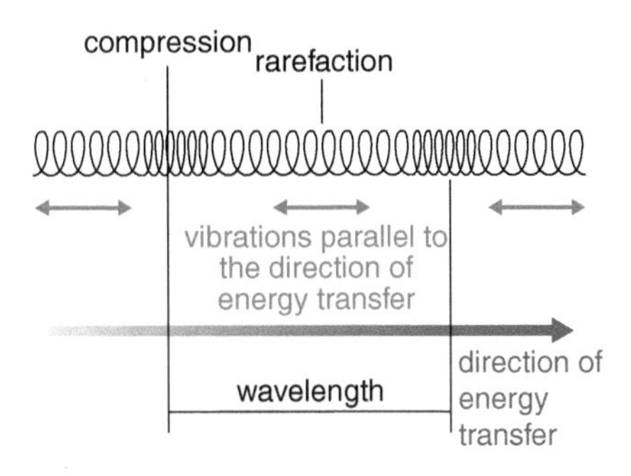

Figure 1.4.6 A longitudinal wave on a Slinky spring

when they spread out. If the particles move from side to side horizontally, the energy carried in the wave moves along the same horizontal direction, away from the energy source.

Sound waves in air are longitudinal waves.

Figure 1.4.7 Hitting the drum makes the drumskin vibrate, moving the air molecules next to it back and forth. The sound travels as the molecules pass on the vibration to their neighbours

3. **How is the longitudinal wave on a Slinky different when you push and pull gently from when you move your hand more vigorously?**
4. **Why wouldn't anyone hear you if you screamed in space?**

The speed of water waves

The speed of a water wave can be found by measuring the time it takes for a water wave to travel a measured distance. For example, make a splash at one end of a 25 m swimming pool and measure, with a stop clock, the time it takes the wave to travel to the other end.

5. a **Is this method an accurate way of measuring the wave speed? Explain your answer.**

   b **Explain how you could improve the accuracy.**
6. **The crest of an ocean wave moves a distance of 20 m in 10 s. Calculate the speed of the ocean wave.**

KEY INFORMATION

A medium (plural media) is a material through which a wave travels.

1.4c Measuring wave speed

Learning objectives:

- describe how to measure the speed of sound waves in air using an echo method
- apply the echo method to waves in water
- apply the relationship between wavelength, frequency and wave velocity.

KEY WORDS

echo
echo sounding

Light and sound are both waves, but in a thunderstorm, you see lightning before you hear thunder. The lightning flash is almost instantaneous but sound travels much more slowly. We can use echoes to measure the speed of sound in air.

Measuring the speed of sound in air

Zoe and Darren measured the speed of sound in air. Sound reflects off a wall in a similar way to light reflecting off a mirror. The reflected sound is called an **echo**.

Zoe stood 50 m from a large wall. She clapped and listened to the echo (Figure 1.4.8).

Zoe tried to clap each time she heard an echo while Darren timed 100 of her claps with a stop clock. He timed 100 claps in 40 s.

The time between claps is $\frac{40\text{s}}{100} = 0.4\text{ s}$

During the time from one clap to the next the sound had time to go to the wall and back, a distance of 100 m.

Speed = distance / time

$$= \frac{100\text{m}}{0.4\text{s}} = 250\text{ m/s}$$

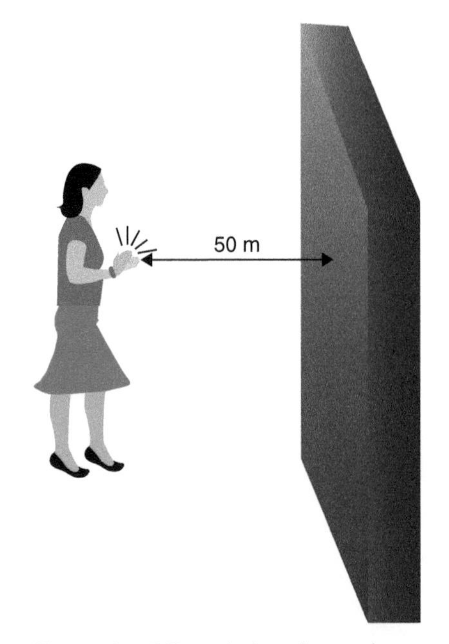

Figure 1.4.8 Zoe tried to clap each time she heard an echo

1. **Jo and Sam also measured the speed of sound in air using the same method. They counted 50 claps in 23 s. Jo also stood 50 m from the wall.**

   What value did they get for the speed of sound?

2. **Suggest why this method is not likely to produce an accurate value for the speed of sound in air.**

Echo sounding

Ships use high-frequency sound waves to find the depth of the seabed or to locate a shoal of fish (Figure 1.4.9). This is **echo sounding**.

ADVICE

In echo sounding remember the wave goes 'there and back', so make sure you use the correct distance in calculations.

Example: A ship sends out a sound wave and receives an echo after 1 second. The speed of sound in water is 1500 m/s. How deep is the water?

Time for sound to reach the seabed = 0.5 s.

$$\text{speed} = \frac{\text{distance}}{\text{time}}$$

distance = speed × time

= 1500 m/s × 0.5 s = 750 m

Figure 1.4.9 The sound waves are reflected off the seabed or from a shoal of fish, coming back as an echo

3 **On a fishing boat, the echo from the shoal of fish is received after 0.1 s. How far below the boat is the shoal?**

4 **A ship is 220 m from a large cliff when it sounds its foghorn.**

- **a** **When the echo is heard on the ship, how far has the sound travelled? (The speed of sound in air is 330 m/s.)**
- **b** **How long is it before the echo is heard?**

Change in speed of sound waves

When a sound wave is transmitted across a boundary from one medium to another, its speed may change. When the speed of a wave changes, there is a change in the wavelength, but there is no change in the frequency. This is because the number of waves leaving the medium each second is the same as the number of waves entering the medium each second.

Medium	Speed of sound (m/s)
air	330
water	1500
steel	5000

5 **Adjacent compressions in a sound wave are 15 cm apart. What is the wavelength of the sound?**

6 **Ann puts her ear to touch an iron railing. Jack hits the iron railing with a stick about 5 m away from Ann. Explain why she hears two sounds, one after the other.**

1.4d A wave equation

Learning objectives:

- describe wave motion in terms of amplitude, wavelength, frequency and period
- describe and apply the relationship between wavelength, frequency and speed
- apply the equation relating period and frequency.

KEY WORDS

amplitude
frequency
hertz (Hz)
period
wavelength

We can describe waves in terms of their wavelength, frequency and speed – and if we know two of these variables, we can work out the other one using the 'wave equation'.

Wavelength, amplitude and frequency

Wavelength (λ) is the distance from a point on one wave to the equivalent point on the adjacent wave. Wavelength is measured in metres (m).

The **amplitude** of a wave is the maximum displacement of a point on a wave away from its undisturbed position.

Frequency (f) is the number of complete waves passing a point in one second. It is measured in **hertz (Hz).** A frequency of 5 Hz means there are five complete waves passing a point in 1 second. Frequencies are also given in kilohertz (kHz) and megahertz (MHz).

1000 Hz = 1 kHz

1000 kHz = 1 MHz

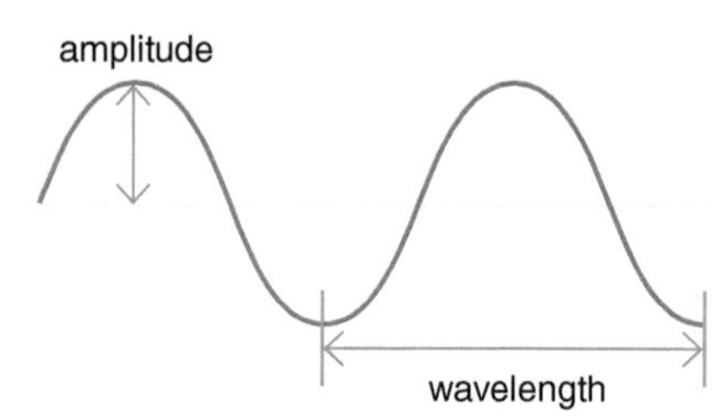

Figure 1.4.10 Amplitude and wavelength of a wave

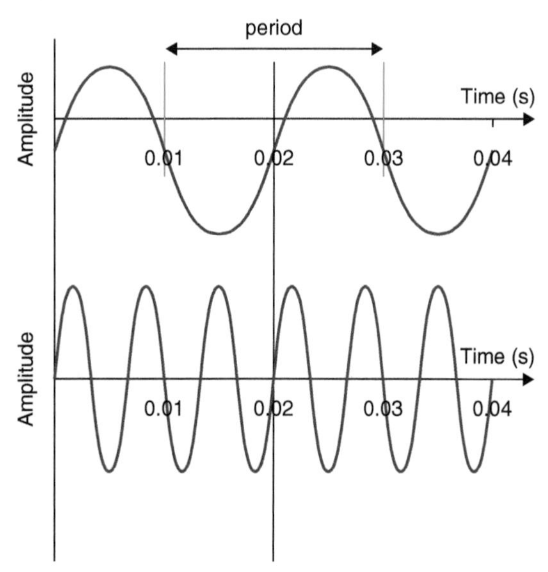

Figure 1.4.11: High-pitched sounds have a higher frequency than low-pitched sounds – the higher the pitch, the more vibrations per second

1. **Define what we mean by**
   a wavelength
   b frequency
   c amplitude.
2. **Suggest how the amount of energy transferred by a wave changes as the amplitude increases.**

Period of a wave motion

The **period (T)** is the time to complete one wavelength.

Period is the reciprocal or inverse of frequency.

Frequency is the number of complete waves passing a point in one second. $T = \frac{1}{f}$,

For example: When the frequency of a wave is 5 Hz there are 5 waves passing a point each second. The time for one wave to pass is $\frac{1}{5}$ s or 0.2 s.

The period, T, $= \frac{1}{f}$

$= \frac{1}{5}$ s or 0.2 s

3 **A wave has a frequency of 2 Hz. How many waves pass a point in 1 second?**

4 **Work out the frequencies of waves with periods of**

a 0.1 s

b 0.25 s.

ADVICE

Always check that you put values into the wave equation using SI units (e.g. wavelengths in metres, frequency in hertz). If the values in the question are **not** in SI units, you first have to convert them.

The wave equation

A wave transfers energy. The wave speed is the speed that the wave transfers energy, or the speed the wave moves at.

All waves obey the **wave equation**:

wave speed = frequency × wavelength

$v = f\lambda$

where the wave speed v is in metres per second (m/s), the frequency f is in hertz (Hz) and the wavelength λ is in metres (m).

If we know two of these three variables we can calculate the third using the wave equation.

Example: A radio station produces waves of frequency 200 kHz and wavelength 1500 m.

(a) Calculate the speed of radio waves.

(b) Another station produces radio waves with a frequency of 600 kHz. What is their wavelength? Assume that the speed of the wave does not change.

(a) $v = f\lambda$ = 200 000 Hz × 1500 m = 300 000 000 m/s

$= 3 \times 10^8$ m/s

(b) wavelength, $\lambda = \frac{v}{f} = \frac{300\,000\,000 \text{ m/s}}{600\,000 \text{ Hz}}$

= 500 m

DID YOU KNOW?

The light that we can see has a wavelength of about $\frac{1}{2000}$ of a millimetre!

5 **A wave has a frequency of 2 Hz and a wavelength of 10 cm. What is the speed of the wave?**

6 **When the frequency of a wave doubles, what happens to its wavelength?**

7 **A TV signal is broadcast at a frequency of 104 kHz. If the wave speed is 3×10^8 m/s, what is the wavelength?**

REQUIRED PRACTICAL

1.4e Measuring the wavelength, frequency and speed of waves in a ripple tank and waves in a solid

These pages are designed to help you think about aspects of the investigation rather than to guide you through it step by step.

Learning objectives:

- develop techniques for making observations of waves
- select suitable apparatus to measure frequency and wavelength
- use data to answer questions.

Frequency and wavelength of waves in a ripple tank

We can use a set of equipment called a ripple tank to explore waves. By careful observation and measurement we can measure and calculate the wavelength and frequency of the waves and then work out their speed. A strobe light can be used to 'freeze' the movement of the waves for making certain measurements.

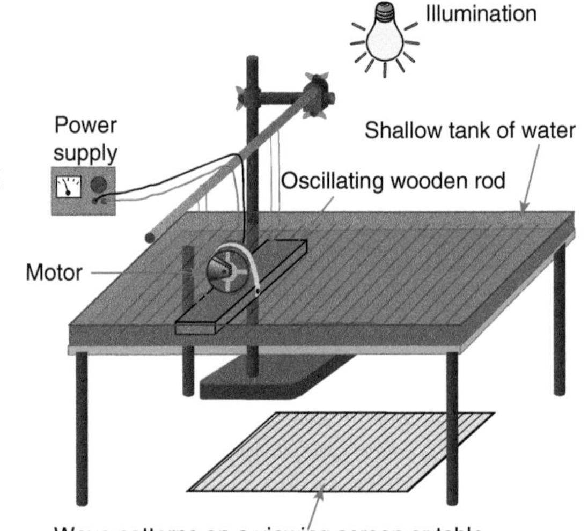

Figure 1.4.12 Ripple tank used for observing waves in water

1. **A motor is attached to the wooden rod. What does this do?**
2. **What are the units of**
   a wavelength?
   b frequency?
3. **Suggest what equipment you could use to measure the wavelength, and how you should set it up.**
4. **When measuring the wavelength, you might measure the length of ten waves on the screen or table and then divide by ten. Explain why this is done.**
5. **Louise is looking at a certain point on the screen and counting how many waves pass that point in ten seconds. How can she then calculate the frequency of the waves?**

SAFETY

A risk assessment is required for this practical work.

When using strobe lights it is necessary to be aware of students with epilepsy, as these lights can induce epileptic fits.

Speed of waves in a ripple tank

Sahil's group are comparing two ways of working out the speed of the waves as they travel through the water. One of these is by using speed equals distance/time and the other is by using speed equals frequency times wavelength.

6 Explain how the group could measure the speed of the waves using the equation speed = distance/time.

7 Now explain how the group could calculate the speed of the waves using the equation speed = frequency × wavelength.

8 Why do scientists sometimes try to use two different methods to measure a quantity

Figure 1.4.13 Shadows of waves travelling across a ripple tank. The wavelength is the distance between two dark patches, which are the peaks (crests) of the waves

Speed of sound in a metal rod

Sound also travels in waves. Unlike water waves we can't see them but we can still measure their speed. The apparatus needs to be different though.

The apparatus in Figure 1.4.14 can be used to measure the speed of sound in an iron bar, about 1 m long. One end of the bar is hit with the hammer. This electrical contact starts the timer running. The sound travels through the bar and is detected by the microphone. An electrical pulse generated by the microphone stops the timer.

This experiment was demonstrated several times and the following values obtained for the time interval: 0.18 milliseconds (ms), 0.21 ms, 0.32 ms, 0.17 ms and 0.22 ms.

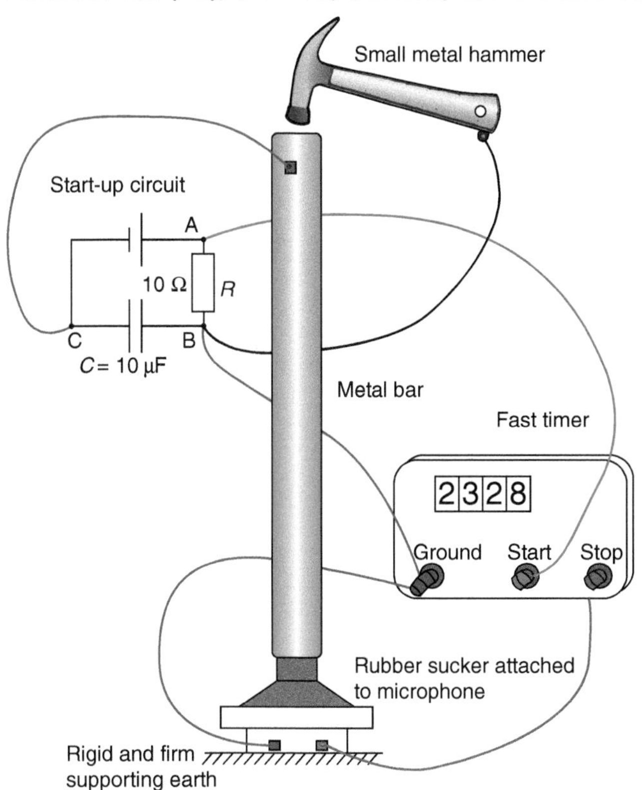

Figure 1.4.14 Measuring the speed of sound in an iron bar

REMEMBER

If you have a set of readings, don't simply average all the results but look at them first to see if there are any anomalies. These should be eliminated before finding the average or they will influence the outcome.

DID YOU KNOW?

Sound travels much faster through solids than through air. It travels through iron over 15 times faster than through air.

9 Suggest how this data should be processed.

10 What calculation should then be done to get a value for the speed?

11 The accepted value for the speed of sound in iron is 5130 m/s. Comment on how well the experimental value compares.

1.4f Electromagnetic waves

Learning objectives:

- recall that electromagnetic waves are transverse waves that can transfer energy through space
- describe the main groupings of the electromagnetic spectrum
- recall and apply the relationship between frequency and wavelength.

KEY WORDS

electromagnetic waves
longitudinal wave
spectrum
transverse wave
visible spectrum

Electromagnetic waves emitted by the Sun, including visible light, travel 149 million kilometres across space. They reach Earth in about 8 minutes.

Transverse and longitudinal waves

As you saw earlier, there are two types of waves: **transverse** and **longitudinal**. In a transverse wave the vibrations are at right angles to the direction of energy transfer and in a longitudinal wave they are parallel to the direction of energy transfer (Figure 1.4.15).

Waves in water (Figure 1.4.16) and a rope or slinky moved from side to side are transverse waves.

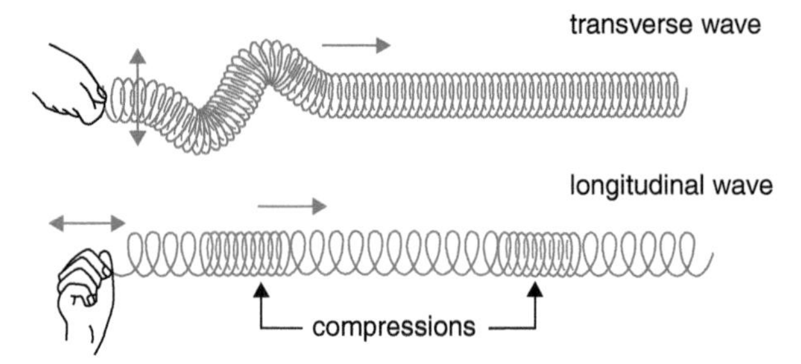

Figure 1.4.15 Comparing transverse and longitudinal waves

1. **Explain what is meant by a) the amplitude and b) the wavelength of a wave.**
2. **Give an example of a longitudinal wave.**

Electromagnetic waves

Light is a type of wave called an **electromagnetic wave**. White light is a mixture of electromagnetic waves with different wavelengths, which can be separated into what we call the **visible spectrum**. When scientists investigate the visible spectrum they can detect invisible waves on both sides, showing that the visible spectrum is really part of a much wider spectrum, which we call the electromagnetic spectrum.

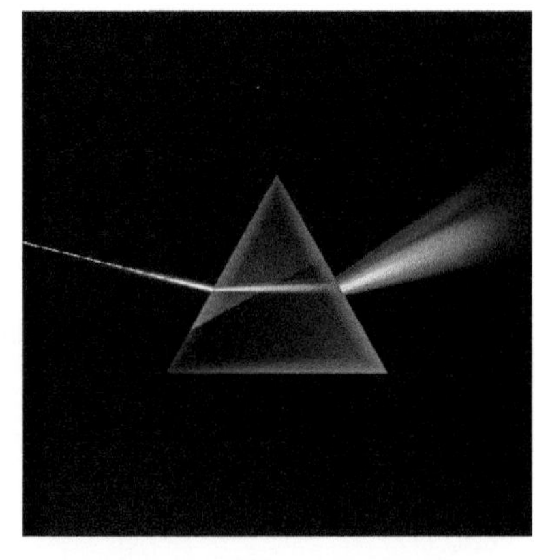

Figure 1.4.16 When white light is passed through a prism, the different wavelengths of the visible spectrum appear to our eyes as different colours

All the waves in the electromagnetic spectrum are transverse waves with many properties in common with visible light.

Just like other waves, all electromagnetic waves transfer energy from one point to another – from a source to an absorber. In electromagnetic waves, electromagnetic fluctuations occur at right angles to the direction in which energy is being transferred by the wave.

Some waves have to travel through a material. Sound waves can travel through air, liquids and solids but not a vacuum. Water ripples travel along the surface of water.

Electromagnetic waves are different from other waves because they do not need a material. They can travel through a vacuum. This is a special property of electromagnetic waves, which enables light and infrared waves to reach us from the Sun. All electromagnetic waves travel at the same speed in a vacuum, 3.0×10^8 m/s.

3 **What are the similarities and differences between transverse and longitudinal waves?**

4 **Explain how waves in the electromagnetic spectrum are different to other waves.**

5 **What properties do all electromagnetic waves have in common?**

ADVICE

Remember the order of electromagnetic wave groupings.

KEY INFORMATION

The shorter the wavelength (the higher the frequency) of an electromagnetic wave the more dangerous the radiation.

The electromagnetic spectrum

Figure 1.4.17 shows that electromagnetic waves span a wide, continuous range of wavelengths and frequencies.

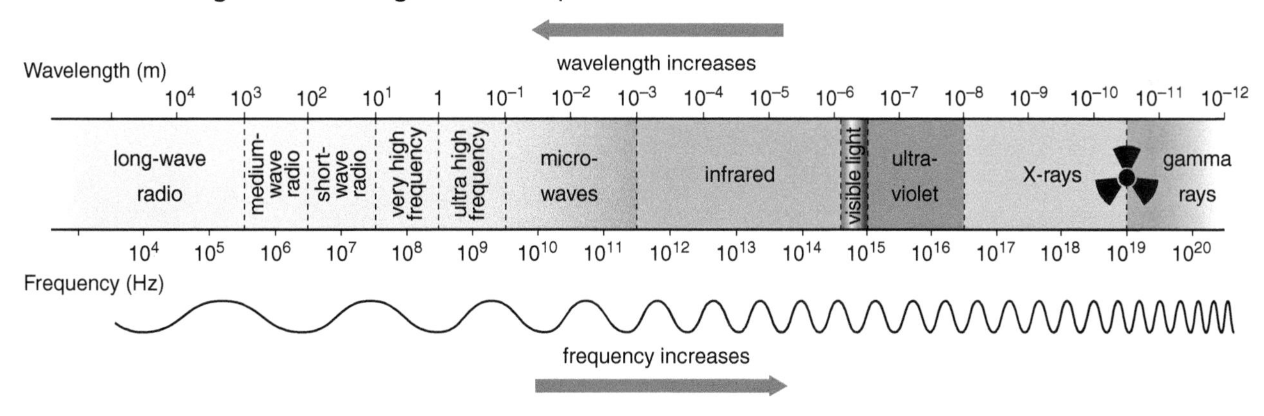

Figure 1.4.17 The wavelengths and frequencies of waves in the electromagnetic spectrum

The groups of waves in the electromagnetic spectrum are shown in the table.

Waves in electromagnetic spectrum	Wavelength (m)
radio, TV and microwaves	above 10^5 m to 10^{-3}
infrared	10^{-3} to 10^{-7}
visible (red to violet)	10^{-7}
ultraviolet	10^{-7} to 10^{-8}
X-rays	10^{-8} to 10^{-10}
gamma rays	10^{-10} to less than 10^{-12}

The shorter the wavelength of the electromagnetic wave, the further it can travel through other materials. The higher the intensity of a wave the more energy it can transfer to another object when the radiation is absorbed.

6 **Which grouping in the electromagnetic spectrum has the highest frequency?**

7 **Ultraviolet light used in a sunbed has a wavelength of 3.5×10^{-7} m. The speed of light is 3×10^8 m.**

Calculate the frequency of this light.

8 **Calculate the frequency of an electromagnetic wave with a wavelength of 20 cm. Use standard form for your answer.**

1.4g Uses of electromagnetic waves

Learning objectives:

- give examples of practical uses of electromagnetic waves
- show that the uses of electromagnetic waves illustrate the transfer of energy from source to absorber
- recall that radio waves can be produced by, or can induce, oscillations in electrical circuits.

KEY WORDS

gamma rays
infrared
microwaves
radio waves
ultraviolet
visible light
X-rays

Electromagnetic radiation has a huge number of uses in medicine, communications and everyday life. These uses illustrate the transfer of energy from a source (something that emits electromagnetic radiation) to an absorber (something that takes in electromagnetic radiation). An absorber is often a detector or a receiver.

Gamma rays and X-rays

Gamma rays have the shortest wavelengths in the electromagnetic spectrum and can be harmful to living cells. In controlled doses, gamma rays are used in radiotherapy to kill cancer cells, and to sterilise surgical instruments.

X-rays pass through soft tissues in the body but are absorbed by bone. This means X-rays can be used to check for broken bones, and also in computerised tomography (CT) scans – multiple images taken at different angles build up a detailed picture of inside a patient's body (Figure 1.4.18).

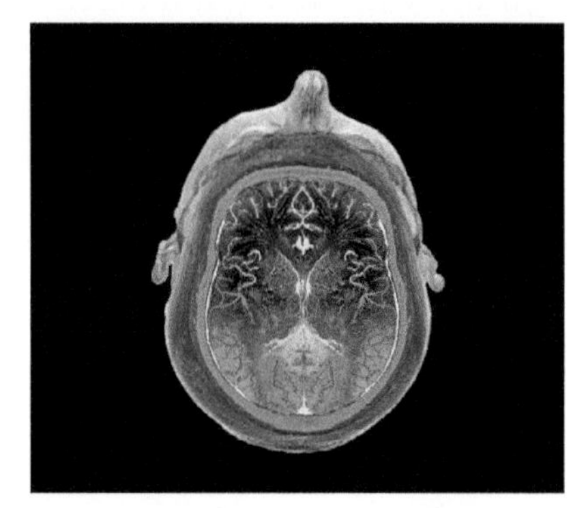

Figure 1.4.18 A CT scan can give good pictures of soft tissue regions but exposes a patient to a much higher radiation dose than a single X-ray

1. **What are the similarities and differences between gamma rays and X-rays?**
2. **Should X-rays used in radiotherapy have longer or shorter wavelengths than those used for medical diagnosis? Explain your answer.**

DID YOU KNOW?

X-ray machines called pedoscopes were used in shoe shops in the 1930s to measure people's feet. Although children loved to watch their toe bones wriggle, people did not realise how dangerous the energy transferred by X-rays can be. Some shop assistants developed cancer because of using them.

Ultraviolet, visible and infrared radiation

Ultraviolet rays have higher frequencies and shorter wavelengths than visible light. One use of ultraviolet is in fluorescent lighting, which is more energy efficient than traditional filament light bulbs.

In small doses, ultraviolet light from sunlight is good for us, as it produces vitamin D in our skin, but larger doses can be harmful to our eyes and may cause skin cancer.

Visible light from lasers is used to send digital data down fibre-optic cables at huge speeds with little loss of signal quality. This has transformed global communications over recent decades.

Infrared radiation has longer wavelengths than visible red light. Electrical heaters and traditional cookers (sources) transfer energy by infrared radiation. Thermal imaging cameras (absorbers) detect low levels of infrared radiation from warm objects.

3. **Suggest why you can't get sunburned through a window, even though you can still see the sunlight.**
4. **Suggest how a thermal image of your house might be useful.**

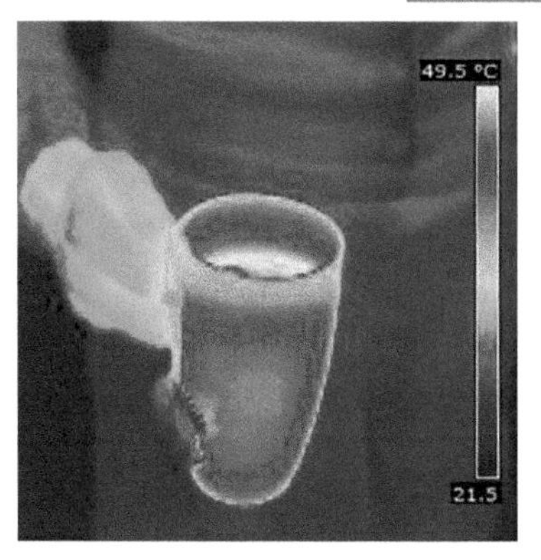

Figure 1.4.19 A thermal imaging camera detects infrared radiation given off by an object or person

Microwaves and radio waves

Microwaves are radio waves with short wavelengths, between 1 mm and 30 cm. Microwaves are used for cooking in microwave ovens, and also to transmit mobile phone signals. Transmitters are placed on high buildings or masts to give better communication over large distances. Sometimes signals are sent from a transmitter (a source) to a receiver (an absorber) via a satellite.

Microwaves used for communication have a longer wavelength than those used for cooking.

Terrestrial radio and TV signals are sent by **radio waves**, travelling at the speed of light. A radio telescope forms images of objects in space by detecting the radio waves they emit. Radio waves have much longer wavelengths than visible light, and a radio telescope must be very large to have the resolution of an optical telescope.

DID YOU KNOW?

Halogen hobs use ring-shaped halogen lamps beneath a glass top. Although you see a bright red light, the glowing filament radiates mostly infrared radiation.

5. **Suggest why satellite TV dishes are placed on the walls or roofs of houses.**
6. **Like many uses of electromagnetic radiation, a radio telescope illustrates the transfer of energy from a source to an absorber. In this example, identify the source, the form in which energy is transferred, and the absorber.**

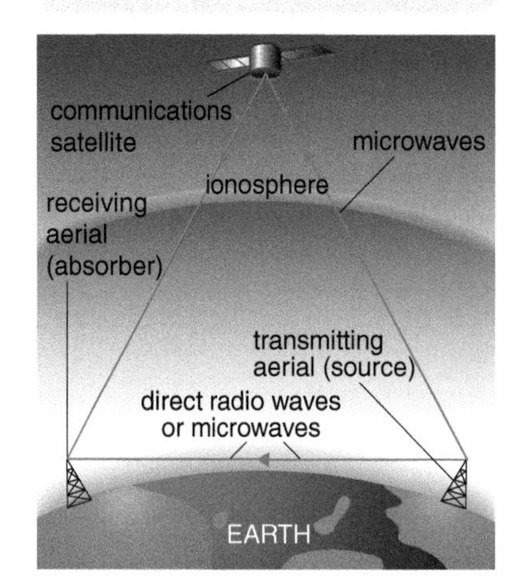

Figure 1.4.20 Direct and satellite wave communication. Microwave satellite communications are used for satellite phones and for satellite TV

HIGHER TIER ONLY

Radio waves are produced by oscillations in electrical circuits. When a current flows through a wire it creates electric and magnetic fields around the wire. When the current changes, fields change, which produces electromagnetic waves of radio frequencies. This is how radio transmitters work.

Radio waves can induce oscillations in an electrical circuit, giving rise to an alternating current with the same frequency as the radio wave itself.

7. **When radio waves are absorbed by an electrical conductor they create an alternating current with the same frequency as the radio wave itself. Suggest how this might be used in a radio receiver.**

DID YOU KNOW?

Bluetooth is a wireless technology for exchanging data over short distances using microwaves.

REQUIRED PRACTICAL

1.4h Investigating infrared absorption and radiation

Learning objectives:

- use appropriate apparatus to observe the interaction of electromagnetic waves with matter
- explain methods and interpret results
- recognise the importance of scientific quantities and understand how they are determined
- use SI units.

KEY WORDS

absorbing
emitting
infrared radiation

Anything that is warm gives off infrared radiation and the warmer it is the better a transmitter it is. However, the colour of an object also affects how much radiation it emits. If we want a radiator to work as well as possible, the colour matters.

Colour also makes a difference to absorbing infrared radiation. If you want to make sure you stay cool on a hot day, choose the right clothing. You'll cook in the wrong colour!

These pages are designed to help you think about aspects of the investigation rather than to guide you through it step by step.

SAFETY

A risk assessment is required for this practical work.

Ensure heater complies with local education authority and CLEAPSS regulations.

Investigating absorption

Alex's teacher is showing the class how to compare different surfaces to see which is better at **absorbing infrared radiation**. She has set up two metal plates, one on either side of a heater. One plate has a shiny surface and the other has been blackened. On the back of each plate a cork stopper has been stuck on with wax. The heater is turned on. After a few minutes one of the stoppers drops off and the other follows several minutes later.

1. **What does this experiment show?**
2. **Which stopper do you predict will drop off first? Explain your reasoning.**
3. **What needs to be done to make sure the experiment is a fair test?**

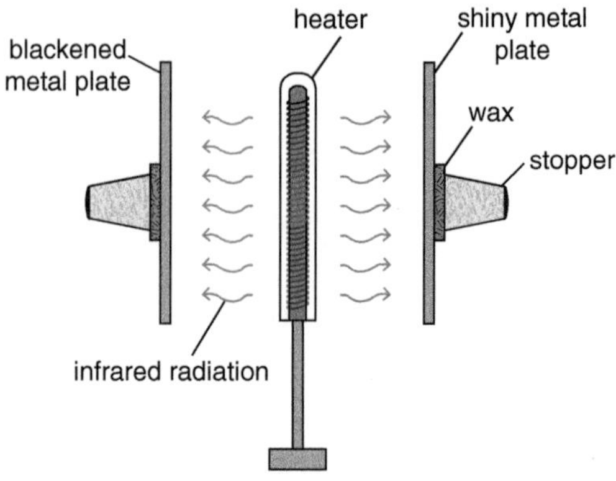

Figure 1.4.21 Experiment to compare absorption of infrared radiation by different surfaces

Commenting on the design of an experiment

Experiments have to be designed well if the results are to be meaningful. The experiment shown in Figure 1.4.22 has been designed to compare how good different surfaces are at **emitting** infrared radiation (radiating thermal energy). The cube is filled with very hot water and an infrared detector is held opposite each of the various faces in turn. The different faces have different finishes on them and it is possible to compare the readings from each.

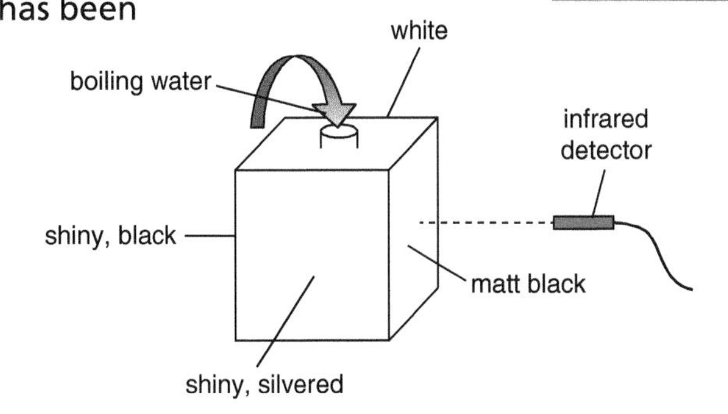

Figure 1.4.22 Experiment to compare infrared radiation emitted by different surfaces

4. **Why does the experiment use one can with four different surfaces rather than four cans, each a different type of surface?**
5. **Why is it designed so that each face has the same width?**
6. **What needs to be true about the positioning of the infrared detector?**
7. **If the detector is held at each face in turn for three 30-second periods, and the average reading for each face calculated, what type of graph could you use to display the data?**

An alternative way of investigating the amount of infrared radiation emitted by different surfaces is to replace the infrared detector with a thermometer with a blackened bulb.

8. **Using what you learned about absorbers of infrared radiation in the first investigation, explain why a thermometer with a blackened bulb is used.**
9. **How does the temperature shown give a measure of the amount of infrared emitted by each surface?**

DID YOU KNOW?

Firefighters use infrared cameras to locate unconscious bodies in smoke-filled buildings. Even though visible light can't penetrate the smoke to enable the body to be seen, the infrared radiation emitted by the body can.

KEY INFORMATION

Make sure you understand about the main features of the design of an experiment. The equipment has been developed in a particular way for good reasons and it is important to understand now what they are.

Applying the ideas

Some ideas in science have an immediate application and ones about infrared affect our everyday lives. We can use the results of these experiments to inform us about the way that various objects should be designed.

Think about the results of the investigation into absorbing infrared radiation.

10. **What does this suggest about the best colour for:**
    - **a** **a firefighter's suit?**
    - **b** **solar panels on the roof of a house, absorbing thermal energy into water to use inside?**

Now think about the experiment on radiating infrared radiation.

11. **What does this suggest about the best colour for:**
    - **a** **a teapot?**
    - **b** **the pipes on the back of a refrigerator, which need to radiate thermal energy from the inside of the fridge to its surroundings?**

1.4i Reflection and refraction of electromagnetic waves

Learning objectives:

- recall that different substances may refract or reflect electromagnetic waves
- construct ray diagrams to illustrate refraction at a boundary
- use wavefront diagrams to explain refraction in terms of change of wave speed in different substances.

KEY WORDS

diffuse
ray diagram
reflection
refraction
specular
wavefront

When an electromagnetic wave meets the surface of another substance, it may be reflected or travel on into the substance. How the wave behaves depends on the nature of the surfaces and the speed of the wave in the different substances.

HIGHER TIER ONLY

Reflection of electromagnetic waves

Most objects you see are visible because they reflect light. If the surface is shiny, it acts like a mirror, and an image is formed. All the light rays from one direction are reflected at the same angle. But at a rough surface, like cloth or paper, the light rays from one direction are scattered in all directions.

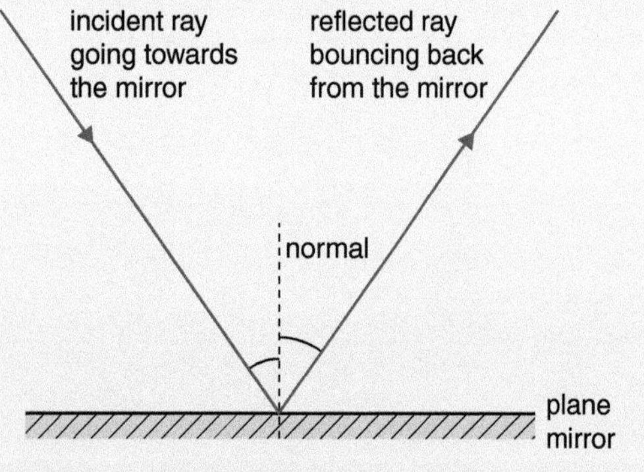

Figure 1.4.23 **Ray diagram** for reflection from a plane mirror. The incident and reflected rays make equal angles with the normal (a line at 90° to the surface of the mirror)

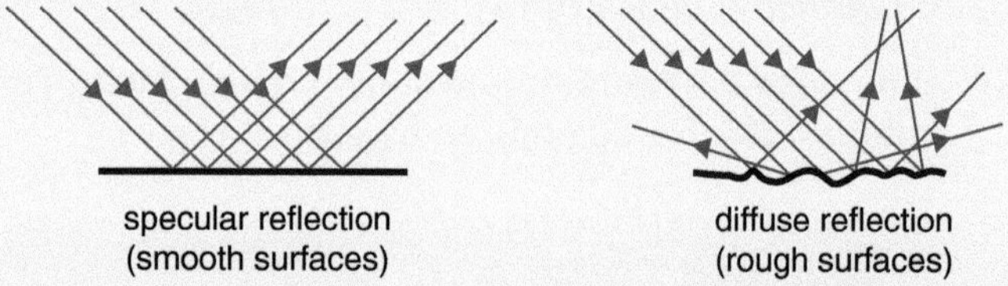

specular reflection (smooth surfaces)

diffuse reflection (rough surfaces)

Figure 1.4.24 **Reflection** from a smooth, shiny surface is called a **specular** reflection. A rough surface scatters light in a **diffuse** reflection

1. **Explain why you can't see a reflection of yourself when you look at a sheet of paper.**
2. **Why do your clothes stay the same colour in a reflection from a mirror?**

KEY INFORMATION

Studying how visible light behaves is relatively easy because we can see what happens. We can then predict, check and understand how other electromagnetic waves behave.

Refraction of electromagnetic waves

If a light ray in air hits another transparent medium, such as glass or water, it may be transmitted. Using a ray of light directed at a glass block, we can see how the ray behaves.

The ray diagram in Figure 1.4.25 shows light bending towards the normal when it enters the glass block and away from the normal when it leaves.

This is called **refraction**. The change of direction happens at the boundary between the two media.

3 **Draw a ray diagram for a ray of light passing from water into air. (Hint: the ray behaves the same as when passing from glass to air.)**

4 **Explain what the 'normal' is in a ray diagram.**

emergent ray
glass block
air
air
normal
incident ray

Figure 1.4.25 Light is refracted towards the normal when it travels from air to glass, and away from the normal when it travels from glass to air

Explaining refraction

A **wavefront** is a line that joins all the points on a wave that are moving up and down together at the same time. The wavefront is at right angles to the direction the wave is travelling.

We can see what happens when waves are refracted by looking at plane water waves in a ripple tank. In Figure 1.4.26 the water wave travels from deep water (Medium 1) into shallow water (Medium 2) at an angle to the boundary. The left-hand part of the wave reaches the boundary first. These wavefronts get closer together because waves travel more slowly in shallower water, and the wavefront changes direction. The wave is refracted towards the normal.

KEY INFORMATION

In a ray diagram, the arrow represents the direction in which the waves is travelling. For simplicity, the wavefronts are not shown, but they would be at right angles to the direction of the ray.

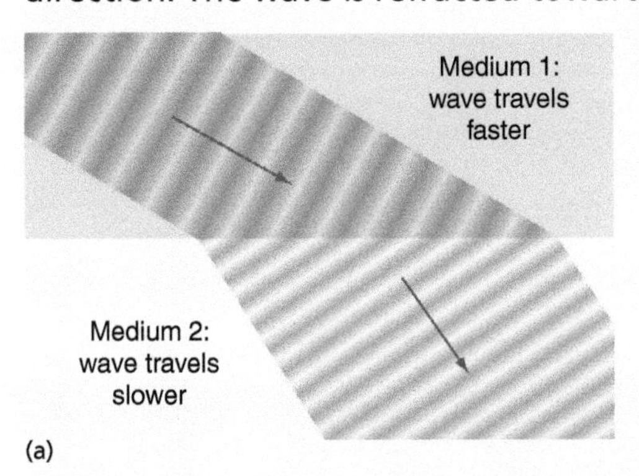

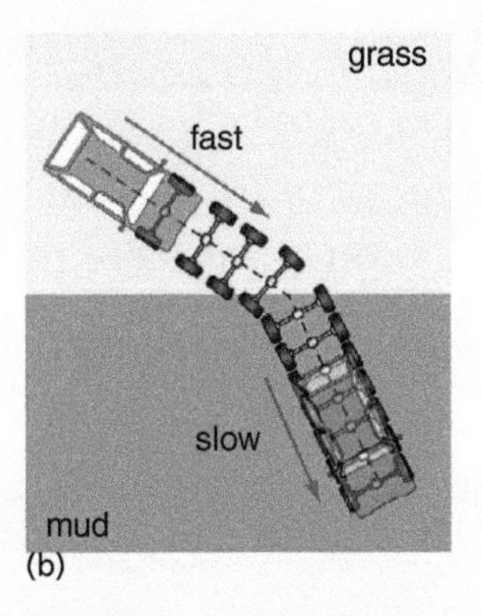

Figure 1.4.26 (a) The direction of travel of the wave, shown by the arrows, bends towards the normal when the wave slows down. (b) When a car drives at an angle into mud, the wheels that reach the mud first slow down first so the car changes direction. (The differential in the axle allows the wheels to turn at different speeds)

Wavefront diagrams like Figure 1.4.26 can be used to show how electromagnetic waves refract at boundaries. The diagram is the same for light travelling from air into water, as light travels more slowly in water than in air.

5 **Draw a wavefront diagram for a wave travelling from water to air.**

6 **The wavefronts get farther apart after a wave leaves a boundary. What has happened to the speed of the wave?**

KEY INFORMATION

From the wave equation, speed = frequency × wavelength. The frequency (number of waves per second) does not change when a wave moves across a boundary, so for a smaller speed the wavelength must also be smaller.

MATHS SKILLS

1.4j Using and rearranging equations

Learning objectives:

- select and apply appropriate equations
- substitute numerical values into equations using appropriate units
- change the subject of an equation.

KEY WORDS

proportional
rearrange an equation
subject of an equation
substitute

When we have values for all the variables in an equation except one, we can calculate the missing variable. It's easier to do this if we first rearrange the equation so the missing variable is the subject.

Period and frequency

Let's use the equation linking period and frequency as an example:

$$T = \frac{1}{f}$$

Remember, frequency, f, is the number of waves passing a point each second. The unit of frequency is the hertz (Hz), which means cycles *per second*. The period, T, is measured in seconds (s).

It's easy to use the equation in this form to calculate the period of a wave, T, if you know the frequency, f.

Example: A wave has a frequency of 5 Hz. What is its period?

Substituting 5 Hz into the equation $T = \frac{1}{f}$ gives $T = \frac{1}{5}$ s. The period is $\frac{1}{5}$ s or 0.2 s.

But what if you know the period, T, and you want to find out the frequency, f?

Example: Calculate the frequency when the period is 4 seconds.

Use the equation $T = \frac{1}{f}$

Rearrange it to make f the **subject of the equation.**

Multiply both sides by f: $Tf = 1$

Divide both sides by T: $f = \frac{1}{T}$

Substitute $T = 4$ s into the rearranged equation.

$$f = \frac{1}{4\,s} = 0.25 \text{ Hz}$$

KEY INFORMATION

You do not need to remember the equations on this page as they will be on the equation sheet. But you need to know when and how to use each equation.
You also need to be able to rearrange an equation. Rearranging an equation means making another variable the subject of the equation. The subject of the equation is on its own, usually on the left-hand side.

1 **Work out the period of a wave when the frequency is:**

a **100 Hz** **b** **1000 Hz**
c **15 000 Hz.**

2 **Work out the frequency of a wave with period:**

a **5 s** **b** **10 s** **c** **150 s.**

Speed, frequency and wavelength

The wave equation links wave speed, frequency and wavelength:

wave speed = frequency × wavelength

$v = f\lambda$

Practise rearranging this equation as required to answer the following questions.

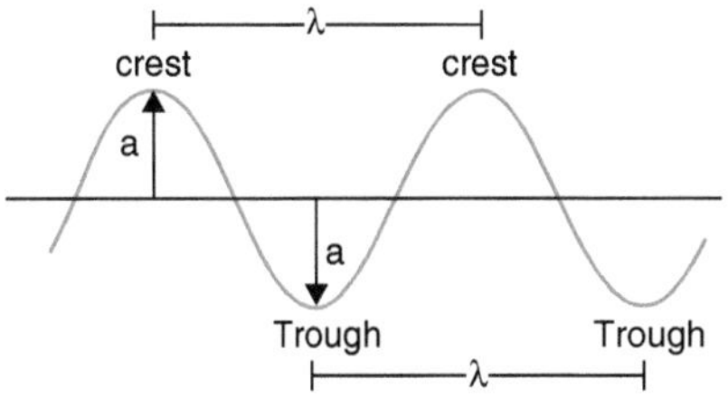

Figure 1.4.27 Wavelength of a transverse wave

3 **Use the equation $v = f\lambda$ to calculate the speed of a wave with:**

a **frequency 100 Hz and wavelength 2 m**

b **frequency 100 Hz and wavelength 2 cm**

c **frequency 100 Hz and wavelength 2 mm.**

4 **Rearrange the wave equation to make frequency the subject.**

5 **Calculate the frequency of a wave with**

a **wavelength 0.5 m and speed 25 m/s**

b **wavelength 0.05 m and speed 250 m/s**

c **wavelength 0.005 m and speed 2500 m/s.**

6 **a** **Rearrange the wave equation to calculate the wavelength given the speed and the frequency.**

b **Calculate the wavelength of a sound wave in air with frequency 500 Hz and speed 330 m/s.**

MATHS

When you rearrange an equation, always do the same operation (such as multiplication or division) to both sides. For example, if you divide one side of the equation by the variable λ, you must divide both sides of the equation by λ.

Changes in velocity, frequency and wavelength

Sound travels at different speeds through different media. One of the factors that contributes to this is density. The denser the material, the closer the atoms or molecules and the quicker the vibrations will travel through that substance.

If speed increases when a sound wave is transmitted from one medium to another, what effect does that have on the wave's frequency and wavelength?

If speed increases then either frequency or wavelength must also increase. This is because in the wave equation, wave speed is **proportional** to frequency and wavelength.

Material	Speed of sound (m/s)
Air	330
Water	1500
Wood	3300
Titanium	6100

7 **Red light has a wavelength of 6.5×10^{-7} m and moves a distance of 3.0 m in 1.0×10^{-8} s. Calculate the frequency of red light.**

8 **A sound wave has a wavelength of 25 cm in air.**

Use the data in the table to calculate its wavelength in titanium. Assume that the frequency of the wave does not change.

Check your progress

You should be able to:

☐ understand that waves transfer energy and information. →	☐ give examples of energy transfer by waves (including electromagnetic waves). →	☐ describe evidence that, e.g. for ripples on a water surface, it is the wave and not the water itself that travels.
☐ know that waves can be transverse or longitudinal. →	☐ give examples of longitudinal and transverse waves. →	☐ explain the difference between transverse and longitudinal waves.
☐ describe how sound waves travel through air or solids. →	☐ describe how to measure the speed of sound waves in air. →	☐ explain how to calculate the depth of water using echo – sounding.
☐ describe the amplitude, wavelength, frequency and period of a wave. →	☐ use the wave equation $v = f\lambda$ to calculate wave speed. →	☐ rearrange and apply the wave equation.
☐ name the main groupings of the electromagnetic spectrum. →	☐ compare the groupings of the electromagnetic spectrum in terms of wavelength and frequency. →	☐ describe how radio waves are produced (HT only).
☐ give examples of the uses of the main groupings of the electromagnetic spectrum. →	☐ describe examples of energy transfer by electromagnetic waves. →	☐ explain why each type of electromagnetic wave is suitable for its applications.
☐ understand that waves can be absorbed, transmitted or reflected at a surface (HT only). →	☐ describe examples of transmission and reflection of waves (including electromagnetic waves) at material interfaces (HT only). →	☐ explain why different substances may refract or reflect electromagnetic waves (HT only).
☐ draw a labelled ray diagram to illustrate reflection of a wave at a boundary (HT only). →	☐ construct ray diagrams to illustrate refraction at a boundary (HT only). →	☐ use wavefront diagrams to explain refraction in terms of a change in wave velocity (HT only).

Worked example

1 **The table below shows the electromagnetic spectrum.**

A	microwave	infrared	visible light	B	X-rays	gamma rays

Write down the names of the waves labelled A and B.

A = radio waves B = ultraviolet waves

Both answers are correct. You could try using a mnemonic to remember the correct order.

2 **X-rays and gamma rays are dangerous to humans. Explain how they can also be used in medical contexts without lasting harm.**

We can use them for X-rays to see our bones and for treating cancer by killing the cancer cells.

This answer is a good start but is incomplete. It doesn't give a full explanation. The answer should also say that we can use them in medical contexts by controlling the exposure dose.

3 **A ship is mapping the seabed using echo sounding. It sends out a sound wave and receives an echo 1.6 s later. If the depth of the water is 1200 m at that position, what is the speed of sound in water?**

$$\text{speed} = \frac{\text{distance}}{\text{time}} = \frac{1200\text{ m}}{1.6\text{ s}} = 750\text{ m / s}$$

The correct equation has been chosen, but the calculation is incorrect. The student has forgotten that the sound wave travels down to the seabed and back in 1.6 s, so the total distance is 2400 m. The speed of sound in water should be 1500 m/s.

4 **A red laser pointer has a wavelength of 650 nm. If the speed of light is 3.0×10^8 m/s, what is the frequency of the laser light waves?**

speed = frequency × wavelength, or

$v = f\lambda$

Rearranging the equation to make frequency the subject:

$f = \frac{v}{\lambda}$

$$f = \frac{3.0 \times 10^8\text{ m / s}}{6.5 \times 10^{-7}\text{ m}} = 0.46 \times 10^{15}\text{ Hz}$$

frequency of red light waves = 4.6×10^{14} Hz

This is a good answer – the calculation is correct and it is clearly set out. The wavelength in nm has been correctly converted to a value in m. Always take extra care when working with standard form.

5 **Complete the wavefront diagram to show the refraction of light from air to water.**

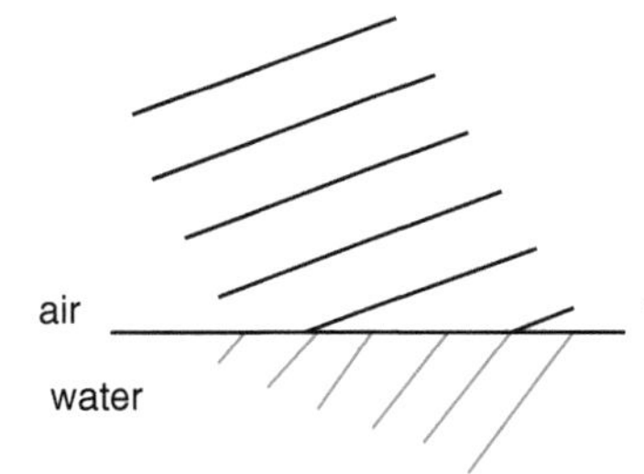

This answer is only partly correct. The wavefronts are closer together in the denser medium, which is correct. But, crucially, when waves go from a less dense medium to a denser medium, they are refracted towards the normal. Here they are refracted *away* from the normal. Also, each wavefront must be continuous from one medium into the other.

End of chapter questions

Getting started

1. Which of the following is a longitudinal wave? **1 Mark**
   A sound wave B water wave C light wave D radio wave

2. A wave is often drawn as a series of peaks and troughs. Use this idea to explain what is meant by the wavelength of the wave. **2 Marks**

3. Which is a correct unit of frequency? **1 Mark**
   A metres B watts C hertz D metres per second

4. A water wave in a ripple tank has a frequency of 4 Hz. Calculate the period (T) of the wave using the formula $T = 1/f$. **1 Mark**

5. A microwave has a wavelength of 1 cm. What type of wave, other than a microwave, might have a wavelength of 1 m? **1 Mark**

6. A ray of light is shone onto a dull, white object but no reflected ray is seen. Explain what type of reflection this is and what happens to the light. **2 Marks**

7. In the diagram, the red lines represent four different waves.
   a Which two waves have the same frequency?
   b Which two waves have the same amplitude?
   c Which wave has the shortest wavelength? **6 Marks**

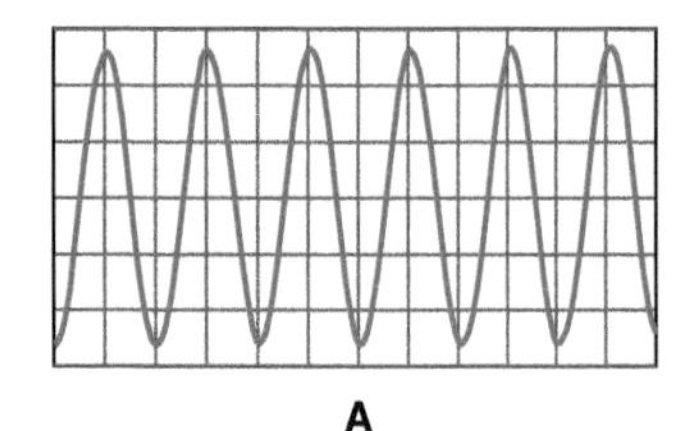

A

B

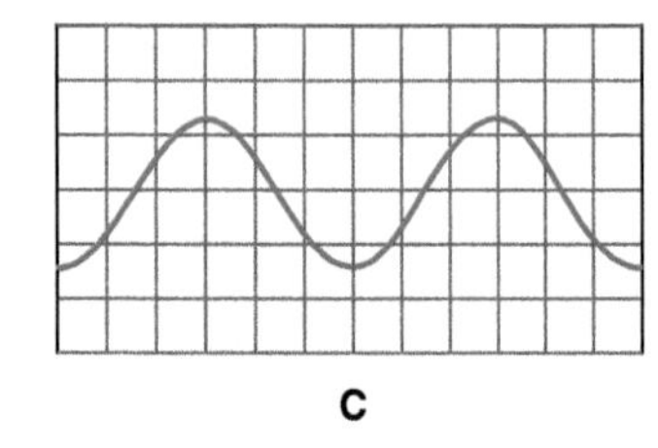

C

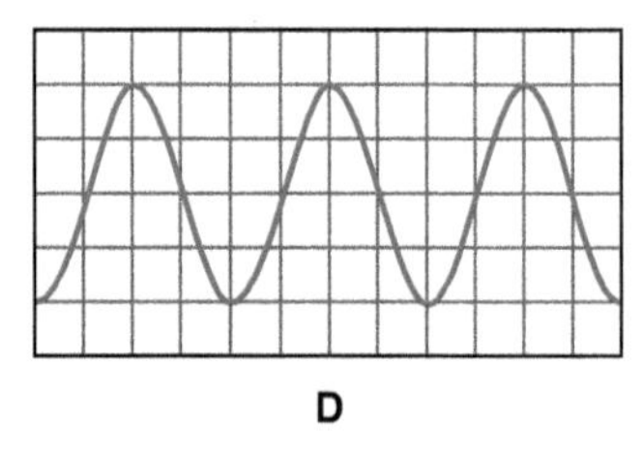

D

Going further

8. Name the parts of the electromagnetic spectrum labelled A and B in the diagram. **1 Mark**

radio waves	microwave	A	visible light	ultraviolet	B	gamma rays

9. Name the two types of electromagnetic radiation that are used to cook food. **1 Mark**

10. Describe one practical application each for
    a microwaves
    b gamma rays. **2 Marks**

11 a Which electromagnetic wave has the shortest wavelength?

b Which electromagnetic wave has the highest frequency?

2 Marks

12 What is the speed of a water wave if it has a wavelength of 8 cm and a frequency of 2 Hz?

2 Marks

More challenging

13 a What type of waves are electromagnetic waves?

b Describe the electromagnetic spectrum.

2 Marks

14 Explain what happens to a ray of light as it goes into a transparent plastic block.

2 Marks

15 Construct a wavefront diagram to show waves in a ripple tank passing into a shallower area created by a block on the bottom of the tank.

3 Marks

Most demanding

16 A survey ship sailing in a straight line at a speed of 4 m/s sends out a sound pulse from its echo-sounder every 10 s. The time between each pulse being sent and its reflection are as follows; 0.25 s, 0.30 s, 0.35 s, 0.40 s. 0.40 s, 0.40 s, 0.40 s, 0.05 s, 0,40 s, 0.35 s, 0.30 s, 0.25 s, 0.25 s and 0.25 s.

Use this data to describe the seabed the ship is sailing over and suggest a reason for the apparent anomalous result.

6 Marks

17 An ultrasound scanner was used to measure the size of a kidney stone. Use the following data to determine the thickness of the stone.

6 Marks

Speed of ultrasound in the stone = 4.00 km/s

Time between emitted ultrasound pulse and reflection from the front of the stone = 0.0500 ms

Time between emitted ultrasound pulse and reflection from the back of the stone = 0.0600 ms

Total: 40 Marks

How the ideas in this topic link together

Large organisms need systems to take in and transport substances to and from every cell in the body. For example, oxygen and glucose must be delivered for cellular respiration, and waste products such as carbon dioxide must be carried away. Specialised surfaces have evolved to import and export such substances, and networks transport them within the organism.

Transport systems in an organism are rather like a bus route – they carry a range of substances, and there are particular places within the organism where materials are taken on board and off-loaded.

In animals like humans, transport and exchange processes are monitored and controlled by nervous and hormonal communications. This is similar to the way that bus and rail transport systems are coordinated by a control centre.

Working Scientifically Focus

- Using models and diagrams to represent processes
- Analysing experimental data

2 Transport over larger distances

Contents

SYSTEMS IN THE HUMAN BODY

IDEAS YOU HAVE MET BEFORE:

ORGANISMS OBTAIN ENERGY BY THE PROCESS OF RESPIRATION.

- The energy released in respiration drives all the processes necessary for life.
- Most organisms respire by aerobic respiration, using oxygen.
- Some cells or organisms can survive without oxygen. They respire anaerobically.

BLOOD IS USED TO TRANSPORT SUBSTANCES TO AND FROM BODY TISSUES.

- The circulatory system moves substances around the body in the blood.
- The function of the heart is to pump blood around the body.
- Blood is made up of red blood cells, white blood cells, plasma and platelets.
- Products of digestion are carried to cells and used to build carbohydrates, lipids and proteins.
- Glucose is required by all cells for respiration; carbon dioxide and water are waste products.
- Oxygen is taken from the lungs to the body. Carbon dioxide is returned from the body to the lungs, where it is excreted.
- Excess protein is converted to urea and transported in the blood to be excreted by the kidneys.

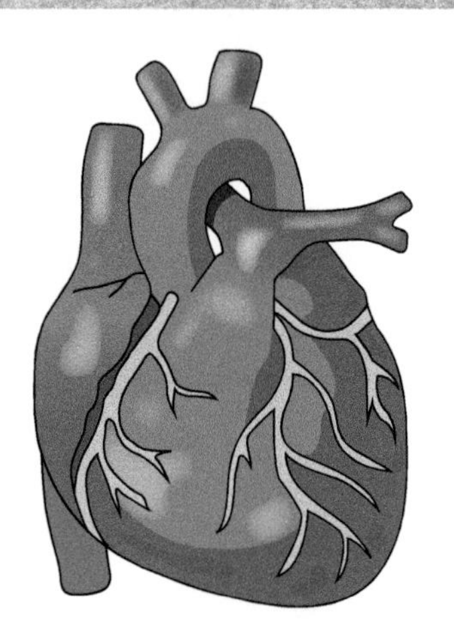

ORGANS WORK TOGETHER AS SYSTEMS.

- Organs are aggregations of tissues.
- Organ systems work together to form organisms.
- Glands in the digestive system produce digestive enzymes.

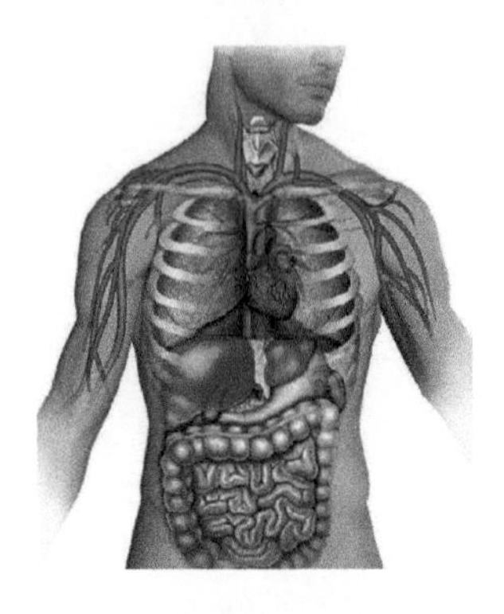

2.1

IN THIS CHAPTER YOU WILL FIND OUT ABOUT:

ANAEROBIC RESPIRATION IS RESPIRATION WITHOUT OXYGEN.

- In animals such as mammals, the muscles can respire anaerobically for short periods during intense activity.
- Anaerobic respiration involves incomplete breakdown of glucose and therefore releases much less energy than aerobic respiration.

WHY DO SOME ORGANISMS NEED A CIRCULATORY SYSTEM?

- Size affects the ability and efficiency of diffusion alone to supply cells with nutrients.
- Small organisms do not have specialised organs for gas exchange or transport of some materials.
- Membrane surfaces and organ systems are specialised for exchanging materials to ensure that all body cells get the nutrients that they need.
- Mammals have transport systems – for example, the heart works as a pump for the transport system in humans.

HOW ARE CONDITIONS IN THE BODY, PROCESSES AND ORGAN SYSTEMS COORDINATED AND CONTROLLED?

- The nervous and endocrine systems are involved in coordination and control.
- The nervous system works using electrical impulses, transmitted using nerves.
- The endocrine system uses chemicals called hormones, which are secreted by endocrine glands.
- The control of hormone secretion by many glands is by negative feedback.

2.1a Cellular respiration

Learning objectives:

- explain the need for energy
- describe aerobic respiration as an exothermic reaction
- write a balanced symbol equation for respiration, given the formula of glucose.

KEY WORDS

aerobic respiration
cellular respiration
exothermic

This runner is using energy to run a marathon. But we all need a continuous supply of energy – 24 hours a day – just to stay alive.

Figure 2.1.1 An average runner uses around 13 000 kJ of energy for a marathon

We need energy to live

Organisms need energy:

- to drive the chemical reactions needed to keep them alive, including building large molecules
- for movement
- to keep warm.

Energy is needed to make our muscles contract and to keep our bodies warm. It's also needed to transport substances around our bodies.

In other sections of the book, you will also find out that energy is needed:

- for cell division
- to maintain a constant environment within our bodies
- for active transport – plants use active transport to take up mineral ions from the soil, and to open and close their stomata
- to transmit nerve impulses.

1 List four uses of energy in animals.

Aerobic respiration

Cellular respiration is the process used by all organisms to release the energy they need from food.

Cellular respiration using oxygen is called **aerobic respiration**. This type of respiration takes place in animal and plant cells, and in many microorganisms.

Glucose is a simple sugar. It is the starting point of cellular respiration in most organisms. The food that organisms take in is, therefore, converted into glucose.

This chemical reaction is **exothermic**. A reaction is described as exothermic when it releases energy. Some of the energy transferred is released as thermal energy.

Figure 2.1.2 Birds and mammals maintain a constant body temperature

2. **What is the purpose of cellular respiration?**
3. **How do birds and mammals make use of the waste thermal energy?**

Bioenergetics

This is the equation for aerobic respiration:

glucose + oxygen → carbon dioxide + water (energy released)

HIGHER TIER ONLY

$$C_6H_{12}O_6 + 6O_2 \longrightarrow 6CO_2 + 6H_2O$$

This equation describes the overall change brought about through each of a series of chemical reactions. A small amount of energy is actually released at each stage in the series.

The first group of steps occurs in the cytoplasm of cells, but most of the energy is transferred by chemical reactions in mitochondria.

4. **When and where does respiration occur?**
5. **Give one characteristic feature of actively respiring cells.**
6. **Why do we often get hot when we exercise?**

DID YOU KNOW?

The muscle an insect uses to fly is the most active tissue found in nature.

Figure 2.1.3 Insect flight muscles have huge numbers of well-developed mitochondria

COMMON MISCONCEPTIONS

Don't forget that *all* organisms respire. The equation is the reverse of photosynthesis, but don't confuse the two. Photosynthesis is the way in which plants make their food.

2.1b Comparing aerobic and anaerobic respiration

Learning objectives:

- describe the process of anaerobic respiration in humans
- compare the processes of aerobic and anaerobic respiration.

KEY WORDS

anaerobic respiration
lactic acid
oxygen debt

Without oxygen, you would die. But when your muscles are actively contracting, they run short of oxygen. When this happens, muscle cells can respire without oxygen for short periods of time. This is anaerobic respiration – respiration without oxygen. Lactic acid is produced.

Anaerobic respiration

If there is not enough oxygen available for aerobic respiration, **anaerobic respiration** takes place in the cytoplasm of muscle cells. It is represented by the equation:

$$\text{glucose} \xrightarrow{\text{energy released}} \text{lactic acid}$$

1. **What is meant by 'anaerobic respiration'?**
2. **When does anaerobic respiration happen in muscle cells?**
3. **Write down the word equation for anaerobic respiration in muscles.**

Incomplete oxidation

Anaerobic respiration in muscles is exothermic (a chemical reaction that releases energy), but it is much less efficient than aerobic respiration. Unlike in aerobic respiration, glucose molecules are not completely broken down (oxidised), which means that much less energy is released – only around 5% of the energy released by aerobic respiration, per molecule of glucose. However, in situations where muscles are running short of oxygen, the amount of energy produced is enough to keep cells running for a while longer. The waste product is **lactic acid** rather than carbon dioxide and water.

4. **Explain why anaerobic respiration releases much less energy than aerobic respiration.**
5. **Why is anaerobic respiration important in muscle cells?**

 Link: 2.1b ⟶ 2.1c

Oxygen debt

During long periods of vigorous activity, our muscles become tired. This is partly because the lactic acid from anaerobic respiration builds up in the muscles and stops them from contracting efficiently.

We say that anaerobic respiration creates an **oxygen debt**. This is the amount of extra oxygen the body needs to react with the accumulated lactic acid and remove it from the cells. The oxygen oxidises the lactic acid to carbon dioxide and water.

Oxygen debt explains why, after exercise, we carry on breathing deeply and quickly for a while – the body is taking in extra oxygen, and transporting it to the muscles in the blood, to remove the build-up of lactic acid from anaerobic respiration.

	Aerobic respiration, in humans	Anaerobic respiration, in humans
Where	in all cells	in muscle cells
When	all the time	during vigorous exercise
By-products	carbon dioxide + water	lactic acid
Energy produced	a great deal	much less – only 5% of that from aerobic respiration
Oxygen needed	yes	no
Oxygen debt produced	no	yes

Figure 2.1.4 This runner is breathing deeply to help clear the lactic acid build-up in her muscles

6 **Using the term 'oxygen debt', explain why a sprinter continues to breathe deeply after running a 100 m race.**

KEY SKILL

You must be able to compare aerobic and anaerobic respiration: the need for oxygen, the products and the amount of energy transferred.

2.1c The need for transport systems in multicellular organisms

Learning objectives:

- explain why multicellular organisms need a transport system, in terms of surface area : volume ratio.

KEY WORDS

diffusion
organs
organ systems
surface area : volume ratio
tissues

A single-celled organism has a relatively large surface area : volume ratio, so substances can diffuse quickly into and out of all parts of the cell. Larger, multicellular organisms have a smaller surface area : volume ratio and cannot rely on diffusion alone. Instead, they need specialised organ systems to carry materials to and from every cell in the body.

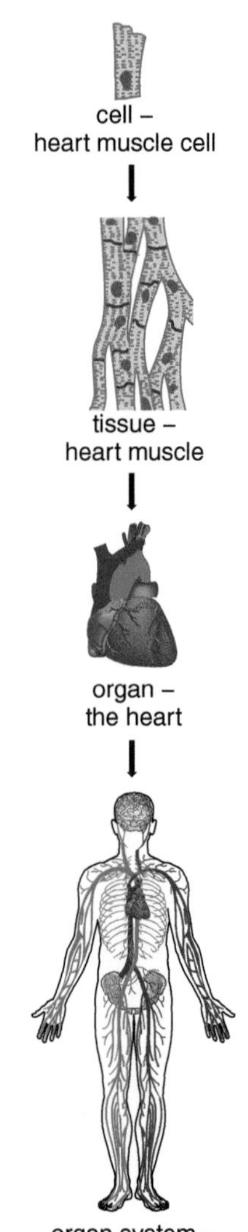

Figure 2.1.5 The organisation of the human circulatory system

Cells, tissues and organ systems

In multicellular organisms, most cells work together in tissues. A **tissue** is a group of cells with a particular function. Many tissues have a number of similar types of cell to enable the tissues to carry out their function.

Tissues are grouped into **organs**. Organs also carry out specific functions.

Different organs are arranged into **organ systems** – for example the circulatory system, digestive system, respiratory system and reproductive system.

Many chemical reactions happen inside living cells. Substances such as nutrients and oxygen must enter the cell to fuel these reactions. The waste products of the reactions, such as carbon dioxide, need to be removed. The circulatory system, digestive system and respiratory system work together to transport these substances to and from all cells in the body.

1. **Arrange the following in ascending order of size:**

   system cell human body organ tissue

2. **Name two types of cell and two types of tissue in the circulatory system.**

Size matters

Most cells are no more than 1 mm in diameter. This is because nutrients, oxygen and waste substances can diffuse quickly in and out of small cells. However, as the volume of a cell increases, the distance increases between the cytoplasm at the centre of the cell and the cell membrane. In addition, larger cells have greater chemical activity, and so need substances to be moved in and out at a greater rate. In cells bigger than 1 mm diameter, the rate of exchange with the surroundings by **diffusion** may be too slow to meet the cell's needs. The cell would probably not survive.

3. **How does the size of a cell affect the chemical activity inside the cell?**
4. **Describe one chemical activity that takes place in cells.**

Looking at surface area : volume ratios

The surface area of a cell affects the rate at which particles can enter and leave the cell. It also affects the rate at which energy is lost or gained.

The volume of the cell affects the rates of chemical reactions within the cell (how quickly materials are used in reactions and how fast the waste products are made).

Look at Figure 2.1.6. The cubes represent cells.

- Cubes have six sides, so the surface area = length × width × 6.
- The volume of a cube = length × width × height.
- **Surface area : volume ratio** = surface area ÷ volume.

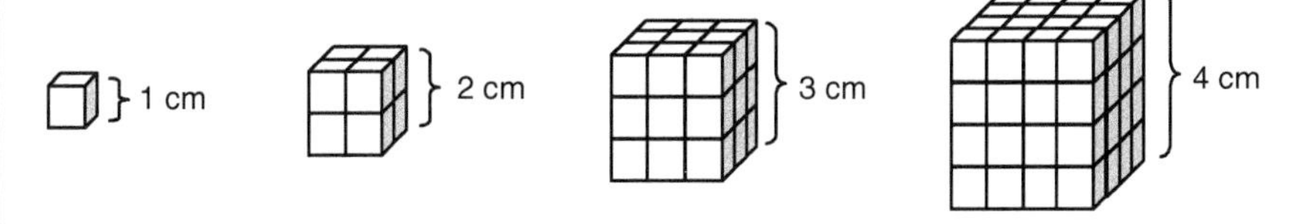

Figure 2.1.6 Cubes of increasing size

5. **What happens to the surface area and volume as size increases?**
6. **What happens to the surface area : volume ratio as size increases?**
7. **What does a decrease in surface area : volume ratio make it more difficult to do?**
8. **Why do multicellular organisms need a transport system?**

MATHS SKILLS

2.1d Surface area : volume ratio

Learning objectives:

- be able to calculate surface area and volume
- calculate and compare surface area : volume ratios
- know how to apply ideas about surface area and volume
- use SI units (e.g. m, mm)
- use prefixes and powers of ten for orders of magnitude (e.g. centi, milli, micro)
- interconvert units.

KEY WORDS

ratio
sphere

The ratio of surface area to volume is very important in living things. In science we use mathematical skills to calculate the surface area:volume ratio, which helps us to understand how organisms work.

Finding the area of a surface

Alveoli provide a large surface area for gas exchange between the air and the bloodstream. If our lungs were smooth on the inside, like balloons, the surface area would be much less and materials would not be exchanged fast enough.

We can calculate surface area in different ways. For a rectangular shape, we multiply the length by the width. For example, an area of skin 4 cm long and 7 cm wide has a surface area of (4 cm × 7 cm) = 28 cm^2.

Living things are not generally made up of regular shapes, such as rectangles, so we have to use other ways of finding the area. One way is to use squared paper and to count how many squares are covered (or are largely covered) by the specimen.

Figure 2.1.7 Elephants have wrinkled skin to increase their surface area

KEY INFORMATION

Note that both the length and the width have to be measured in the same units and that the answer is in those units, squared.

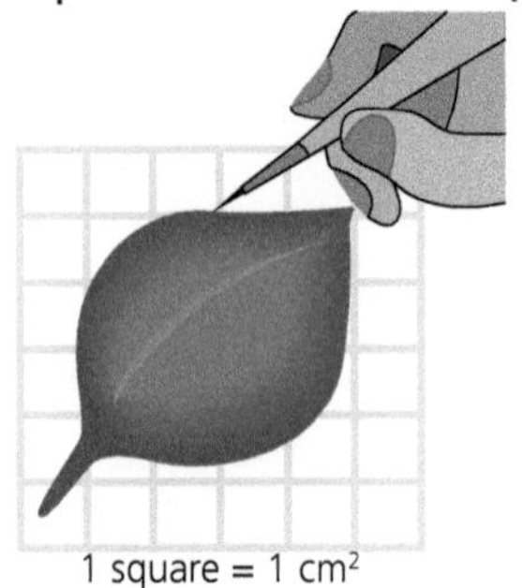

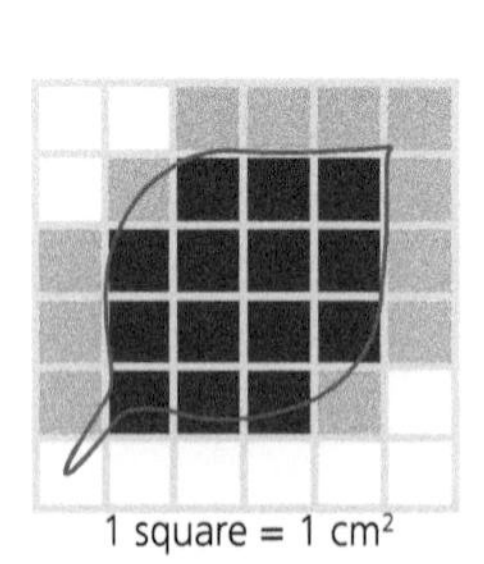

Figure 2.1.8 Counting squares that are largely covered is (approximately) balanced by not counting squares that are slightly covered

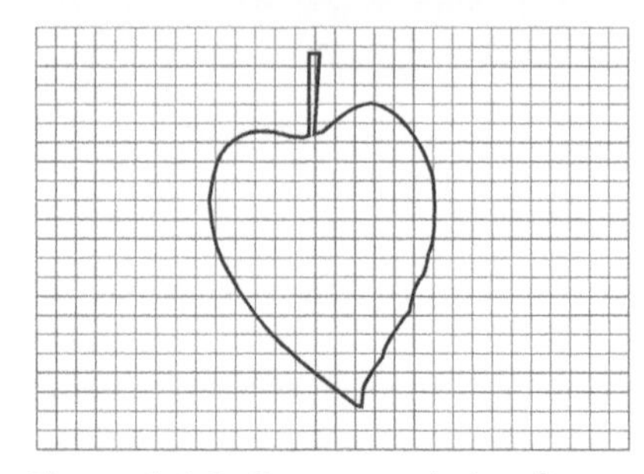

Figure 2.1.9 One square is 1 cm^2

1. **Estimate the surface area of the leaf in Figure 2.1.9.**
2. **Calculate the surface area of:**
   a a piece of tree bark that is 30 cm long and 3 cm wide
   b a razor shell that is 50 mm long and 8 mm wide.

Working out the volume

It is easier for warm-blooded animals to keep warm on cold days if their volume is large, but then it is harder for these same animals to lose heat on a hot day. We calculate the volume of a cube by multiplying length by width by height. A die with a side of 2 cm has a volume of (2 cm × 2 cm × 2 cm) = 8 cm^3. Again, all the distances need to be in the same units and the volume is also in those units, cubed.

3 What is the volume of:
a a science laboratory 10 m wide, 15 m long and 3 m high?
b a block of wood 2 cm wide, 3 cm long and 4 cm high?

4 Finding the volume of a tree branch is tricky, but one way is to immerse the branch in a tank full of water. Suggest how the volume is measured.

Surface area : volume ratio

In science it is useful to compare the surface area with the volume. We do this by finding the ratio of one compared with the other. To find the **ratio**, divide the surface area by the volume. For example, for a cube with sides 2 cm long:

surface area = 2 cm × 2 cm × 6 = 24 cm^2;
volume = 2 cm × 2 cm × 2 = 8 cm^3;
surface area : volume ratio = 24 : 8 = 3 : 1

The shape of an organism also affects its surface area : volume ratio. **Spheres** have the smallest surface area compared with their volumes. Many small mammals have a shape that is almost spherical – for example, a mouse – and puppies and kittens curl up into a ball to sleep. As small animals, they want to minimise their surface area : volume ratio so as to minimise thermal energy loss.

5 Compare how surface area, volume, and surface area : volume ratio change as the size of a cube increases. Calculate values for cubes with sides of 1, 2, 3, 4, 6 and 8 cm.

6 Imagine that, the shapes A, B and C in Figure 2.1.10 are animals.

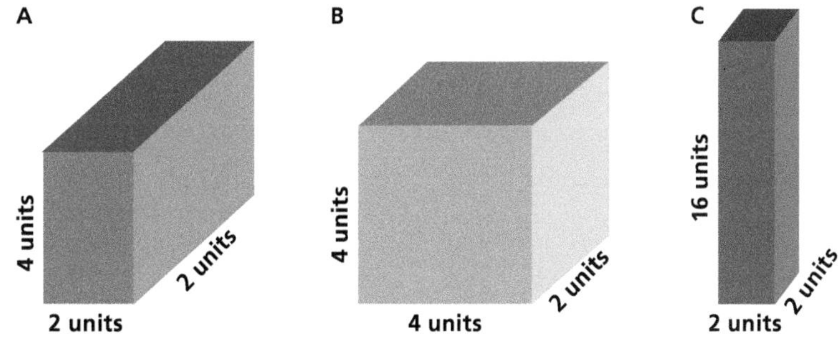

Figure 2.1.10 What do you notice about the surface area : volume ratios of these 'animals'?

Which animal will have problems keeping:
a cool? Explain your answer.
b warm? Explain your answer.

7 Devise a method to measure the volume of the air you breathe out in one breath.

DID YOU KNOW?

Arctic foxes have much smaller ears than desert-dwelling fennec foxes. Arctic foxes must reduce thermal energy loss but fennec foxes must increase thermal energy loss to their environment.

2.1e Exchange surfaces

Learning objectives:

- explain how efficient exchange surfaces are adapted to carry out their function
- calculate and compare surface area : volume ratios.

KEY WORDS

alveolus (plural: alveoli)
concentration gradient
diffusion
exchange surface
gas exchange
ventilation

A single-celled organism, with a relatively large surface area : volume ratio, can rely on diffusion alone to exchange substances with the environment. Multicellular organisms have much smaller surface area : volume ratios, and therefore need surfaces and organ systems that are specialised for exchanging materials efficiently with the surroundings.

Organism		Surface area : volume ratio
bacterium		6 000 000
amoeba		60 000
fly		600
dog		6
whale		0.06

Transport systems and exchange surfaces

When an organism is multicellular and has several layers of cells, oxygen and nutrients take longer to diffuse in and are all used up by the outer layers of cells. Exchange systems allow transport to and from all cells for the organism's needs. As well as exchanging substances, cells also have to lose thermal energy fast enough to prevent overheating.

An efficient **exchange surface** has a:

- large surface area to maximise rate of exchange
- thin membrane to provide a short diffusion path
- method of transporting substances to and from the exchange surface.

1. **How are efficient exchange surfaces adapted to carry out their function?**
2. **Explain why large organisms need transport systems but small organisms do not.**

Gas exchange

Gas exchange involves:

- taking in oxygen, which reacts with chemicals in food during respiration to give out energy
- removing carbon dioxide, which is a waste product of respiration in cells.

Gas exchange happens at a respiratory surface – a membrane separating the inside of the body from the external environment. Single-celled organisms use their cell surface membrane for gas exchange. In large organisms, the exchange

COMMON MISCONCEPTIONS

Some students think that breathing and respiration are the same thing, but they are not. Breathing is ventilation. Respiration is the release of energy from food inside each cell.

surface is part of specialised organs, such as lungs and gills in animals and leaves in plants.

In humans, gas exchange happens in the **alveoli** of the lungs. Blood transports the gases to and from the surface of each alveolus through the capillaries that cover it.

Air entering the alveoli has a greater oxygen concentration than the deoxygenated blood flowing through the lungs. This causes a steep **concentration gradient**, from high to low concentration, and allows efficient diffusion of oxygen in the air to the blood.

Deoxygenated blood has a greater carbon dioxide concentration than air in the alveoli, so carbon dioxide diffuses from the blood into the alveoli before being breathed out.

All gas exchange surfaces have a large surface area, a thin permeable membrane and a moist exchange surface. Many also have a **ventilation** system and an efficient transport system to keep the diffusion gradients as high as possible, to maximise the rate of gas exchange.

3 Explain how gas exchange happens in the alveoli.

4 What is the difference between breathing and respiration?

Adaptations of your lungs

Your lungs are efficient exchange surfaces.

- Alveoli are spherical, and there are millions in each lung, which provides a vast total surface area.
- The exchange surface at the alveolus wall is very thin (one cell thick), so diffusion distance for gases is very short.
- Each alveolus is closely surrounded by a network of capillaries to ensure a good blood supply. Oxygen is constantly taken away in the blood, and carbon dioxide is constantly brought to the lungs to be removed. So gas exchange happens at a steep concentration gradient.
- Constant ventilation continually brings oxygen-rich air into the lungs and expels carbon dioxide, which also helps maintain steep concentration gradients of gases.
- The alveoli surfaces are moist. Gases dissolve to allow efficient diffusion across the exchange surface.

5 Why are lungs efficient exchange surfaces?

6 Explain how your breathing, together with the efficient blood supply to your lungs, maintains steep concentration gradients of oxygen and carbon dioxide across the gas exchange surface.

7 Explain why it is important to maintain concentration gradients across an exchange surface.

MAKING LINKS

Look back to topic 1.3e to remind yourself about diffusion into and out of cells.

DID YOU KNOW?

The lungs have over 300 million alveoli and the left lung is slightly smaller than the right, so there is room for your heart.

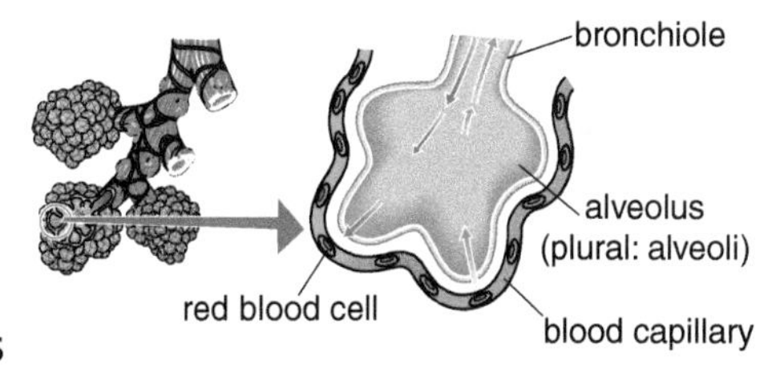

Figure 2.1.11 Gas exchange in an alveolus. Oxygen diffuses into the capillary from the alveolus (red arrows) and carbon dioxide diffuses into the alveolus from the capillary (blue arrows)

KEY INFORMATION

All exchange surfaces have a large surface area, thin membranes and are moist. Many have an efficient transport system.

2.1f The human heart

Learning objectives:

- describe the structure and function of the heart
- explain how the structure of the heart is adapted to its function
- explain the movement of blood around the heart
- use simple compound measures such as heart rate
- carry out calculations of heart rate.

KEY WORDS

aorta
atrium (plural: atria)
coronary artery
pacemaker
vena cava
ventricle

Larger animals like humans need specialised exchange surfaces and a transport system to carry important substances to every cell throughout the body, and to remove waste. To keep substances moving, the transport system needs a pump, called the heart.

The heart

The heart is made of muscle. Heart muscle continually contracts and relaxes. It uses a lot of energy. Heart muscle receives oxygen and glucose for respiration from the blood brought by the **coronary artery**.

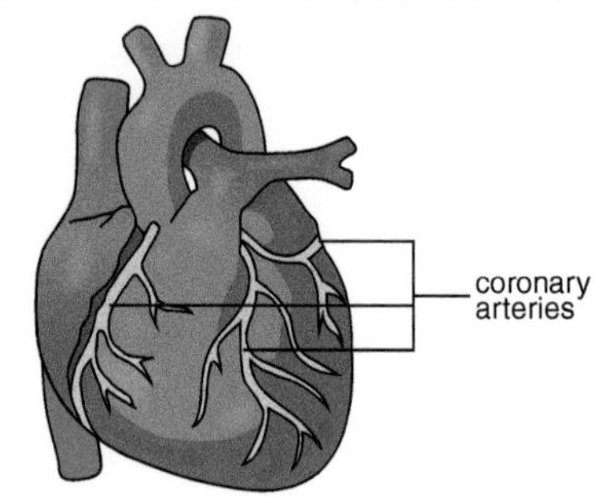

Figure 2.1.12 The coronary arteries supply the heart muscle, bringing oxygen and nutrients and carrying away waste substances

The heart has two pumps (a dual circulation) that beat together about 70 times every minute of every day. This is a heart rate of 70 beats per minute (bpm).

Each pump has an upper chamber (**atrium**) that receives blood and a lower chamber (**ventricle**) that pumps blood out. Both **atria** fill and pump blood out at the same time, as do both ventricles.

Blood from the lungs contains oxygen and enters the heart at the left atrium. It passes into the left ventricle and is pumped out to the body.

Blood from the body contains very little oxygen and enters the heart at the right atrium, passes into the right ventricle and is pumped to the lungs, where gas exchange takes place.

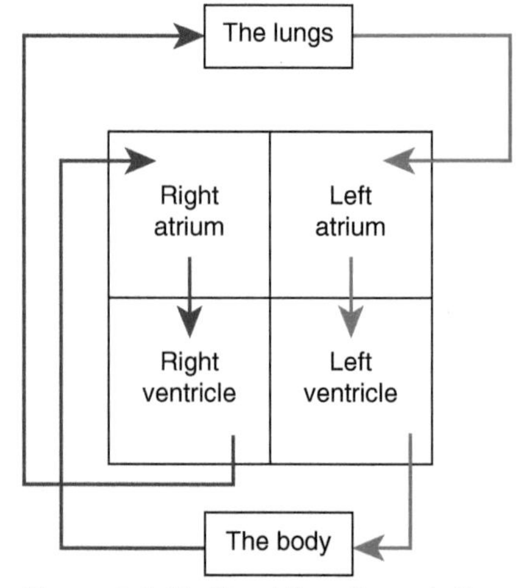

Figure 2.1.13 Blood flow through the heart

The natural resting heart rate is controlled by a group of cells located in the right atrium that act as a **pacemaker**. If a person has an irregular heart rate – because of disease or injury to the heart, for example – an artificial pacemaker can be implanted. This electrical device sends small impulses to the heart muscle to correct irregularities in the heart rate.

1. **What is the function of the heart?**
2. **A doctor took a patient's pulse, and counted 19 beats in 15 seconds. Calculate the patient's heart rate, in beats per minute (bpm).**

REMEMBER!

In diagrams of the heart, the right side of the heart is shown on the left of the diagram, and the left side of the heart is shown on the right. This is because you are looking at it from the front, as if examining the heart of someone standing facing you.

The parts of the heart

The heart has four main blood vessels:

- The pulmonary vein transports oxygenated blood from the lungs to the left atrium.

- The **aorta** (main artery) transports oxygenated blood from the left ventricle to the body.
- The **vena cava** (main vein) transports blood from the body to the right atrium.
- The pulmonary artery transports deoxygenated blood from the right ventricle to the lungs.

Ventricles have thicker walls than atria because they pump blood further. The left ventricle pumps blood around the body. It has a thicker wall than the right ventricle, which only pumps blood to the lungs. Valves between the atria and the ventricles prevent the backflow of blood. They open to let blood through and then shut.

3 **Describe the functions of the different chambers of the heart.**

4 **Describe how the atria and ventricles move blood through the heart.**

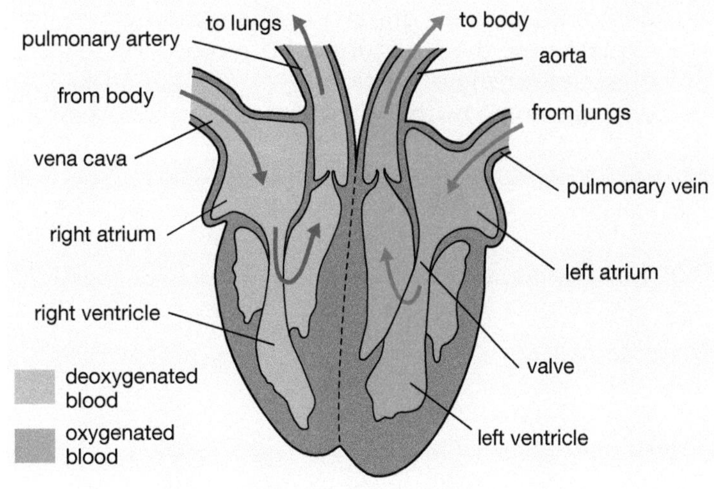

Figure 2.1.14 Why does the heart need two pumps?

Explaining blood flow

Figure 2.1.15 summarises the sequence of events in a heart beat, called the cardiac cycle.

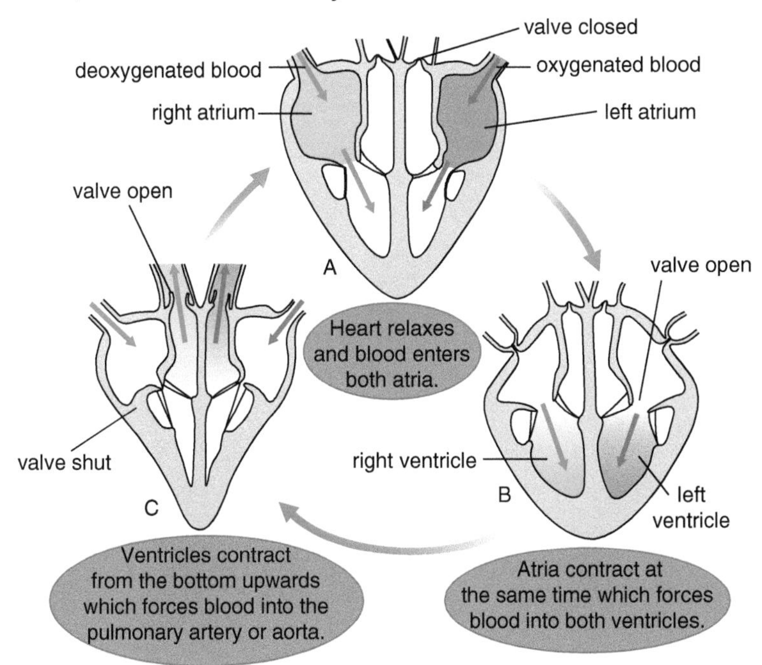

Figure 2.1.15 The cardiac cycle

KEY INFORMATION

Atria receive blood, ventricles pump it out.

DID YOU KNOW?

The complete cardiac cycle normally takes 0.8 seconds.

5 **Explain the sequence of contractions and valve openings as blood passes through the heart.**

6 **If a coronary artery supplying the left ventricle becomes blocked, what effect does this have on the functioning of the heart? Explain your answer.**

2.1g The human circulatory system

Learning objectives:

- describe the human circulatory system, including its relationship with the gaseous exchange system
- describe functions of parts of the circulatory system
- explain how the structures of the blood vessels are adapted for their functions
- describe some of the substances transported into and out of organisms.

KEY WORDS

artery
capillary
dual circulatory system
vein

The heart pumps blood through a vast system of blood vessels, which carry it close to every cell in the body, and then back to the heart. As the blood moves through this circulatory system, it delivers useful substances such as oxygen to respiring cells, and carries away waste products such as carbon dioxide.

Transport systems

Humans have a **dual circulatory system**. This means the blood flows in two circuits round the body:

- from the heart to the lungs, and back to the heart
- from the heart to the body, and back to the heart.

Blood is pumped out of the heart into vessels called **arteries**, under high pressure. As the blood flows from the arteries, through tiny **capillaries** in body tissues, and then back to the heart in **veins**, the pressure decreases.

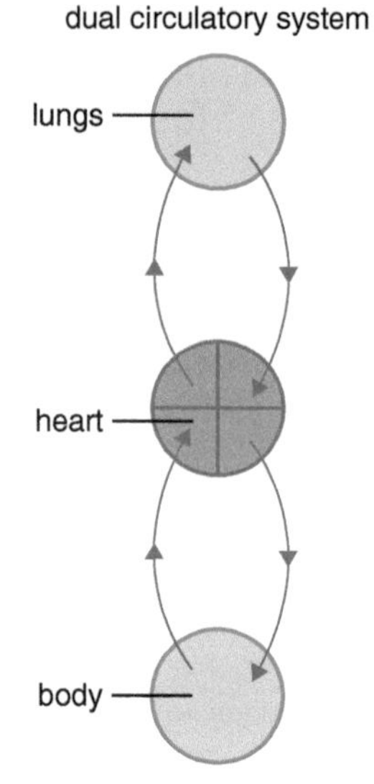

Figure 2.1.16 Why does blood travel to the lungs?

1. **What is a dual circulatory system?**
2. **Which vessels carry blood away from the heart?**

How does the system work?

As well as dissolved food, your body cells need oxygen for cellular respiration, and need to get rid of waste carbon dioxide. Your circulatory system, working with your respiratory system, delivers oxygen to and picks up carbon dioxide from every cell.

Deoxygenated blood from the body is pumped from the right ventricle of the heart, through the pulmonary artery, into the capillary networks around the alveoli of the lungs. Here, at the specialised surfaces of the alveoli, oxygen from the air diffuses into the blood, and carbon dioxide from the blood diffuses into the air.

As you breathe, you ventilate your lungs, bringing in fresh oxygen-rich air and exhaling air rich in carbon dioxide. This, with the efficient blood supply, maintains steep concentration gradients for diffusion of these gases across the exchange surface of the alveoli.

Oxygenated blood goes to the left side of your heart in the pulmonary vein. Then it is pumped from the left ventricle, through the main artery (aorta), to the body. It enters smaller arteries (arterioles) supplying your organs and tissues. Then it passes into networks of tiny capillaries, which lie very close to every body cell. Here, oxygen diffuses from the blood into the cells, and carbon dioxide diffuses from the cells into the blood, each down its concentration gradient. Next, deoxygenated blood moves into venules, and then into veins, which carry it back to the right side of the heart, and the circulation then begins again.

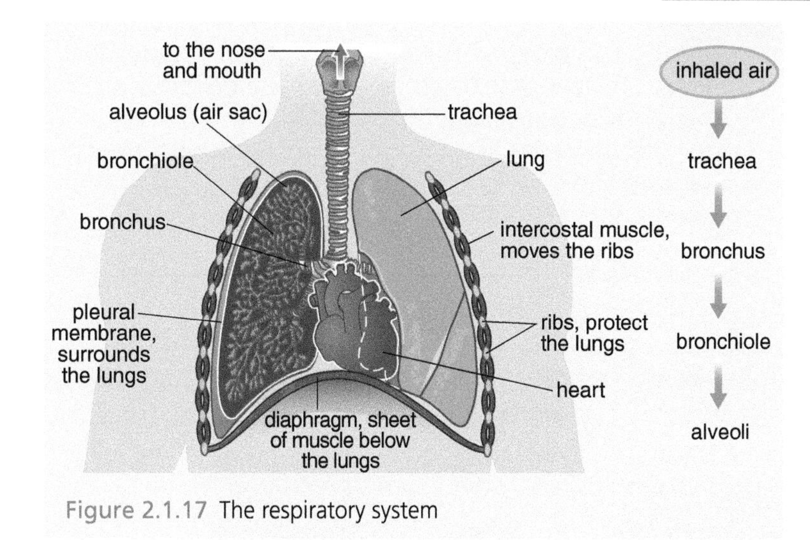

Figure 2.1.17 The respiratory system

3. **Describe how air reaches the alveoli.**
4. **Describe how cells deep in your body can receive oxygen and dispose of carbon dioxide efficiently.**

Adaptations of your circulatory system

Blood flows from the heart to arteries, arterioles (small arteries), capillaries, venules (small veins), veins and then back to the heart.

Each type of blood vessel is adapted to carry out different functions. The table summarises the adaptations.

Arteries	Capillaries	Veins
thick, elastic wall small lumen	single cell wall	thin wall large lumen valve
carry blood from the heart, with a pulse from the heart beat	carry blood from arteries to veins, with no pulse	carry blood to the heart, with smooth flow and no pulse
carry oxygenated blood (except the pulmonary artery)	blood slowly loses its oxygen	carry deoxygenated blood (except the pulmonary vein)
have thick, impermeable, elastic walls with small lumen, to carry blood under high pressure	have one-cell-thick, permeable walls, with huge total surface area, for efficient exchange with body cells	have thinner, impermeable walls with a large lumen, providing less flow resistance, for carrying blood under low pressure
do not have valves	do not have valves	have valves to prevent backflow

5. **Explain how each blood vessel is adapted for its function.**
6. **Why are the walls of arteries and veins impermeable?**

2.1h Blood cells

Learning objectives:

- identify the parts of the blood and their functions
- explain how the different parts of the blood are adapted to their functions
- identify different types of blood cell in a photograph or diagram.

KEY WORDS

haemoglobin
oxyhaemoglobin
plasma
platelets
red blood cells
white blood cells

The blood is a liquid medium adapted to carry substances efficiently to and from every cell of the body, in the circulatory system. Blood is made up of different parts, each with its own function.

Figure 2.1.18 An illustration to show the three main types of blood cell that are transported in the plasma

What is blood?

Blood is a tissue. It is a mixture of cells, solutes and a liquid. The straw-coloured liquid part of blood is **plasma**.

Red blood cells, **white blood cells** and **platelets** are suspended in the plasma. Each part of the blood has a specific function:

- Plasma transports substances around the body, for example carbon dioxide.
- Platelets are cell fragments which help the clotting process at wound sites.
- Red blood cells carry oxygen from the lungs to body cells.
- White blood cells help to protect the body against infection.

There are millions of red blood cells in the plasma. This is why blood looks red.

1 **Look at Figure 2.1.18. Identify the red blood cell, white blood cell and platelet.**

2 **What is the function of:**
a **red blood cells** **b** **white blood cells?**

Looking closer

The blood parts can be separated by spinning them very fast in a machine called a centrifuge. About 55% of blood is plasma; plasma consists of roughly 90% water and 10% solutes. It is very important because it transports many substances, for example:

- hormones
- antibodies
- nutrients, such as glucose, amino acids (proteins), minerals and vitamins
- waste substances, like carbon dioxide and urea.

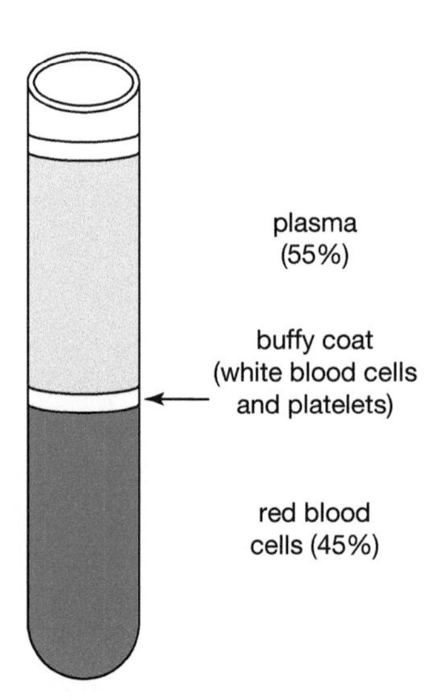

Figure 2.1.19 Blood looks like this if it is separated in a centrifuge

Red blood cells transport oxygen from the lungs to tissues all over the body. Blood is able to transport oxygen efficiently because in 1 mm^3 of blood there are about 5 million red blood cells. Red blood cells:

- are tiny, allowing them to pass through the narrow capillaries
- have a biconcave disc shape, giving them a large surface area : volume ratio, which increases the efficiency of diffusion of oxygen into and out of the cell, and reduces the diffusion distance to the centre of the cell
- contain **haemoglobin,** which binds to oxygen to transport it from the lungs to the body tissues
- have no nucleus, increasing the space available for haemoglobin.

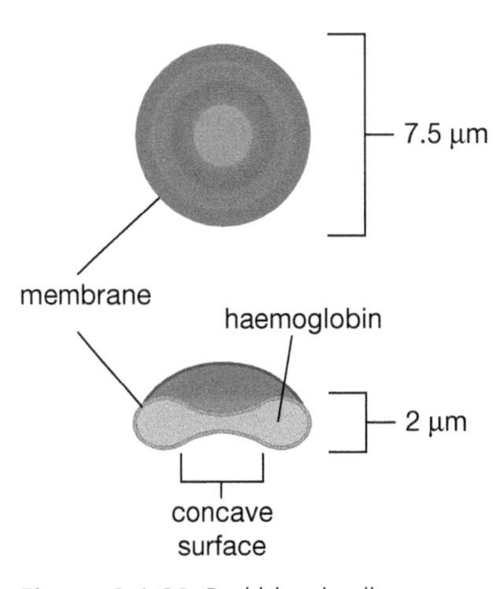

Figure 2.1.20 Red blood cells are adapted to carry oxygen

White blood cells are adapted to fight infections. There are several types. Some are able to produce proteins called antibodies, which help destroy invaders. Others can change shape to engulf bacteria, and can even squeeze out of blood vessels to get to an infection site.

Platelets are adapted to help seal wounds and prevent excessive bleeding. They are tiny cell fragments without a nucleus that become 'sticky' when activated at a wound site, and help form a clot.

KEY INFORMATION

Red blood cells transport oxygen; plasma transports many substances including carbon dioxide.

3 **Describe the role of plasma, and how it is adapted for this role.**

4 **How are red blood cells, white blood cells and platelets adapted to their different functions?**

How the blood carries oxygen

Haemoglobin binds with oxygen at high concentration to form a bright red compound called **oxyhaemoglobin**. The bonds between the haemoglobin and oxygen are weak, and oxyhaemoglobin dissociates to haemoglobin and oxygen in low oxygen concentrations.

oxygen + haemoglobin $\rightleftharpoons$ oxyhaemoglobin

5 **Describe how haemoglobin transports oxygen.**

6 **Use surface area : volume ratio to explain how red blood cells are adapted to their function.**

7 **Sickle cell anaemia is a serious inherited blood disorder in which the red blood cells develop abnormally and contain defective haemoglobin. Explain why somebody with sickle cell anaemia is likely to feel very tired and breathless while exercising.**

DID YOU KNOW?

Crabs have blue blood, earthworms and leeches have green blood and starfish have clear or pale yellow blood!

2.1i The human digestive system

KEY WORDS

absorption
carbohydrase
digestion
enzyme
lipase
protease

Learning objectives:

- explain how large insoluble food molecules are broken down by digestion into small soluble molecules
- explain how the products of digestion can be used in cells
- describe some of the substances transported into and out of organisms, including dissolved food molecules and urea.

Your body breaks down the food you eat. Large, insoluble food molecules are digested into much smaller, soluble molecules, which are absorbed into the blood from the small intestine, and carried to every body cell.

The digestive system

The digestive system is a long tube that runs from the mouth to the anus. It consists of several organs working together to break down and absorb food. Each organ is adapted to perform a different function.

The digestive system uses **enzymes** to break down, or digest, large insoluble food molecules into small soluble molecules. **Digestion** is completed in the small intestine, where the products – the small, soluble food molecules – then pass through the gut wall into the blood. This is called **absorption**. The blood carries the soluble molecules to the body cells, where they can be used for respiration or to make new large molecules that cells need as energy reserves or for growth and repair.

1. **Why do we digest food?**
2. **What is absorption?**

Digestion and synthesis

Digestion is the breakdown of large molecules into smaller ones, while synthesis is the process of building large molecules from smaller units. Enzymes catalyse both types of reaction in the body.

Most enzymes, including those that help synthesise large molecules, work inside cells. But digestive enzymes work outside cells. They are produced by cells in glands and in the lining of the gut, and pass into the gut to mix with the food. There are three groups of enzymes in digestion: carbohydrases, proteases and lipases.

- **Carbohydrases** break down carbohydrates into simple sugars. For example, amylase breaks down insoluble starch to water-soluble sugars, which are further broken down into glucose. Glucose is used in cells for respiration, and also to synthesise new carbohydrates.

- **Proteases** break down insoluble proteins to soluble amino acids. Cells use amino acids to synthesise new proteins. Any unwanted amino acids are converted by the liver to urea, which is carried by the blood to the kidneys and excreted in urine.
- **Lipases** break down lipids (fats and oils) to glycerol and fatty acids, which are used for energy, to build cell membranes and to make hormones. Cells can also use these building blocks to re-synthesise fats as a store of energy because cells can break them down and use them in respiration.

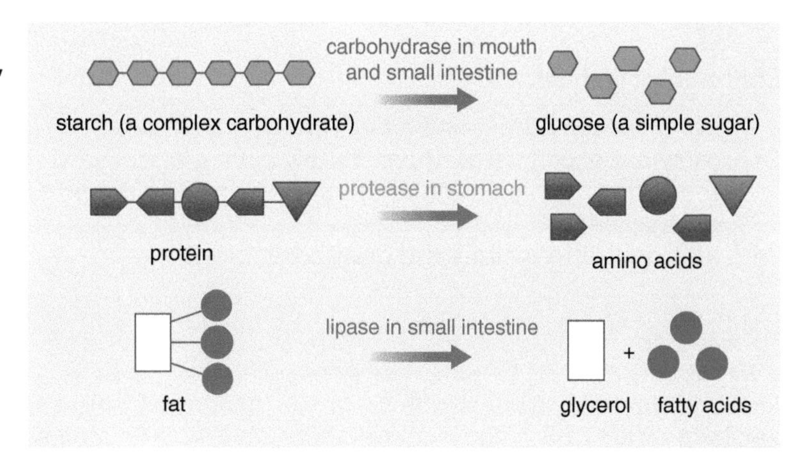

Figure 2.1.21 Why is digestion by enzymes important?

3 **What reaction do proteases catalyse?**

4 **What are the products when a lipase breaks down fats?**

5 **Suggest why digestive enzymes do not work inside cells.**

A special exchange surface

The soluble products of digestion pass through the wall of the small intestine, and are carried away in the blood capillaries. Fatty acids are also carried away in other vessels called lacteals.

The small intestine is an effective exchange surface because:

- it is about 7 m long, so there is time for absorption of soluble molecules as food travels along
- it has a very thin, permeable membrane for easy diffusion
- the cells lining the small intestine have many small projections called villi, each with even tinier projections called microvilli, increasing the surface area for absorption
- blood capillaries transport molecules away, maintaining the concentration gradient for diffusion
- lacteals carry fatty acids away through the lymphatic system, which eventually returns them to the blood.

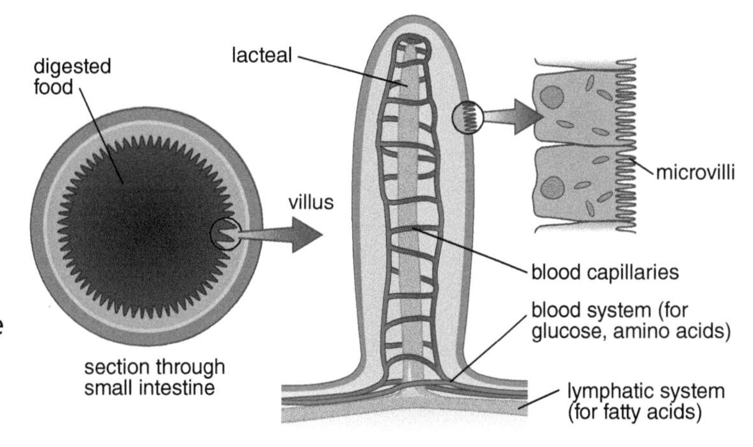

Figure 2.1.22 Villi help make the small intestine an effective exchange surface

Between meals, the concentration of dissolved food molecules in the blood can be higher than in the intestine. At this time, the molecules are moved into the blood using active transport.

6 **Explain why the small intestine is an effective exchange surface.**

7 **Suggest why fatty acids are not absorbed directly into blood.**

DID YOU KNOW?

If flattened out, the surface of the small intestine would cover an area of about 250 m² (the size of a tennis court).

REQUIRED PRACTICAL

2.1j Food tests

Learning objectives:

- use a Bunsen burner and a boiling water bath safely
- carry out experiments appropriately having due regard for the correct manipulation of apparatus, and health and safety considerations
- interpret observations and draw conclusions.

KEY WORDS

Benedict's test
carbohydrates
qualitative
reagents

Food samples can be analysed using qualitative tests to see if they contain carbohydrates (starch and sugars), proteins and fats.

These pages are designed to help you think about aspects of the investigation rather than to guide you through it step by step.

In this practical, eye protection should be worn, and a risk assessment is required.

Describing how apparatus is used and working safely

Qualitative tests can be used to test for the presence of different food groups using ground up food. The food is added to distilled water, stirred and filtered.

The **Benedict's test** is used to test for sugars: the filtrate is transferred to a test tube, Benedict's reagent is added and the tube is placed in a water bath of boiling water for 5 minutes to see if a colour change occurs.

To test for lipids, the filtrate is added to a test tube and shaken gently with Sudan III stain.

The Biuret test is used to test for protein.

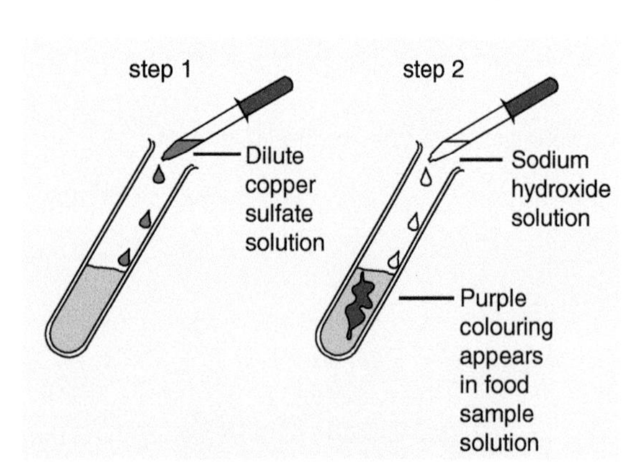

Figure 2.1.24 Test for protein (Biuret test)

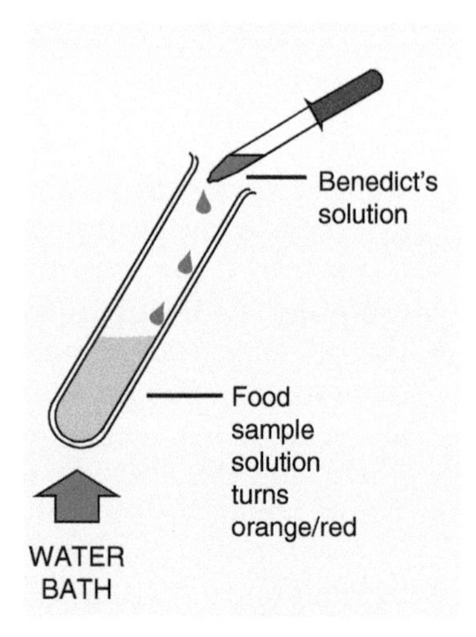

Figure 2.1.23 Test for glucose

1 Which piece of apparatus could be used to:

a grind the food up
b add drops of indicator solution
c measure small volumes of liquids?

2. **Why is the ground up food stirred with the distilled water and then filtered?**
3. **Describe the test that could be carried out to show if protein was present in the food.**
4. **State two safety precautions that should be taken for each test.**

Making and recording observations

Ravi is going to test three different foods to see which food groups, including starch, are present. She needs to design a table for the results that all the data will fit into. It needs to be fully labelled and include units. Ravi is going to test each food for each food group twice and is also going to use a control tube, using water.

5. **Why is Ravi going to repeat the test for each food group?**
6. **Why is a control tube set up?**
7. **Construct a table that Ravi can use that will fit all the data she is going to collect.**

Interpreting observations

Food samples A, B and C were tested for different food groups.

Complete this table to show the initial colour of the reagent and the colour change of a positive test.

Food group	Colour change of a positive test	
	Initial colour	Final colour
Glucose		
Protein		
Fat		

In the first test, the foods were boiled with Benedict's reagent, in the second test Biuret reagent was added and in the third test Sudan III reagent was added.

Food sample	Colour with Benedict's reagent	Colour with Biuret reagent	Colour with Sudan III reagent
A	brick red	blue	red layer at top of tube
B	blue	purple-pink	no red layer
C	brick red	purple-pink	no red layer

8. **Which food/foods contained these different food groups? Explain your answers.**

   a **No protein**
   b **Glucose**
   c **Lipids**

2.1k The human nervous system

Learning objectives:

- describe the structure of neurones and of the nervous system
- explain how the nervous system is adapted to its functions.

KEY WORDS

central nervous system
myelin sheath
neurone
receptor

Cells called neurones in the nervous system communicate with each other and with muscles, glands and other structures. This enables humans to react to their surroundings and to coordinate both their internal functions and their behaviour.

The structure of the nervous system

Your nervous system enables you to detect your surroundings, and coordinate your body and behaviour.

The structure of the nervous system is well adapted to these functions.

It is made up of:

- the **central nervous system** (CNS) – the brain and spinal cord
- the peripheral nervous system (PNS) – all the nerves extending to and from the body tissues and organs.

1. **What is the function of the nervous system?**
2. **What are the two parts of the nervous system?**

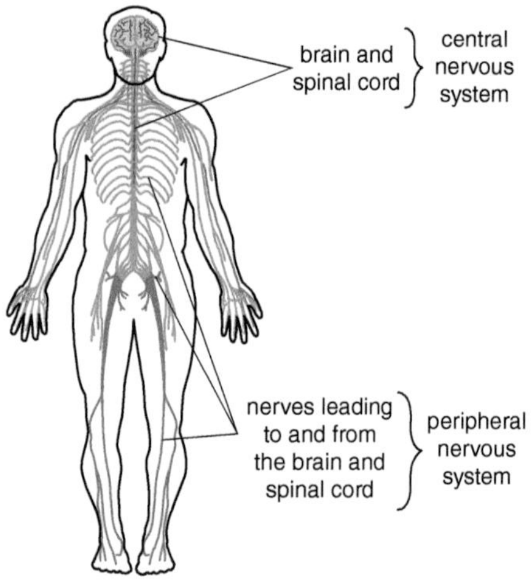

brain and spinal cord } central nervous system
nerves leading to and from the brain and spinal cord } peripheral nervous system

Figure 2.1.25 The human nervous system

Neurones

The nervous system consists of nerve cells or **neurones**. Neurones are specialised for transmitting messages in the form of an electrical impulse.

The part of the cell containing the nucleus is called the cell body. The cell body of all neurones is found in the CNS. Neurones, however, have an extended shape so that they can carry nerve impulses from one part of the body to another. They also have fine branches at their tips to communicate with other neurones.

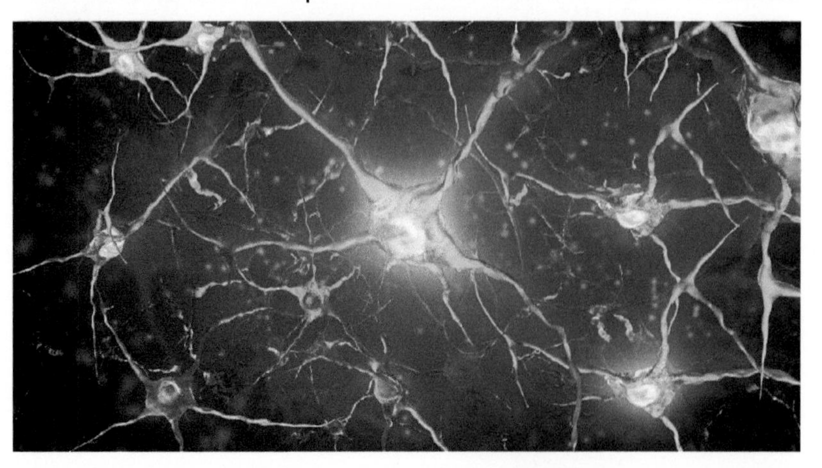

Figure 2.1.26 A computer-generated illustration to show that projections that extend from a nerve cell communicate with other nerve cells

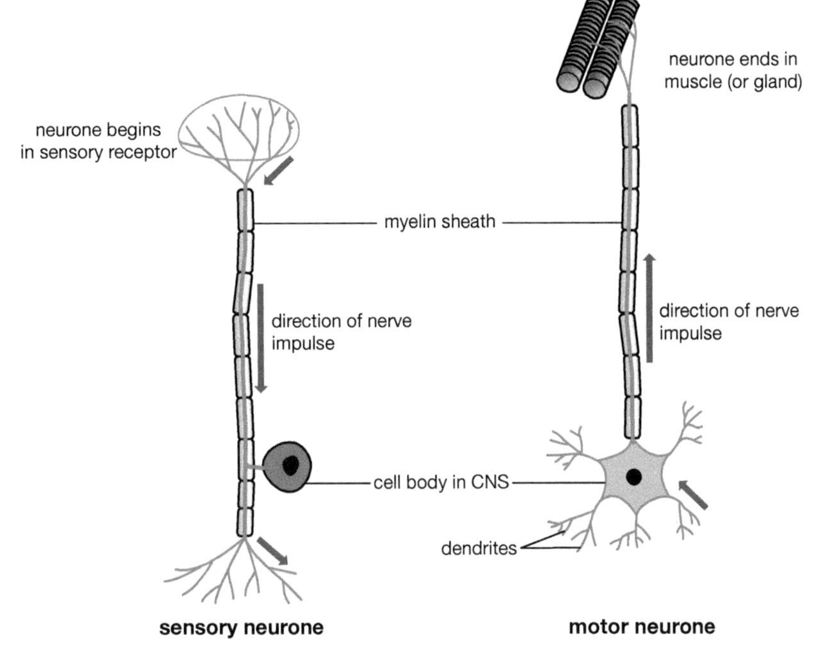

Figure 2.1.27 Two types of neurone

Receptors are cells that detect any changes in the environment. Receptors are sometimes grouped to form sense organs. Sensory neurones relay nerve impulses from these receptors to the CNS.

The CNS processes the information and coordinates how the body should respond.

Motor neurones relay impulses from the CNS to the effector – for instance, a muscle may respond by contracting or a gland by secreting a hormone.

The sequence of events is:

stimulus → receptor → coordinator → effector → response

3. **What is the scientific name for a nerve cell?**
4. **Describe the pathway of a nerve impulse, beginning with the stimulus and ending with the response.**

The transmission of a nerve impulse

The nerve impulse is electrical. This means that it can be transmitted quickly.

In vertebrates, neurones are covered with a fatty layer called the **myelin sheath**. The myelin sheath acts as an insulator and speeds up the transmission of the impulse. But the myelin sheath is not continuous. Periodically, there are small gaps in it – around 1 μm in size. These allow the nerve impulse to jump from one gap to the next, further increasing the speed of transmission.

5. **What is the outer, fatty layer around a neurone called?**
6. **Explain how the nervous system is adapted to its function.**

MAKING LINKS

Look back to topic 1.3l to remind yourself about the size of neurones and nerves.

DID YOU KNOW?

If the myelin sheath does not develop properly, or becomes or inflamed or damaged – in conditions such as multiple sclerosis, and diseases such as leprosy – the transmission of nerve impulses can be seriously affected.

MAKING LINKS

Link the information given here back to topic 1.6, and how nerve cells are adapted to their functions.

2.1l Reflex actions

Learning objectives:

- explain the importance of reflex actions
- describe the path of the pain withdrawal reflex arc
- explain how the structures in the reflex arc relate to their function.

KEY WORDS

reflex action
reflex arc
relay neurone
synapse

Many of the body's responses that are coordinated by the nervous system are not under conscious control. A reflex action is an example. An impulse from a receptor travels along a sensory neurone to the spinal cord, where it is passed to a relay neurone and then directly to a motor neurone, without passing through the brain. The impulse travels down the motor neurone and causes a response in an effector.

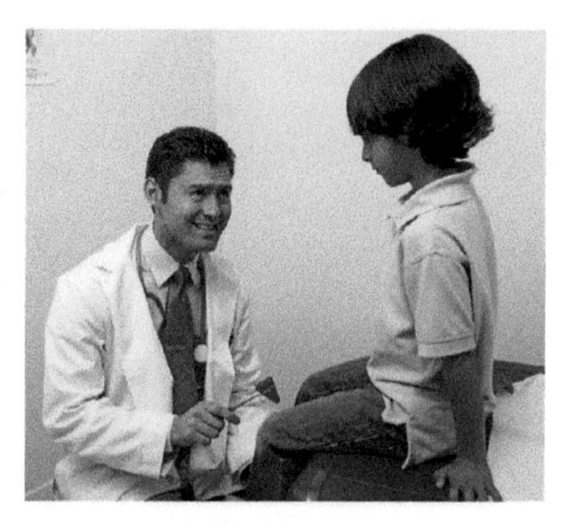

Figure 2.1.28 Nasim's doctor taps the tendon just below the knee cap. Nasim's leg kicks up. This is the knee-jerk reflex. A normal reaction time is around 50 ms

Reflexes are related to survival

Reflex actions are rapid, automatic responses to a stimulus. We do not have to think about them. Reflex actions form the basis of behaviour in simpler organisms. In humans, they prevent us from getting hurt. In other animals, our human ancestors and babies they are also related to survival.

Some reflex actions include:

- the pain withdrawal reflex, in which you remove your hand from a hot or sharp object
- the grasping reflex, in which a baby grips a finger
- blinking our eyes if an object approaches rapidly
- the pupil reflex, whereby the pupil gets wider in dim light and narrower in bright light.

1. **What is a reflex action?**
2. **Why are reflex actions important?**

The pain withdrawal reflex arc

Our spinal reflexes do not involve the brain. Or, at least, not to begin with. In a reflex action, the nerve impulse follows a pathway called the **reflex arc**.

Figure 2.1.29 shows the pathway taken by a nerve impulse when a person puts a hand on a hot object.

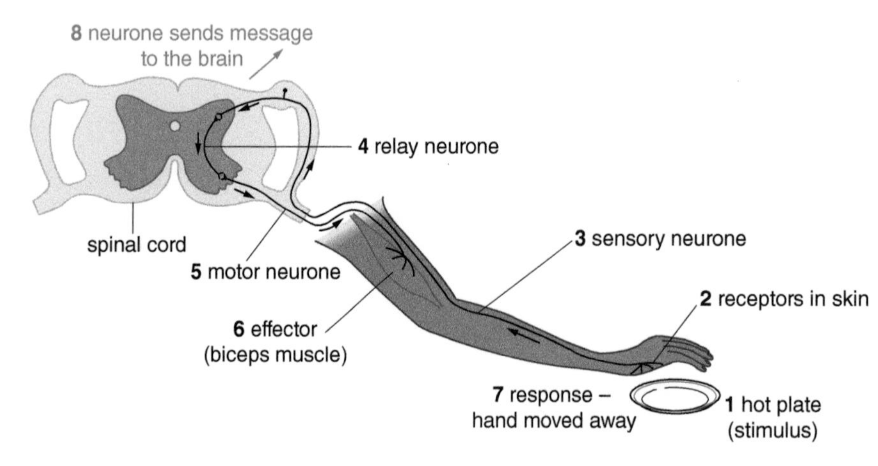

Figure 2.1.29 The pain withdrawal reflex arc. Follow the numbers 1–7. The pathway is through the spinal cord

The pathway includes:

- a sensory neurone – transmits nerve impulses from the receptor to the CNS
- a **relay neurone**, in the spinal cord – transmits the impulses from the sensory to the motor neurone
- a motor neurone – sends impulses from the CNS to the effector.

In this case, the effector is the biceps muscle, which moves the arm. The hand is moved away from the hot object.

Because other neurones in the spinal cord link via a synapses with those of the reflex arc, a message is sent to the brain *after* the hand has been removed (number 8 in the figure). It tells us that the plate was hot.

DID YOU KNOW?

Over 100 different types of transmitter molecule, or neurotransmitter, have been identified. Many medical drugs, recreational drugs and poisons work by affecting transmitters.

3 **Name the nervous pathway that a nerve impulse takes during a reflex action.**

4 **Explain how the parts of this pathway relate to their function.**

REMEMBER!

Be able to apply your knowledge and describe the pathway taken by the reflex arc in another type of reflex action.

Linking nerves

The three neurones in the reflex arc don't link together *physically*. There's a gap – called a **synapse** – between each pair. This means that *many* nerves can connect with each other. In the brain, neurones can link up with up to 10 000 others.

Nerve impulses pass across a synapse with the help of chemical transmitter molecules, which diffuse rapidly across the gap.

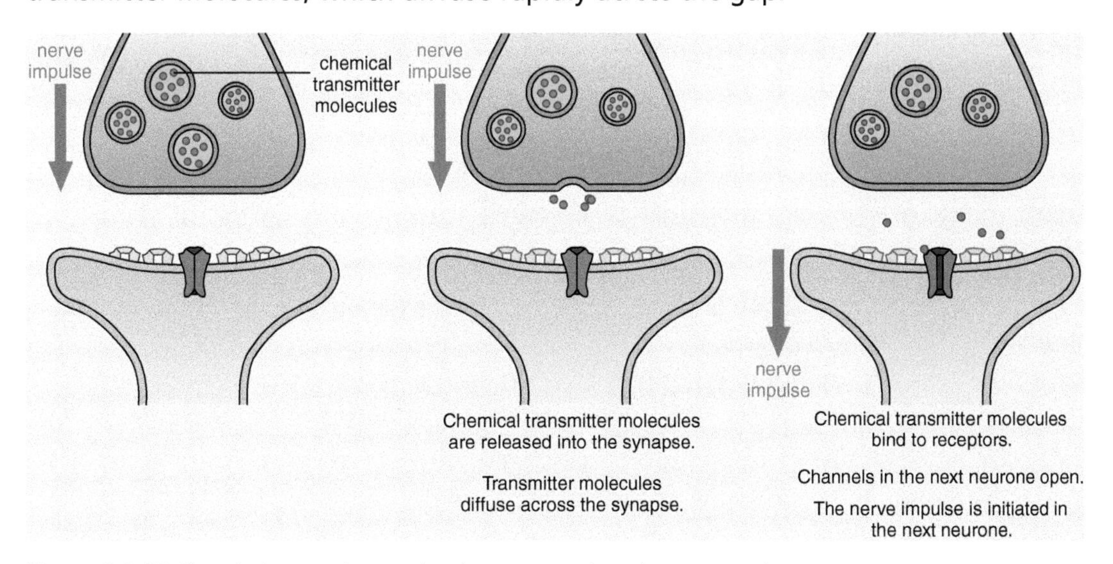

Figure 2.1.30 Chemical transmitter molecules cause an impulse to move from one neurone to the next

5 **What is the gap between neurones called?**

6 **How does a nerve impulse travel from one neurone to the next?**

7 **Compose a flow diagram to describe and explain the sequence of events in a reflex arc.**

REQUIRED PRACTICAL

2.1m Investigating reaction time

Learning objectives:

- select appropriate apparatus and techniques to measure the physiological function of reaction time
- carry out physiological experiments safely
- translate information between numerical and graphical form.

KEY WORDS

reaction time
valid

Our reflex actions protect us from harm. Quick reactions are also important in sport. Reaction times vary from person to person, but typical values range from 0.3 s to 0.9 s.

These pages are designed to help you think about aspects of the investigation rather than to guide you through it step by step.

Measuring reaction time

A group of students worked in pairs to find their **reaction times** using the ruler drop test. One student dropped a 30 cm ruler while another student caught it between their outstretched thumb and index finger of their dominant hand.

In the test, the release of the ruler is detected by our eyes and a message is sent to the sensory region of our brain. A message is then sent to another part, the motor region, which instructs muscles in our hand to contract.

After the student has carried out the test 10 times, they leave the lab to drink a cup of coffee. The student returns and takes the test again to investigate the effect of coffee on their performance.

Figure 2.1.31 In cricket, fielders need fast reaction times

Test number	Experiment 1: Normal	Experiment 2: After coffee
	Distance the ruler dropped (mm)	
1	119	98
2	116	98
3	117	92
4	113	91
5	150	92
6	113	93
7	108	92
8	109	92
9	108	91
10	107	91

The student's results for the ruler drop test

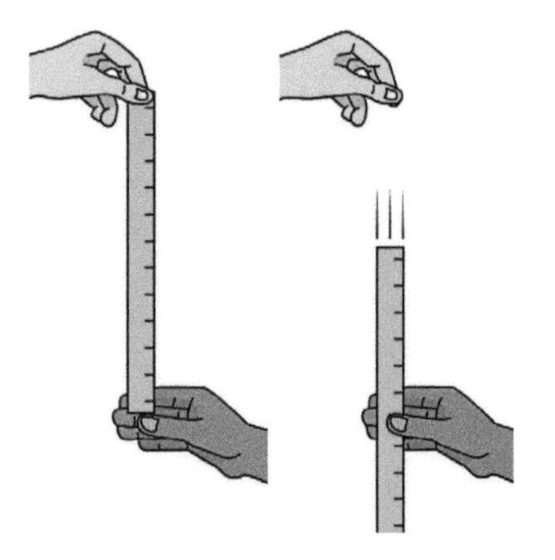

Figure 2.1.32 The ruler drop test

1. **Write a risk assessment for this experiment.**
2. **Identify any anomalous results.**
3. **Calculate the average distance fallen by the ruler before and after drinking coffee.**

Calculating reaction time

The distance travelled by the ruler before it is caught gives an indication of a student's reaction time, but not the reaction time itself.

Another group of students finds a formula on the Internet that is used to calculate reaction time:

$$t = \sqrt{\frac{2d}{a}}$$

where t = time in seconds

d = distance in metres

a = acceleration due to gravity = 9.81 m/s^2

They use this formula to calculate their reaction times.

4. **Calculate the mean reaction time for the student before and after coffee.**
5. **The ingredient in coffee that affects our nervous system is caffeine. When testing its effect on other students, explain why the experiment must be carefully controlled to produce valid results.**

Pooling class results

All the students in the year group measured their minimum reaction time. They used an alternative test on the computer. They had to click on the mouse when the screen changed colour. The reaction time was measured by the computer timer.

The students produced a tally of those falling into different ranges of reaction time.

Mean reaction time (ms)	101–200	201–250	251–275	276–300	301–325	326–350	351–375	376–400	401–500
Number of students within range	1	3	16	24	33	14	6	2	1

The tally of student reaction times across the year group

6. **What type of chart would be best suited to displaying the data shown above? Draw a chart of the results.**
7. **Determine the median and modal reaction time categories.**
8. **Suggest why the computer method for measuring reaction time may be better than the ruler drop method.**

DID YOU KNOW?

Caffeine works by affecting chemical transmitter molecules. It binds to one type that makes us sleepy. Fewer receptors are, therefore, available and nervous activity speeds up.

KEY SKILL

Selecting equipment, carrying out a risk assessment and processing data appropriately are important skills.

KEY INFORMATION

You will revisit reaction times in Unit 2 when you learn about drivers' stopping distances.

2.1n The endocrine system

KEY WORDS

endocrine gland
endocrine system
hormone
target organ

Learning objectives:

- recall that the endocrine system is made up of glands that secrete hormones into the blood
- understand why the pituitary gland is the 'master gland'
- describe the principles of hormonal coordination and control by the human endocrine system.

The endocrine system is composed of glands that secrete hormones directly into the bloodstream. It works alongside the nervous system to coordinate and control the body's activities. Compared to the nervous system, the effects of the endocrine system are slower, but they act for longer.

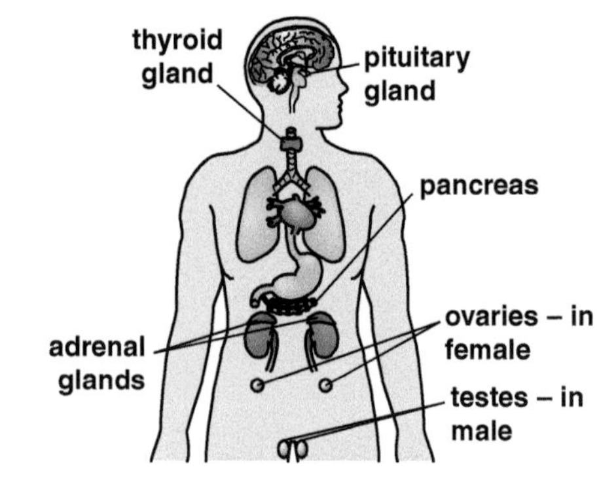

Figure 2.1.33 The location of the major endocrine glands in the body

The endocrine system

Hormones are produced by glands of the **endocrine system**. **Endocrine glands** secrete hormones directly into the blood. Some examples include the adrenal glands just above the kidneys, which secrete adrenaline, and the thyroid gland in the neck, which secretes thyroxine.

Hormones are often described as *chemical messengers*. They circulate in the blood and produce an effect on **target organs**. Many hormones are large molecules.

Like the nervous system, hormones work on effectors. But, unlike the nervous system, the effects they produce don't take milliseconds. With the exception of adrenaline, effects of hormones take minutes, hours or, in the case of hormones involved in our development, they act for years.

1. **What is the function of the endocrine system?**
2. **On what does the endocrine system act?**

The master gland

The pituitary gland is an outgrowth from the base of the brain.

Some of the hormones that it secretes, such as growth hormone, have a direct effect on their target organs. Other hormones have an indirect effect; they cause other glands to secrete hormones. It is, therefore, called the master gland as it regulates the secretion of other endocrine glands.

3. **Explain why the pituitary gland is called the master gland.**
4. **Name one hormone that exerts its effects over the whole body.**

 Link: 2.1n ⟶ 2.1o

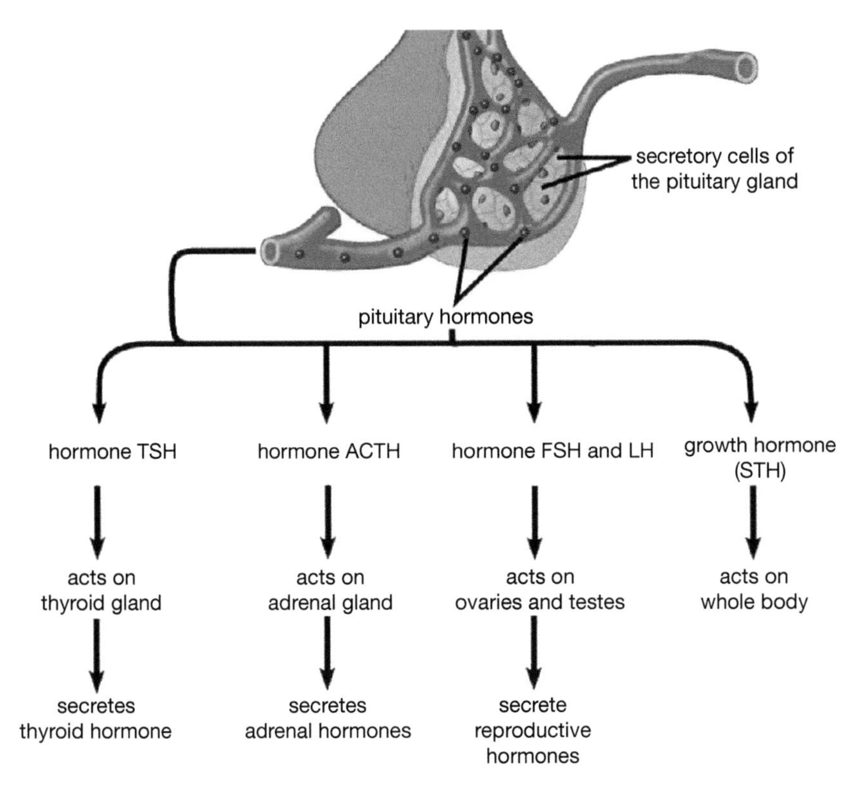

Figure 2.1.34 Some of the hormones secreted by the pituitary gland

Control systems

All control systems in the body, whether nervous or endocrine, have the same model.

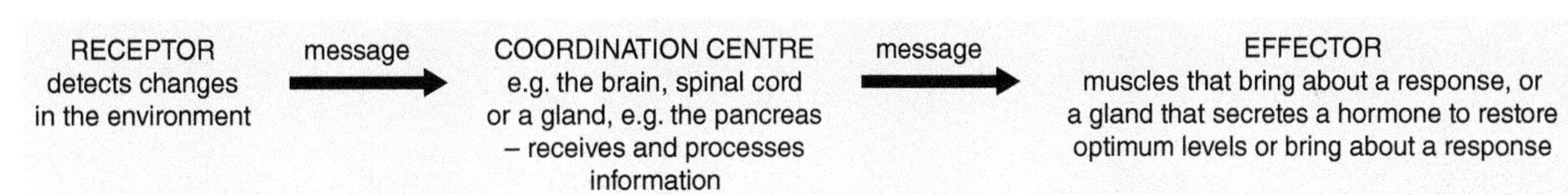

Figure 2.1.35 A model showing the components of the body systems that are responsible for coordination and control

The nervous system and endocrine system are different in nature but, in practice, the two systems interact with and regulate each other.

	Nervous system	Endocrine system
response	rapid and precise	slower but acts for longer
nature of message	nerve impulse – electrical and chemical	a hormone – chemical
action	carried in nerves to specific location, e.g. muscle	carried in blood to all organs, but affects the target organ only

5 **How are changes detected by the body?**

6 **Compare and contrast the nervous system and the endocrine system.**

DID YOU KNOW?

The squid has giant nerve cells, up to 1 mm in diameter. Study of these has been key to our understanding of how nerve impulses are transmitted.

REMEMBER!

Remember the sequence of receptor → coordination centre → effector.

2.1o Negative feedback

Learning objectives:

- describe the effects of adrenaline
- explain the role of thyroxine in the body
- understand the principles of negative feedback, as applied to thyroxine.

KEY WORDS

adrenal gland
adrenaline
basal metabolic rate
negative feedback
pituitary gland
thyroxine

Hormones maintain conditions in the body within narrow limits using negative feedback. This is important to ensure your body cells always have the optimum environment, even when external conditions change.

HIGHER TIER ONLY

Adrenaline

Harry is frightened of having of having a flu vaccination. His heart begins to race. His skin goes pale. He feels 'butterflies' in his stomach. These effects are the result of a hormone called **adrenaline**, secreted from the **adrenal glands.**

Adrenaline is called the 'flight-or-fight' hormone, because when you get a fright it prepares you very quickly to run, or maybe to fight. It causes blood flow to be diverted away from your skin and your digestive system toward your muscles and brain, carrying much needed oxygen and glucose for emergency action. At the same time, your heart rate rises, further boosting blood flow to your muscles and brain.

Once the danger is over (you've escaped!) and your brain and muscles no longer need increased oxygen and glucose, your adrenal glands stop secreting adrenaline, and your body returns to its normal state.

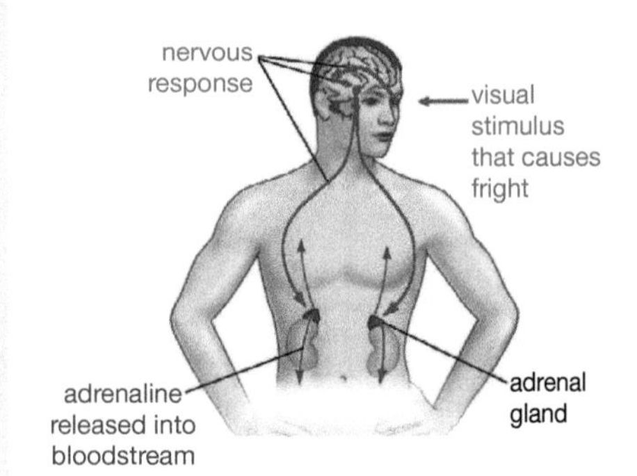

Figure 2.1.36 When something causes us to be frightened, the hormone adrenaline is responsible for preparing our body for 'flight or fight'

1. **Which hormone is released when we become frightened?**
2. **There is only a set volume of blood in the body. Explain how more blood can be pumped to the brain and muscles.**

Thyroxine

The thyroid gland affects a person's activity by producing the hormone thyroxine, which stimulates the body's **basal metabolic rate** – it increases the metabolism of all the body's cells. It also plays an important role in growth in children.

The **pituitary gland** controls secretions from the thyroid gland. When the pituitary gland secretes thyroid-stimulating

DID YOU KNOW?

Thyroxine is also involved in the growth and development of other animals. It is responsible for metamorphosis in amphibians – the process that transforms them from tadpoles to adults.

hormone (TSH); look back at Figure 2.1.34, the thyroid gland responds by secreting **thyroxine**. If levels of thyroxine become too high, TSH secretion is blocked, which stops secretion of more thyroxine. This is an example of **negative feedback**, which acts to stabilise a system – if a level becomes too high or too low, it stimulates a response that acts to bring the level back toward normal.

Here's another example of negative feedback control. If the body temperature falls, a thermoregulatory centre in the brain detects this and causes the pituitary gland to release TSH, which stimulates the thyroid to secrete more thyroxine. Because thyroxine increases the respiration rate of cells, it causes more thermal energy to be released, raising the body temperature back toward normal. If the body temperature becomes too high, the thermoregulatory centre detects the temperature rise, and the hormone's secretion is blocked.

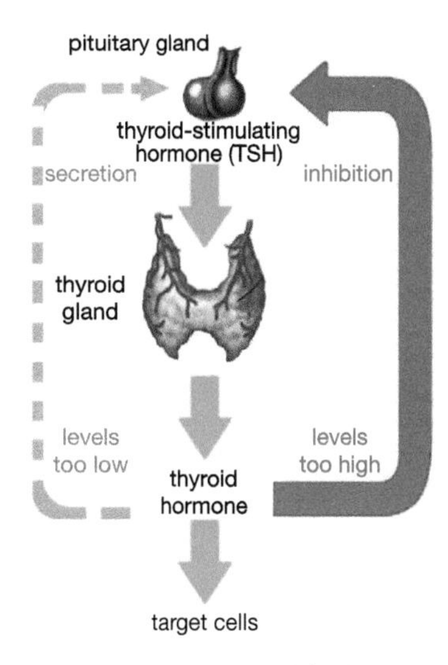

Figure 2.1.37 A simplified version of the negative-feedback system that controls thyroxine secretion

3. **Suggest a symptom of an overactive thyroid.**
4. **What is the function of thyroxine?**
5. **Which hormone regulates thyroxine production?**

The principles of negative feedback

The endocrine system keeps the conditions in the body constant using feedback systems. A simple negative-feedback system is the central-heating system in your home. If the temperature falls in your living room, the thermostat detects this and switches on the heating. The room warms up.

When the thermostat temperature is reached, it detects this, and turns off the heating. This is negative feedback because when the desired effect is reached, the system is switched off.

Negative feedback within the endocrine system tends to reverse a change. This prevents a system from becoming overactive. The system is stabilised by its own products.

REMEMBER!

You should aim to be able to explain how negative feedback mechanisms involving thyroxine operate.

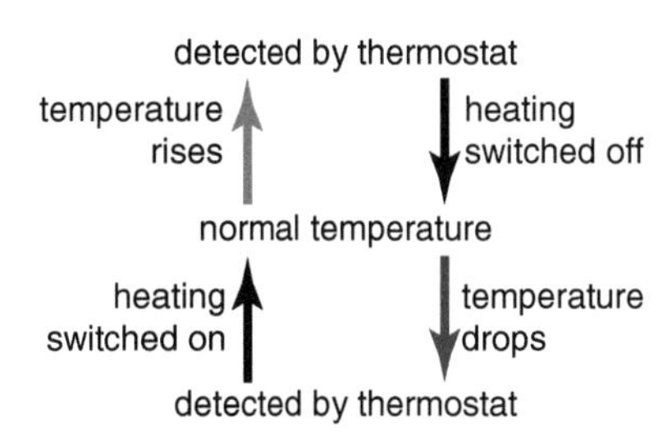

Figure 2.1.38 Negative feedback in a central-heating system

6. **Give a definition of a negative-feedback system.**
7. **Figure 2.1.38 illustrates a simple negative feedback system, but it's not a perfect comparison with thyroid secretion. Explain why.**

Check your progress

You should be able to:

recall that organisms can respire with oxygen (aerobic respiration) or without oxygen (anaerobic respiration). →	use word equations to describe the processes of aerobic and anaerobic respiration. →	use a symbol equation for aerobic respiration and be able to compare the two processes.
describe the functions of different parts of the circulatory system. →	describe how the circulatory system transports substances. →	explain how the circulatory system is adapted to its function.
describe the effect of surface area : volume ratio on the diffusion of substances. →	describe the features of a range of exchange surfaces in plants and animals. →	explain the features of exchange surfaces.
know that digested food is transported from the small intestine to body cells. recall how to test for carbohydrates, lipids and proteins. →	describe the adaptations of the intestine as an exchange surface. →	explain how the small intestine is adapted for efficient food absorption.
describe the structure of the nervous system. →	describe how neurones are adapted to their role and the transmission of an impulse in a reflex arc. →	explain how nerve impulses are transmitted across synapses.
recall typical results for factors affecting reaction time. →	plan methods of measuring human reaction times. →	explain and evaluate methods of measuring human reaction times.
describe the endocrine system as being composed of glands that secrete hormones to target organs around the body. →	compare the endocrine and nervous systems. →	explain how negative feedback is involved in control of responses, using thyroxine as an example.

Worked example

The diagram shows the human heart.

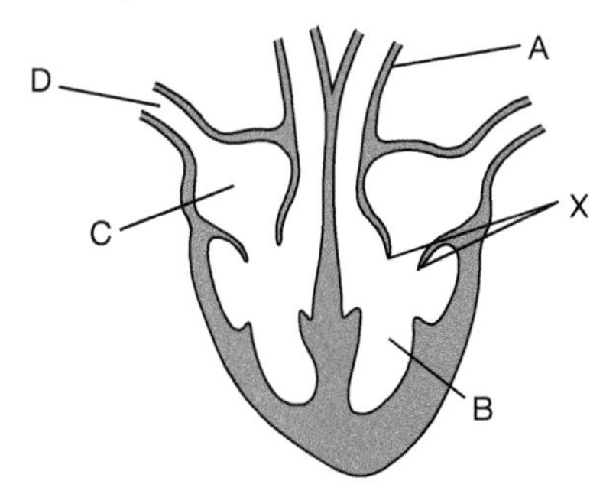

1 Label parts A–D on the diagram.

A Artery

B Left ventricle

C Right atrium

D Vena cava

> B–D correctly identified. Giving 'artery' for A is insufficient.
>
> 'Aorta' is more accurate.

2 a What is the name of part X?

Valve

> Valve is correctly identified.

b What is the function of part X?

Valves prevent the backflow of blood.

> Correct. A full answer would also explain that this valve opens as the atrium contracts, so that blood passes into the ventricle. Then it closes as the ventricle contracts so that blood is pumped out through the artery and not back into the atrium.

c Describe a difference between blood in part B and blood in part C.

Blood in part C has returned from the body and is deoxygenated. Blood in part B has come from the lungs and is oxygenated.

> A correct and full answer.

d Why are the walls of the ventricles of the heart thicker than those of the atria?

Ventricles have thicker walls because they have to pump blood further. The left ventricle pumps blood around the body. The right ventricle pumps blood to the lungs. The atria only have to pump to the ventricles.

> Again, a good, full answer.

3 The heart pumps blood to the lungs so that gas exchange can happen.

Why are alveoli an efficient exchange surface?

They have a large surface area compared to volume to give a large area for gas exchange. Good blood supply to take oxygen to body cells where it is used for respiration for energy for growth.

> Two correct answers are identified. Then the student forgets the question being asked and moves away from the topic of alveoli.

> Overall, this student shows a good understanding of the topic but does not always give complete answers. Careful reading of the questions, and consideration of exactly what is being asked for, in addition to more detailed responses, would help improve this answer.

End of chapter questions

Getting started

1. Give two features of an efficient exchange surface. **2 Marks**
2. Name the main blood vessels in the circulatory system. **2 Marks**
3. What is the function of the **1 Mark**
   a heart? **1 Mark**
   b coronary arteries? **1 Mark**
4. How are hormones transported to their target organs? **1 Mark**
5. Use words from the box to complete the following sentences.

proteases	glycogen	simple sugars	carbohydrases	starch	fatty acids

Fats are digested by lipases into .... and glycerol. .... break down proteins into amino acids. Carbohydrates are digested by .... into .... **2 Marks**

6. Use words from the box to complete the word equation for aerobic respiration.

alcohol	carbon dioxide	glucose	lactic acid

.... + oxygen .... + water (+energy) **1 Mark**

Going further

7. Give two ways in which the small intestine is adapted for efficient absorption. **2 Marks**
8. Compare the nervous and endocrine systems according to their:
   a type of message
   b speed of response. **2 Marks**
9. a Which gland is referred to as the 'master gland'? **1 Mark**
   b Why is it called the 'master gland'? **1 Mark**
10. For each of the following blood vessels, state whether they carry oxygenated or deoxygenated blood:
    a vena cava **1 Mark**
    b pulmonary vein **1 Mark**
    c vena cava **1 Mark**

More challenging

11 **A teenage girl had her heart beat measured at 74 beats per minute. Each beat pumped 70 cm^3 of blood. Calculate how much blood will be pumped in 10 minutes. Give your answer in litres.** 1 Mark

12 **a Copy and complete the table comparing aerobic and anaerobic respiration in humans.** 3 Marks

	Aerobic respiration	Anaerobic respiration
Where does it occur?	in all cells	
When does it occur?		during vigorous exercise
What are the by-products?	carbon dioxide + water	
How much energy is energy produced?	a great deal	
Oxygen needed?	yes	
Oxygen debt produced?		yes

b During periods of vigorous activity, our muscles become tired. Why is this? 1 Mark

c During vigorous exercise, we may build up an oxygen debt. Explain what is meant by 'oxygen debt' and how the body deals with it. 2 Marks

13 **The diagram shows a single-celled organism that lives in an aquatic environment. Explain why it does not need a specialised respiratory system, digestive system or transport system.** 3 Marks

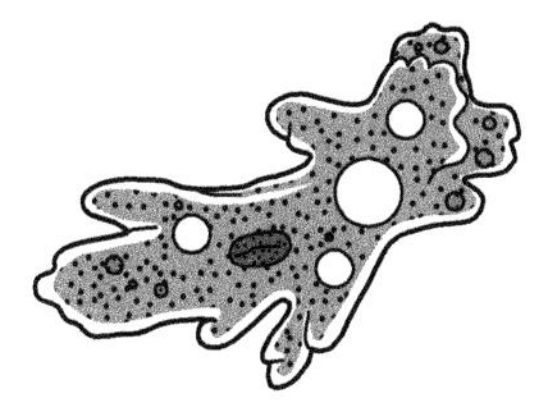

Most demanding

14 **Explain how thyroxine levels affect the temperature of the body.** 2 Marks

15 **Coeliac disease destroys the villi in the small intestine. Explain the effect of having many fewer villi in the small intestine.** 4 Marks

16 **A mother places her finger on her baby daughter's palm. The baby grasps it. It is a reflex action. Describe the process by which this reflex action occurs. You can use a diagram to help with your description.** 4 Marks

Total: 40 Marks

PLANTS AND PHOTOSYNTHESIS

IDEAS YOU HAVE MET BEFORE:

PLANT TISSUES ARE ADAPTED FOR THEIR FUNCTIONS

- Plants have adaptations that allow them to survive and grow, such as air pores in the leaves.
- Cells have specific adaptations so that they can carry out a specialised function in the plant.
- Photosynthesis happens in cells in leaves. Water is absorbed through root hair cells.
- Diffusion is the movement of molecules from a higher concentration to a lower concentration until they are equally distributed.
- Substances that plants need diffuse into the plant. Other substances pass out by plants by diffusion.

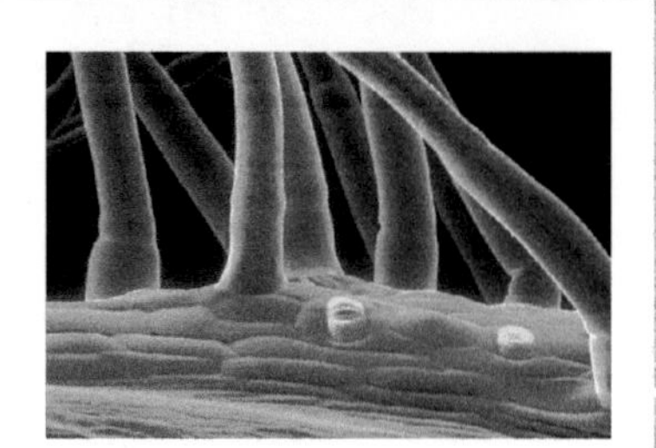

PLANTS MAKE THEIR FOOD WITH THE HELP OF SUNLIGHT

- Green plants use light, water and carbon dioxide to produce glucose in their leaves by photosynthesis.
- The rate of photosynthesis can be affected by different factors.
- Plants can be damaged by a range of ion deficiency conditions.

PLANTS CAN MOVE NUTRIENTS AND WATER TO WHERE IT IS NEEDED

- Plants have a network of vessels that transport water and minerals to their leaves and flowers.
- All cells contain watery cytoplasm. Plants need water so that the chemical reactions needed to sustain life can occur.

2.2

IN THIS CHAPTER YOU WILL FIND OUT ABOUT:

HOW DO THE ADAPTATIONS OF PLANTS HELP THEM TO SURVIVE?

- Adaptations of cells and tissues in leaves allow them to photosynthesise efficiently.
- Stomata are adapted to control the exchange of gases.
- Cells and tissues in leaves, stems and roots are designed for the maximum exchange of substances in and out of the plant.
- Different factors affect the rate of diffusion in plant systems.
- Substances move in and out of the leaf during different processes: for example, photosynthesis, respiration and transpiration.

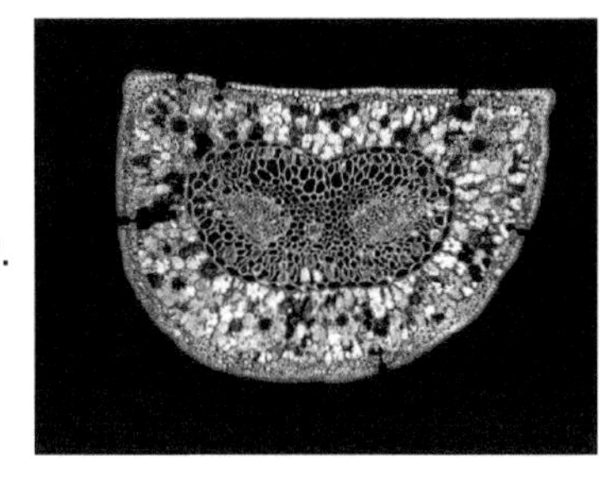

HOW DO DIFFERENT FACTORS AFFECT PHOTOSYNTHESIS?

- The useful products of photosynthesis are simple carbohydrates, such as glucose and sucrose. Oxygen production can be used as a measure of photosynthesis in some water plants.
- Different environmental factors interact to limit the rate of photosynthesis in different habitats at different times.
- The environment in which plants are grown can be artificially manipulated.
- Viral, bacterial and fungal pathogens cause communicable plant diseases, which limit photosynthesis.
- Plant diseases in crop and garden plants can be controlled to a certain extent by human interventions.

HOW IS WATER UPTAKE BY A PLANT AFFECTED BY ENVIRONMENTAL CONDITIONS?

- There are two transport systems in plants: xylem transports water from the roots to the leaves, and phloem transports substances in all directions around the plant.
- Water movement through the plant by transpiration is affected by different environmental factors.
- Water loss in plants is a consequence of adaptations for photosynthesis.

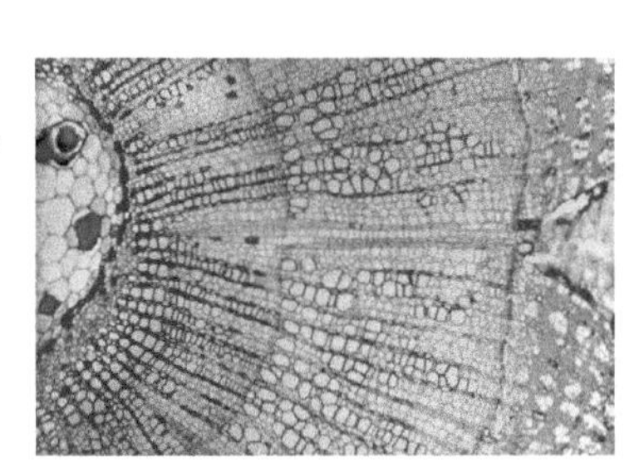

2.2a Meristems

Learning objectives:

- describe the function of meristems in plants
- describe and explain the use of stem cells from meristems to produce clones of plants quickly and economically.

KEY WORDS

clone
meristem
tissue culture

Like animals, plants have undifferentiated cells that can divide to produce new cells. In plants, these cells are found in regions called meristems. Growth – in response to environmental conditions, for example – is controlled and coordinated by the division of meristem cells.

MAKING LINKS

Look back to topic 1.3l to remind yourself about cell differentiation.

Plant stem cells

Meristems are found at the growing tips of shoots and roots in a plant, and contain the cells that divide as the plant grows. Meristem cells are undifferentiated, like stem cells in animals. As they divide, the daughter cells differentiate into different types of cells depending on where they are in the plant.

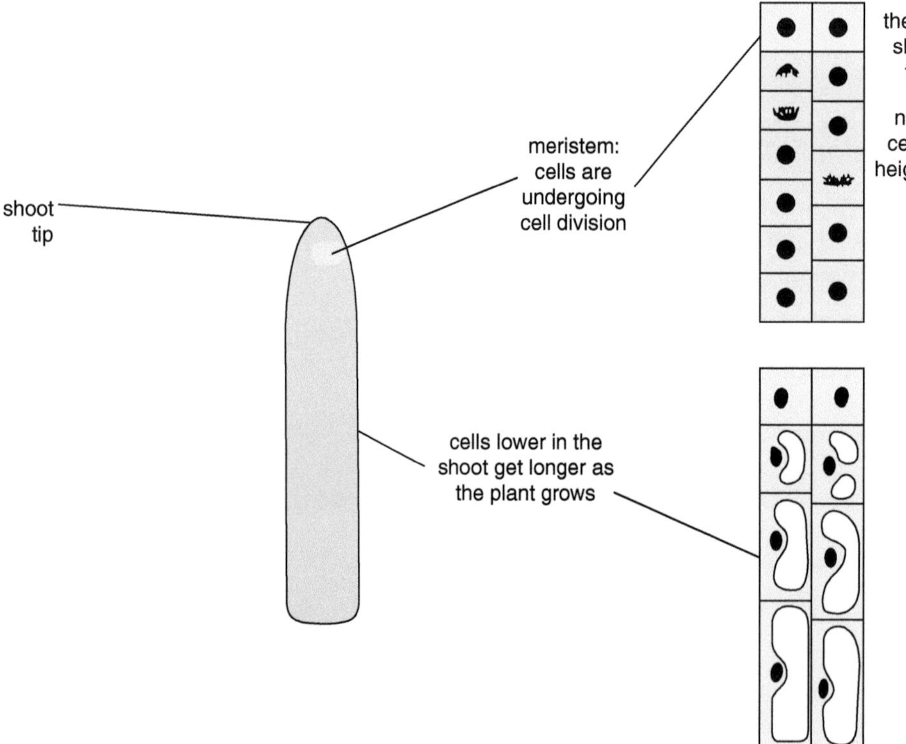

Figure 2.2.1 Plant stem cells are found in meristem tissue, towards the tip of each shoot, and at the tip of each root

1. **Where are stem cells found in plants?**
2. **What happens to some of the daughter cells as the meristem cells divide?**

Cloning plants

When cells from the meristems of plants are removed, scientists are able to grow them in **tissue culture**. Tissue culture is used to produce **clones** of plants. Clones are groups of cells or organisms, all produced from one parent, which are genetically identical to each other and to the parent. This technique means that plants that are useful to us – such as crop varieties or garden plants – can be grown quickly and economically.

When plants are grown in tissue culture, the culture medium contains nutrients, along with agar to support the growth of the plantlets. Plant hormones are also added to the medium to help stimulate the meristem cells to divide and the daughter cells to grow.

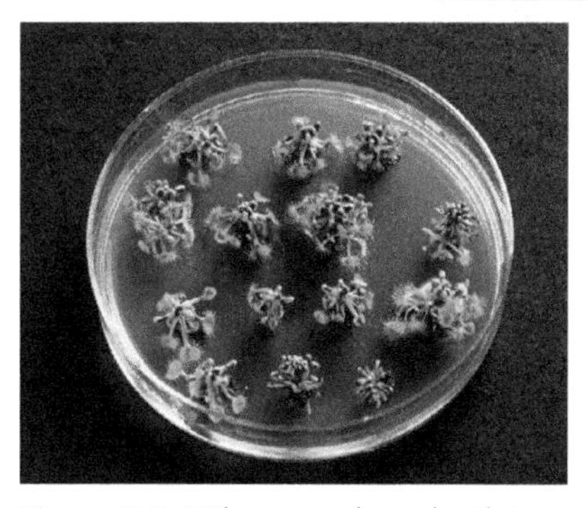

Figure 2.2.2 These sundew plantlets are being grown in tissue culture. They are all genetically identical

3. **All the plantlets grown in tissue culture from the meristem tissue of one plant are genetically identical. Why is this?**
4. **What are the contents of a tissue culture medium?**

Plant cell differentiation

Meristem cells all look about the same. As they divide, the process of differentiation begins. Some cells, at the surface of the plant tissue, become specialised to protect the plant from water loss and from invasion by harmful microorganisms, while at the same time allowing gas exchange. Some cells inside the plant differentiate into xylem and phloem tissue (see topic 2.2b) to carry water and nutrients around the plant. Other cells form more general tissues with functions such as photosynthesis, storage and support.

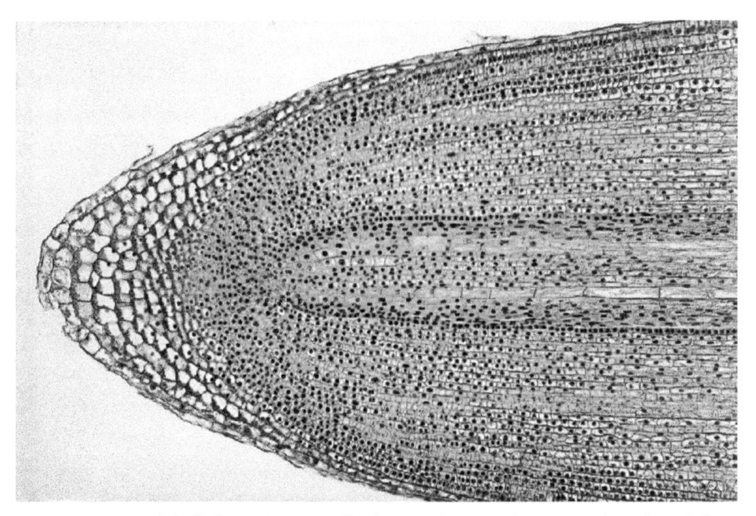

Figure 2.2.3 This light micrograph shows the meristem at the tip of the root of a lily. The cells have been stained with colour so that they show up better and viewed with a high-power lens (×400 for school microscopes)

DID YOU KNOW?

There are about a dozen basic cell types in plants, which is far fewer than in animals.

5. **Describe the function of plant meristems.**
6. **Explain why differentiation is important in multicellular organisms such as plants.**

2.2b Plant structures

Learning objectives:

- describe some of the substances transported into and out of plants
- understand that the roots, stem and leaves form a transport system in plants.

KEY WORDS

phloem
stomata
xylem

Plants do not have a heart or blood, but they still need a transport system to move food, water and minerals to every cell. Specialised cells in the roots, stem and leaves form a plant organ system for the transport of substances.

Transporting food and water

A growing seedling develops roots, a stem and leaves. Each part has its own function:

- the leaves make food for the plant by photosynthesis from water and carbon dioxide
- the roots take in water from the soil with dissolved ions, and help to anchor the plant in the soil
- the stem supports the plant and transports substances up and down the plant.

Glucose made in the leaves has to be transported to every cell in the plant for respiration. Veins in the leaves carry the glucose to the main stem, so it can be transported around the plant. The tissue in the veins and stem that transports glucose is called **phloem**.

Water and mineral ions enter the plant through the roots. Water-carrying tissue in the stem and in leaf veins transports the water around the plant, especially to the leaves. This tissue is called **xylem**.

1. **Why does a plant need a transport system to move glucose away from the leaves?**
2. **Why is water needed in the leaves?**

Transporting gases

Leaves have tiny pores called **stomata** that allow gases to enter and leave. Most leaves are very thin, so gas molecules can diffuse into the cells once they have entered the stomata.

The leaf takes in carbon dioxide gas from the air for photosynthesis. The oxygen gas produced in the photosynthesis reaction is released, and diffuses out of the leaf.

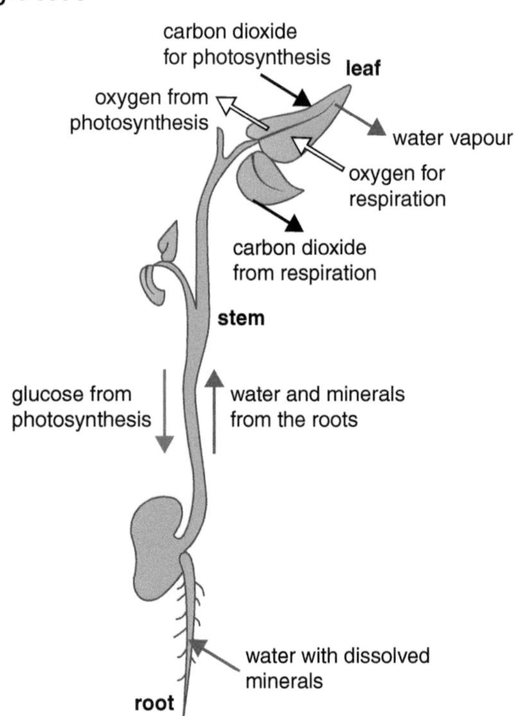

Figure 2.2.4 Transport of substances in a young bean plant

All cells need to respire. Respiration needs oxygen, as well as glucose. The stomata and thin structure of leaves means that oxygen in the air can easily diffuse into cells for respiration, when required. The carbon dioxide produced in respiration diffuses out of the leaf.

Water also leaves the plant through the stomata, as water vapour.

3. **How does the structure of a leaf help gases to enter and leave the plant?**
4. **In daylight, when plants are photosynthesising, less oxygen diffuses from the air into the cells for respiration. Suggest why this is.**

Roots and transport

Plants take in water from the soil, with dissolved ions that are important for the healthy functioning of the plant. For example, nitrate ions are needed to make proteins and magnesium ions are used to make the green pigment chlorophyll, which is essential for photosynthesis.

The table lists some of the mineral ions that you might hear about. You only need to remember about nitrates and magnesium for your examinations.

Mineral ions	Use in the plant
nitrates, containing nitrogen (N)	to make amino acids for protein synthesis
magnesium (Mg)	to make chlorophyll for photosynthesis
phosphates, containing phosphorus (P)	in respiration to make DNA and new cell membranes
potassium (K)	in respiration in photosynthesis

5. **How do water and mineral ions reach all parts of the plant?**
6. **If a plant was growing in soil that was deficient in magnesium ions, how do you think it might look?**

MAKING LINKS

See topic 5.1d for a discussion of ions.

COMMON MISCONCEPTIONS

Some students think that water is absorbed through the leaves, but this is wrong. Water is only absorbed through the roots.

2.2c Transpiration

Learning objectives:

- explain the need for exchange surfaces and a transport system in multicellular organisms
- explain how the structure of the root hair cells in plants relates to their function
- explain how the structure of xylem is adapted to its functions in the plant.

KEY WORDS

active transport
diffuse
lignin
osmosis
root hair cells
transpiration
xylem

All living organisms need water for life processes. Green plants also need water for photosynthesis. Specialised tissues in the plant are adapted to obtain water and to transport it to all the cells that need it.

DID YOU KNOW?

Plant transport systems are found in the bark of trees. Some animals chew this away, causing the tree to die because it cannot transport water and glucose to where they are needed.

Water movement in plants

Most plants are multicellular organisms. They need to take in important materials such as water from their surroundings and transport them to every cell.

Plants absorb water from the soil into their roots, through specialised cells called root hairs. The water passes into the root, up the stem in the **xylem** tissue and into the leaves, where it is needed for photosynthesis. Water on the surfaces of the cells inside the leaves evaporates and **diffuses** out of leaves through the stomata.

This movement of water through the plant, in through the roots and out through the leaves, is called **transpiration**.

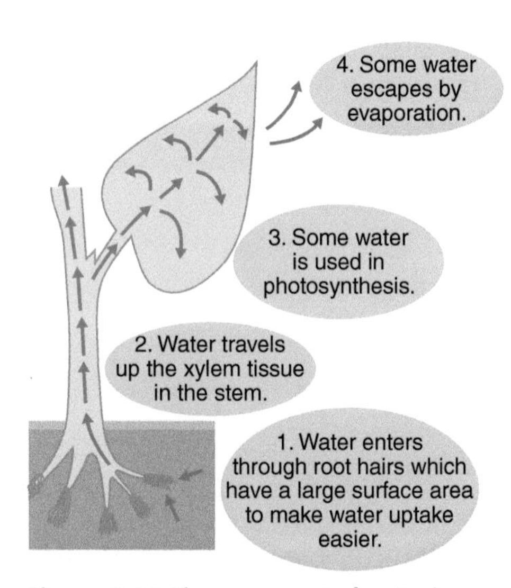

Figure 2.2.5 The movement of water in a plant

1. **What is transpiration?**
2. **Describe the path of water through a plant.**

Adaptations for water absorption

Roots branch out into the soil, forming a network of tubes that absorb water and dissolved minerals. **Root hair cells,** found just behind each root tip, increase the surface area for absorption, and also help to anchor the plant. Each root hair cell has a long, thin extension that reaches out between the soil particles. Water moves into the root hairs by **osmosis**.

The concentration of mineral ions is lower in the soil than in the root hair cells, so **active transport** is needed to move the mineral ions into the root hairs. Root hair cells contain many mitochondria to provide the energy for this.

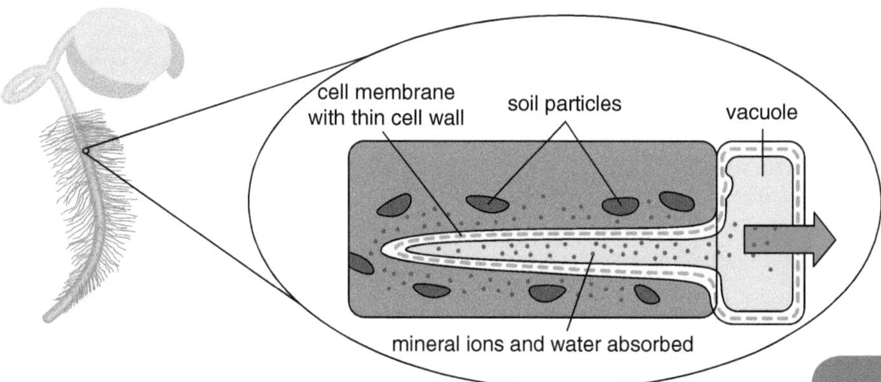

Figure 2.2.6 Root hair cells are specialised exchange surfaces

Root hair cells are an efficient exchange surface for the absorption of water and dissolved minerals because they:

- provide a large surface area
- have only a thin outer membrane to allow rapid absorption
- have a thin cell wall to reduce the distance that materials must move across
- have a large permanent vacuole to absorb as much water as possible
- are close to the xylem vessels in the root, so that materials only have a short distance to travel to enter the xylem and then be moved up and around the plant.

These adaptations mean that a lot of water and mineral ions can get into the cells very quickly and then pass around the plant efficiently.

3 How are root hair cells adapted for their function?

4 How do plants absorb the minerals they need?

MAKING LINKS

Look back at topics 1.3e, 1.3f and 1.3i to remind yourself about diffusion, osmosis and active transport.

Figure 2.2.7 The network of roots takes in water for this bean seedling

Adaptations for water transport

Water flows in the xylem vessels, from the plant roots, through the stem, to the leaves. Xylem vessels are tubes made from many long cells with thick, reinforced walls containing **lignin**, lying end to end. Lignin makes the cell walls strong, and also waterproof, which means the cells die. The cell contents and end walls break down, so the 'chain' of cells forms a continuous tube with a hollow centre. Water and dissolved minerals flow through the tubes. Lignified xylem cells form the wood in a tree.

5 Explain how xylem vessels are adapted to their function.

6 Why is it important for the function of xylem tissue that the xylem cells die?

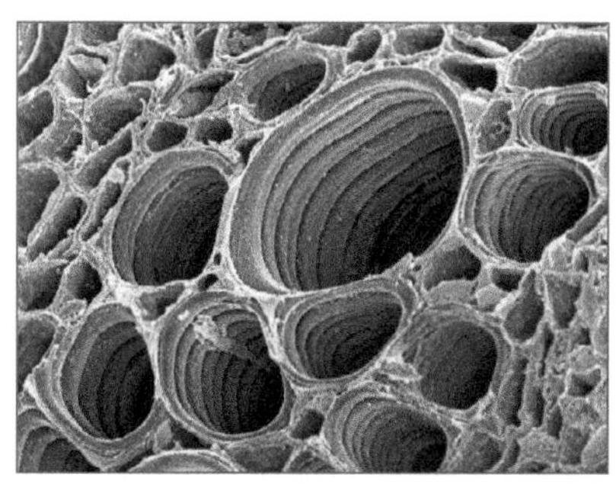

Figure 2.2.8 A scanning electron micrograph of xylem cells from a plant leaf. Colours have been added to the image: lignin is shown in orange

2.2d Looking at stomata

Learning objectives:

- describe the process of transpiration
- explain the relationship between transpiration and leaf structure
- explain the structure and function of stomata.

KEY WORDS

flaccid
guard cell
stomata
turgid
xylem

Gases continually pass in and out of leaves through tiny openings called stomata, found on the lower surface of each leaf. They control the exchange of gases in plants, and also the flow of water in transpiration.

Leaf structure and transpiration

Water evaporates from the cells inside the leaves of a plant. The water vapour diffuses out through tiny holes in the surface of the leaves called **stomata** (singular: stoma). Evaporation from the leaf causes water to be drawn up the plant through the **xylem** tissue. This is what causes transpiration.

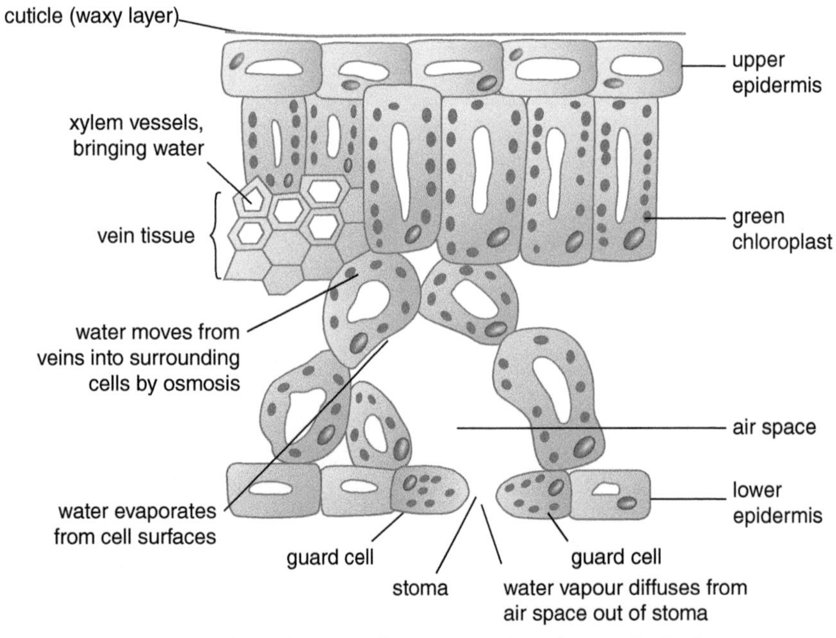

Figure 2.2.9 A generalised drawing of a cross-section of part of a leaf

KEY INFORMATION

Water evaporates from the cell surfaces and the water vapour diffuses out of the leaf.

Transpiration is important because:

- water is needed in the leaves for photosynthesis
- water carries dissolved materials around the plant
- cells full of water help support the plant
- evaporation of water cools leaves.

The rate of transpiration increases when the temperature

is higher, the humidity is lower or the wind speed is greater, because these conditions make evaporation faster. Transpiration is also faster when there is more light available, because the stomata open up in the light, letting carbon dioxide in for photosynthesis.

1. **Why are stomata important in transpiration?**
2. **Why do plants need water?**

Opening and closing of stomata

If too much water is lost through the stomata, plants may wilt and die. **Guard cells** in the leaf open and close the stomata to control the rate of transpiration.

The curved guard cells are attached to each other at both ends and surround the stoma. When it is light, the stomata open so carbon dioxide gas can diffuse into the leaf for photosynthesis. The guard cells take in water by osmosis and become **turgid** (swollen). The thick inner walls can't stretch, so as the cell expands the inner walls bend apart and the stoma opens.

In the dark, when photosynthesis stops and carbon dioxide is no longer needed, the guard cells lose water by osmosis and become **flaccid** (limp). The inner walls move together and close the pore.

3. **Suggest why some plant species have more stomata than others.**

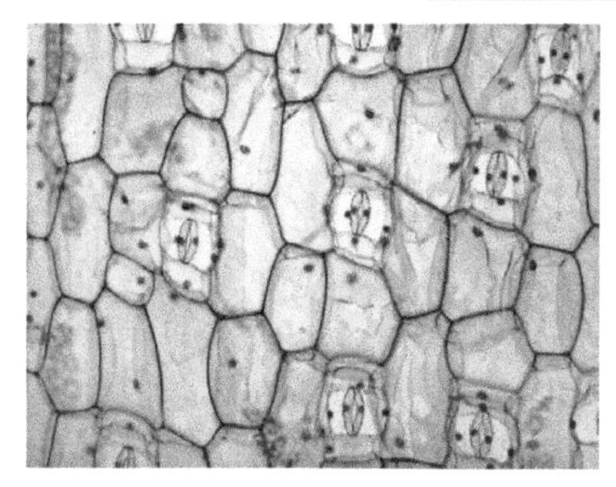

Figure 2.2.10 This micrograph of the underside of a leaf clearly shows the pairs of guard cells, each surrounding a stoma

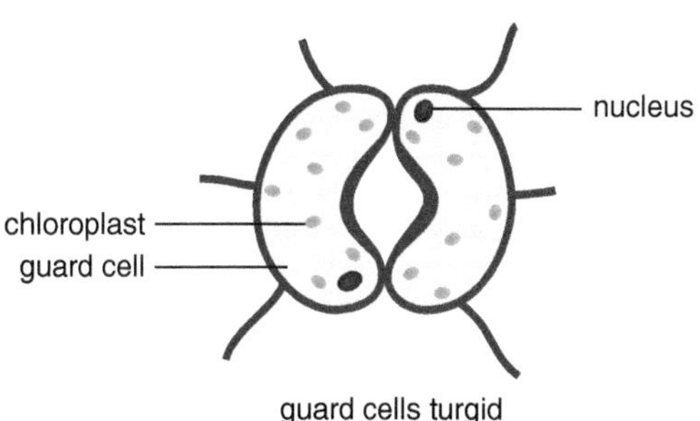

Figure 2.2.11 Stomata open in daylight and close in darkness

Water balance

Healthy plants balance water uptake and loss. If plants lose water faster than it is replaced, they wilt. Wilting protects the plant. Leaves droop down to reduce the surface area for water loss by evaporation. The stomata close to prevent water loss, but this means that photosynthesis stops and the plant may overheat.

Some plants have adaptations to reduce water loss while keeping the stomata open so photosynthesis can carry on. Marram grass grows on sand dunes near beaches. It has rolled leaves with a thick, waxy outer surface (cuticle) and stomata sunk into pits. Interlocking hairs hold water vapour. These features all reduce water loss.

4. **Explain how marram grass is adapted to prevent water loss.**
5. **Explain how wilting protects a plant.**

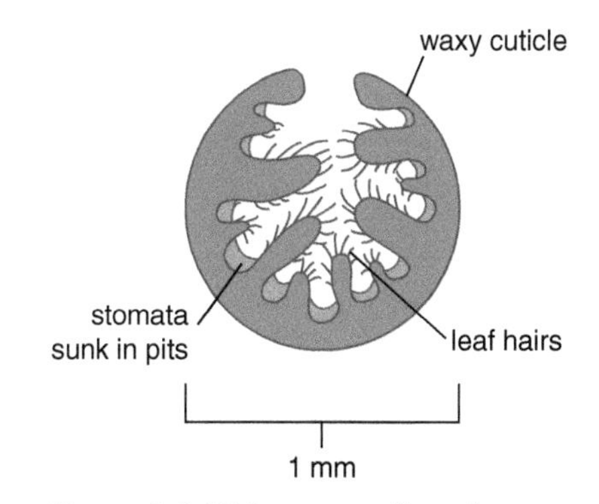

Figure 2.2.12 A cross-section of a rolled marram grass leaf, which grows on exposed sand dunes

2.2e Rate of transpiration

Learning objectives:

- describe how transpiration is affected by different factors
- understand and use simple compound measures such as rate of transpiration
- draw and interpret appropriate graphs, charts and tables.

KEY WORD

potometer

The rate of transpiration in plants depends on several environmental factors, including the light intensity, the temperature and how windy it is. We can investigate how different conditions affect transpiration rate.

Environmental factors and transpiration

During transpiration, water is transported from the roots to the leaves in xylem vessels. Water diffuses out of the leaf cells and forms a film over the cell surfaces. This water evaporates into the air spaces between cells. The water vapour moves down the concentration gradient towards the stomata, and escapes from the leaf.

Factors that increase the rate of transpiration are:

- an increase in temperature
- an increase in wind speed
- a decrease in humidity.

When plants photosynthesise, the stomata are open to allow carbon dioxide to diffuse into the leaf. So factors that increase the rate of photosynthesis will also increase transpiration rate.

Some students set up the experiment shown below to investigate how light intensity affects the rate of transpiration.

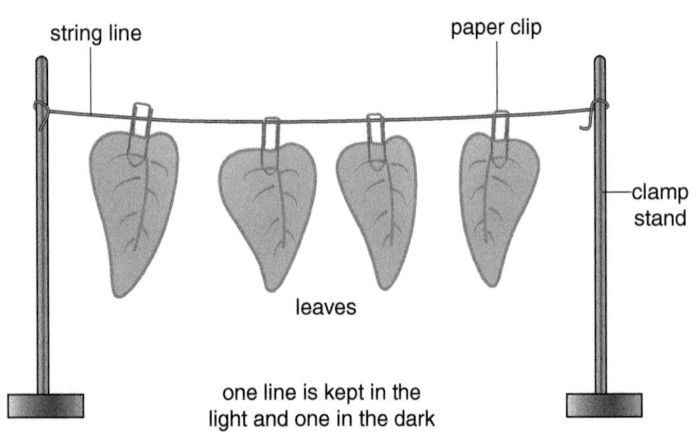

Figure 2.2.14 Experiment to investigate the effect of light on water loss from leaves

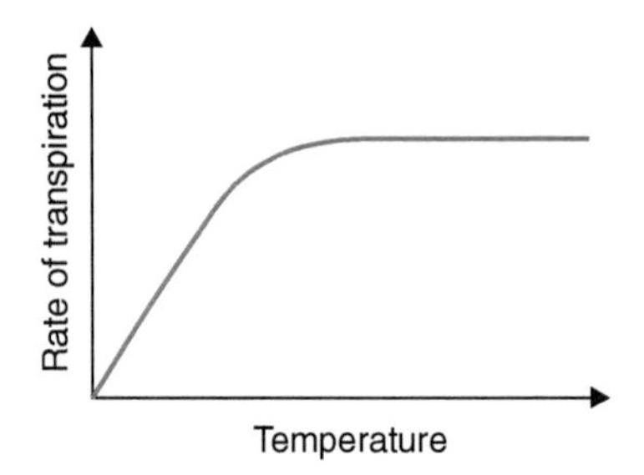

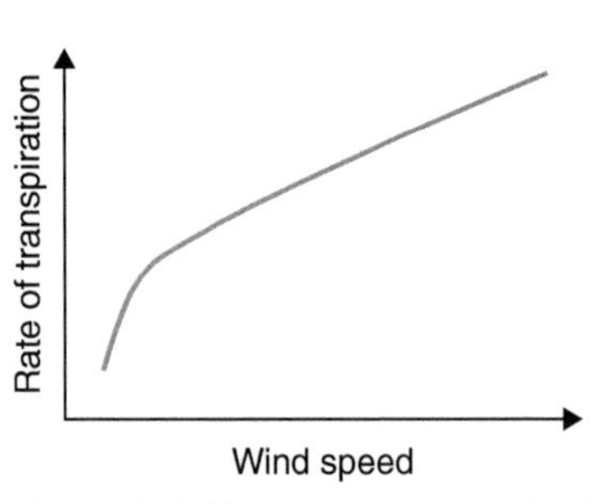

Figure 2.2.13 Temperature and wind speed affect transpiration rate. Why do you think the graphs have these shapes?

1. **In Figure 2.2.14, why were some leaves put in the dark and some in the light?**
2. **What measurements could the students take?**
3. **Why is this experiment not measuring transpiration directly?**

KEY INFORMATION

Remember that water diffuses out of cells and evaporates to form water vapour, which can escape from the leaf.

Measuring water uptake

A **potometer** is used to measure water uptake by plant shoots in different conditions. The rate of movement of the air bubble in the tube is used as a measure of the rate at which water is taken up.

On still days, there is no wind to move the water vapour away from the leaf surface as it diffuses out. The concentration of water vapour outside the leaf builds up and the concentration gradient decreases. When the water vapour concentration outside the leaf equals that inside the leaf, transpiration stops.

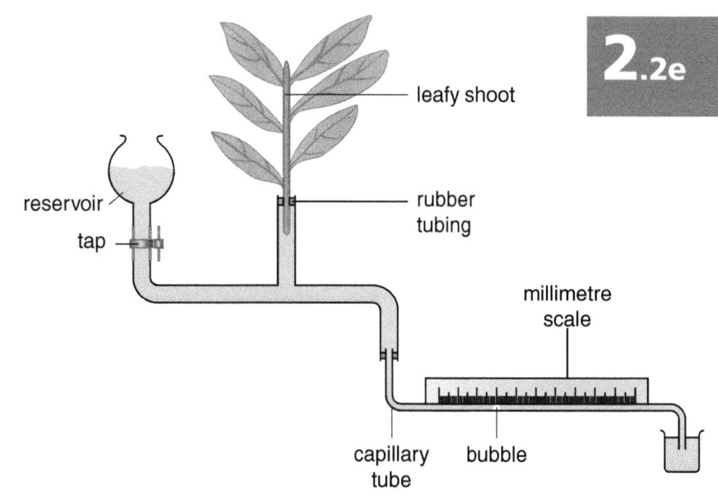

Figure 2.2.15 A potometer

4. **Describe an experiment using a potometer, and any other apparatus you think is needed, to investigate how transpiration rate is affected by air movement.**
5. **Why is the rate of water uptake not exactly the same as transpiration rate?**

Analysing changes in water uptake

The table shows the results of a student's potometer experiment investigating the effect of light intensity on the rate of transpiration, in a leafy shoot of a geranium plant.

Light intensity (arbitrary units)	Rate of uptake of water (mm/min)			
	Trial 1	Trial 2	Trial 3	Mean
10	5.0	7.0	5.0	
20	5.0	7.0	5.0	
30	12.0	12.0	11.0	
40	24.0	23.0	26.0	
50	32.0	33.0	32.0	

6. **Why did the student carry out three trials of the experiment?**
7. **What factors should the student have kept the same during this experiment?**
8. **Calculate the mean transpiration rate for each light intensity, and plot a graph. How does transpiration rate vary with light intensity?**
9. **Suggest why there was no change in transpiration rate between light intensities of 20 and 10 units.**
10. **How would you expect the transpiration rate to change at even higher light intensities? Explain your answer.**

DID YOU KNOW?

Transpiration keeps plants cool. Some trees lose hundreds of thousands of litres of water in one day through transpiration.

2.2f Chlorophyll and other plant pigments

Learning objectives:

- explain how to set up paper chromatography
- distinguish pure from impure substances
- interpret chromatograms and determine R_f values
- carry out and represent mathematical and statistical analysis.

KEY WORDS

chromatogram
chromatography
mobile phase
R_f value
stationary phase

Plants use chlorophyll and other pigments in their leaves to carry out photosynthesis. These pigments can be separated and identified using chromatography.

Chromatography

Chromatography can be used to separate substances in a mixture, and to help identify the substances. If you put a spot of dye onto chromatography paper and put the edge in water, the colours in the dye move up through the paper as the water rises.

In this simple example, the chromatography paper is called the **stationary phase**, and the water – the solvent – is the **mobile phase**.

The different coloured compounds in the dye move at different speeds up the paper. This means they get to different heights on the paper and appear as separate patches.

The compounds in a mixture may separate into different spots with different solvents, but a pure compound produces a single spot in all solvents.

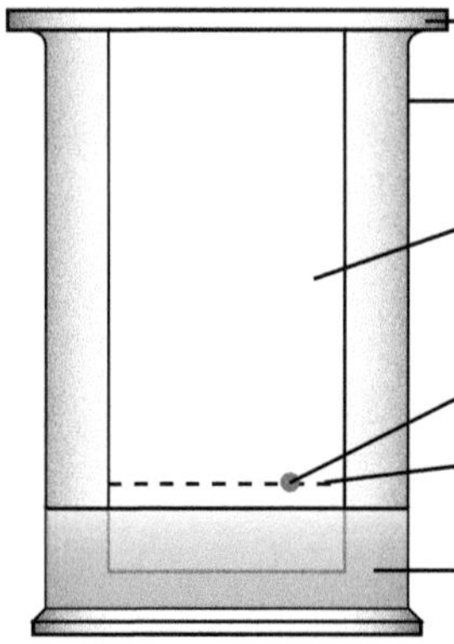

Figure 2.2.16 Carrying out chromatography – the solvent here is water

KEY INFORMATION

The edge of the paper needs to *just* dip into the water. The spot must not be in the water.

1. **Why is the paper called the 'stationary' phase and the water called the 'mobile' phase?**
2. **Why shouldn't the spot of mixture be under the water surface?**

Separating pigments

Chromatography can be used to separate and identify the different pigments in plant leaves. First, a leaf extract is made by grinding up several leaves. The extract is a liquid containing all the pigments that were in the leaves.

A spot of this 'unknown mixture' is placed on a line at the bottom of a sheet of chromatography paper. Next to this, spots of various known plant pigments are also placed on the line.

Link: 2.2f ⟶ 2.2g

The sheet is held in the solvent and left for separation to take place. Figure 2.2.17 could be the resulting chromatogram. For each of the five known pigments, one spot has travelled up the paper a certain distance. This distance can be measured. The leaf mixture has four colours: the first spot has moved the same distance as pigment 3, the second spot is at the same height as chlorophyll B, the third spot matches the level of chlorophyll A and the top spot is aligned with pigment 1. So the leaf mixture contains pigments 1 and 3 and chlorophylls A and B.

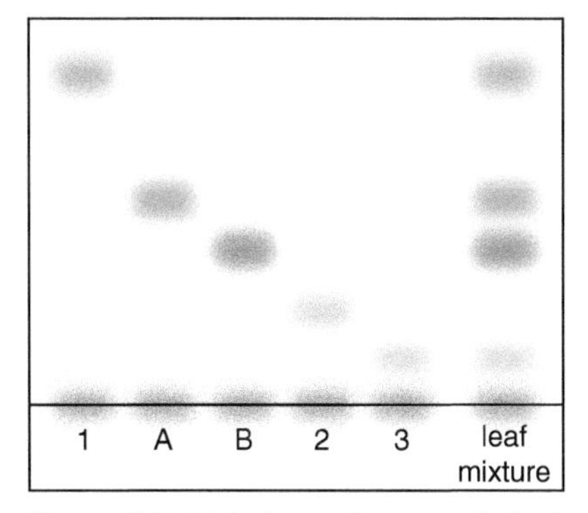

Figure 2.2.17 A chromatogram of plant pigments: pigments 1, 2 and 3 and chlorophyll A (A) and chlorophyll B (B)

3 **Explain how you know that the leaves tested did not contain pigment 2.**

4 **The pigment samples are pure substances. The leaf extract is impure (it is a mixture). Explain how the chromatogram confirms this.**

Calculating R_f values

In our example, it is clear that the pigments have moved different distances as they are all on the same chromatogram. We need a standard measure to compare separation in any situation. This standard is the R_f **value**, where:

$$R_f = \frac{\text{distance moved by substance}}{\text{distance moved by solvent}}$$

This is the ratio of the distance moved by a substance (from the centre of the original spot) to the distance moved by the solvent.

For example, in Figure 2.2.18 the substance has moved 45 mm and the solvent has moved 60 mm.

$$R_f = \frac{45\text{ mm}}{60\text{ mm}} = 0.75$$

Different substances have different R_f values in different solvents, which can be used to help identify each substance.

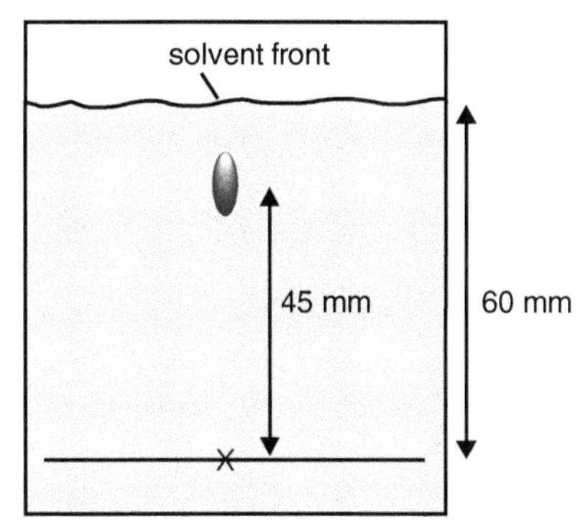

Figure 2.2.18 Measuring R_f values

5 **What is the R_f value for a dye that moves 56 mm in a solvent that moves 70 mm?**

6 **If the R_f value for a blue dye is 0.68 and in a chromatogram the solvent front moved 90 mm, how far would a blue dye spot move?**

REQUIRED PRACTICAL

2.2g Paper chromatography

Learning objectives:

- safely and accurately use a range of appropriate apparatus to separate and distinguish plant pigments by chromatography
- extract and interpret information from charts and tables
- determine R_f values.

KEY WORDS

solvent front

Paper chromatography can be used to separate the photosynthetic pigments in plants. From the resulting chromatogram, we can calculate an R_f value for each pigment, which allows us to identify it. Here, we look at the skills needed to carry out this kind of identification using chromatography.

These pages are designed to help you think about aspects of the investigation rather than to guide you through it step by step.

Safe and correct use of apparatus

A chromatogram is set up – spots of the mixture are placed along a start line on a strip of chromatography paper, and the edge of this is held in a small volume of solvent.

Sam, Alex and Jo were each given some spinach leaves and asked to identify some of the pigments that spinach uses for photosynthesis.

First they crushed the leaves with a little clean sand and a few drops of a solvent called propanone in a pestle and mortar. This made a dark green liquid. They put a spot of liquid on the start line of a strip of chromatography paper, by dipping a tooth pick in the liquid, touching the paper and letting it dry. They repeated this many times. Then they each developed their chromatogram, using propanone as the solvent.

Once the solvent had risen up through the chromatography paper, but before it had reached the end of the paper, the chromatograms were dried and measured.

In this practical, eye protection should be worn, and a risk assessment is required.

1. **The start line is drawn 2 cm from the bottom of the paper. Why is the line drawn in pencil and not ink?**
2. **Why did the students repeatedly put drops of liquid onto the same spot on the paper, letting it dry each time?**
3. **How far into the solvent should the paper be held in position?**
4. **Until when should the developing chromatogram be left in the solvent?**

Making and recording results

Figure 2.2.19 shows Sam's chromatogram. Alex and Jo's results were similar, but not identical.

From the start line, the students measured:

- the distance from the start line to the final level of the **solvent front** (Figure 2.2.19)
- the distance each spot of pigment had travelled.

They recorded these distances in a table.

Chromatogram	Distance from start line (mm)				
	Solvent front	Pigment A	Pigment B	Pigment C	Pigment D
Sam's	75	72	48	21	13
Alex's	76	74	60	20	14
Jo's	65	63	42	18	11

5 **Suggest why the solvent front of Jo's chromatogram is a shorter distance from the start line than the others.**

6 **Alex has one anomalous result.**

a **What do we mean by an 'anomalous result'?**

b **Suggest an explanation for this result.**

c **What should Alex do to check the result?**

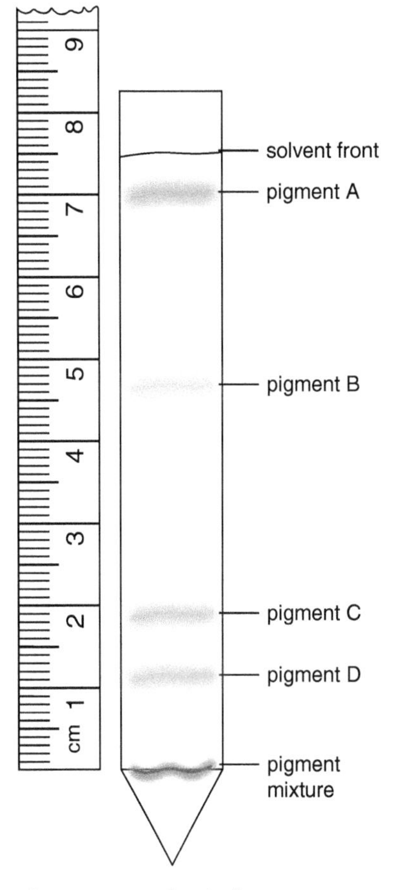

Figure 2.2.19 Sam's chromatogram from spinach leaves

Using R_f values

As the solvent front may not move the same distance on each chromatogram, it important to calculate the ratio of distances travelled by solvent and pigment. This ratio is called the R_f value:

$$R_f = \frac{\text{distance moved by substance}}{\text{distance moved by solvent}}$$

R_f values allow us to check if a pigment is the same in different chromatograms.

Each pigment has a particular R_f value for a certain solvent. So by calculating R_f values from Sam's chromatogram, for example, and comparing these with known values for some plant pigments, we can identify the pigments in the spinach leaves Sam tested.

7 **Calculate an R_f value for each pigment in Sam's chromatogram. Repeat this for Alex's and Jo's chromatograms. (Ignore Alex's anomalous result.)**

8 **The table shows reference R_f values for some plant pigments. Use these values to identify each of the pigments in the spinach leaves.**

Pigment	Colour	R_f value
Chlorophyll A	blue-green	0.28
Chlorophyll B	green-yellow	0.18
Carotene	yellow-orange	0.96
Xanthophyll	yellow	0.64

DID YOU KNOW?

Paper chromatography is only one chromatographic technique that can be used. Other techniques are thin layer chromatography and gas-liquid chromatography.

2.2h Photosynthesis

Learning objectives:

- describe the process of photosynthesis as an endothermic reaction
- write a word equation for photosynthesis
- write a balanced symbol equation for photosynthesis given the formula of glucose.

KEY WORDS

chlorophyll
chloroplast
endothermic
photosynthesis

Plants and algae are amazing organisms. In a process called photosynthesis, they use pigments such as chlorophyll to harness energy from outer space and transfer it into chemical energy in glucose. Without plants, there would be no life!

Describing photosynthesis

Photosynthesis is the process in green plants that makes food using light energy from the Sun. ('Photo' means 'light', and 'synthesis' means 'making'.) Photosynthesis takes place in the **chloroplasts** in the cells of the leaves of plants. The chloroplasts contain **chlorophyll**, which absorbs sunlight.

To carry out photosynthesis, plants also need two raw materials from the environment:

- carbon dioxide, absorbed through the leaves
- water, absorbed through the roots.

The products of photosynthesis are glucose and oxygen. We can describe the reaction using a word equation:

$$\text{carbon dioxide} + \text{water} \xrightarrow{\text{light}} \text{glucose} + \text{oxygen}$$

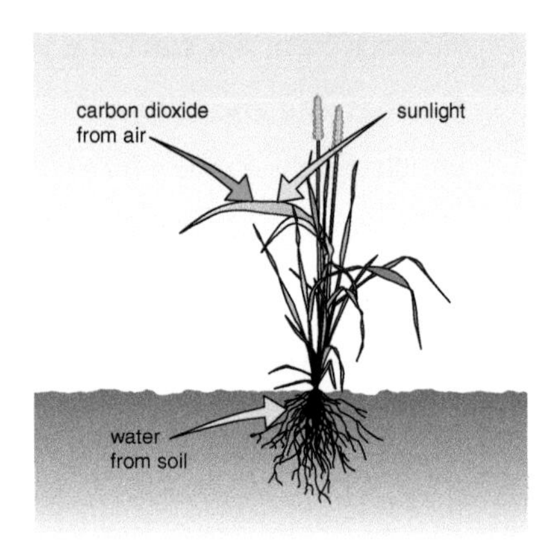

Figure 2.2.20 A plant needs light, chlorophyll, water and carbon dioxide for photosynthesis

The symbol equation for photosynthesis is:

HIGHER TIER ONLY

$$6CO_2 + 6H_2O \xrightarrow{\text{light}} C_6H_{12}O_6 + 6O_2$$

Photosynthesis is an **endothermic** reaction. This means it needs an input of energy. The energy is transferred to the chloroplasts by light.

1. **What raw materials are used in photosynthesis?**
2. **What else is needed for photosynthesis, apart from the raw materials?**
3. **What are the products of the reaction?**
4. **What is an endothermic reaction?**

DID YOU KNOW?

Some leaves have green parts and yellow or white parts. They are called variegated leaves. Variegated leaves only photosynthesise in the green parts, where the chlorophyll is found.

Using glucose for respiration

Dissolved sugars are transported around the plant in the phloem vessels to every respiring cell.

Plants cannot make glucose at night when it's dark, but their cells still need to respire. So during daylight some glucose is converted into an insoluble carbohydrate called starch, to be stored until it is needed.

Starch is the main energy store in plants. It is found in every cell, where it is used for respiration in low light levels or in the dark.

Plant cells use the energy they gain from respiration for chemical reactions.

REMEMBER!

In Chapter 2.1 you learned that respiration is an exothermic reaction that occurs continuously in all living cells. As well as releasing energy, the reaction produces carbon dioxide and water.

5 When and why do plants respire?

6 Why do plants need to have a store of energy?

Building with glucose molecules

Starches made by plants are large molecules, made from many glucose molecules joined together.

Plants also use the small glucose molecules as building blocks for other substances. The glucose may be:

- used to build a carbohydrate called cellulose, which strengthens the cell wall
- used to produce fats and oils for storage
- combined with nitrate ions and other ions taken in from the soil through the roots, to make amino acids for protein synthesis.

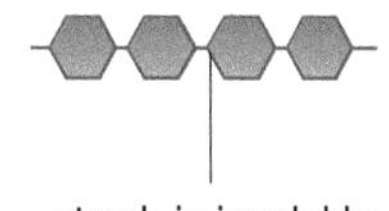
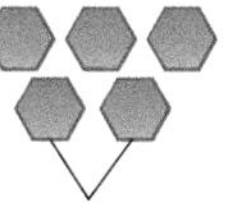

Figure 2.2.21 Starch is a large molecule made of many glucose molecules linked together. A glucose molecule is a ring of atoms with the formula $C_6H_{12}O_6$

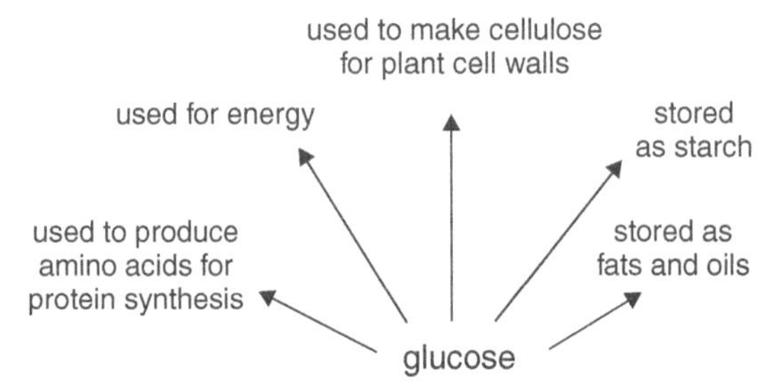

Figure 2.2.22 How plants use the glucose produced by photosynthesis

7 List five ways in which plants use glucose.

8 Explain the relationship between photosynthesis and respiration, in leaves.

2.2i Factors affecting the rate of photosynthesis

KEY WORDS

limiting factor

Learning objectives:

- identify factors that affect the rate of photosynthesis
- interpret graphs relating different factors to the rate of photosynthesis
- explain the interaction of factors in limiting the rate of photosynthesis and relate to the cost effectiveness of controlling conditions in greenhouses.

Plants grow faster in summer than in winter. They can produce food more quickly in warm, bright conditions. Temperature and light are factors that affect the rate of photosynthesis, as well as the availability of raw materials such as carbon dioxide.

Factors that affect the rate of photosynthesis

To make food by photosynthesis, green plants need the raw materials carbon dioxide and water. They also need light. The rate of photosynthesis increases when:

- the light intensity is greater, so more energy is available
- the concentration of carbon dioxide in the air is higher
- the temperature is warmer.

Figure 2.2.23 Conditions are very different in the lush rainforest and the arid desert

1. **Why do plants generally grow faster in the summer than in winter?**
2. **Suggest how the environmental conditions in the habitats shown in Figure 2.2.23 affect photosynthesis in each habitat.**

Greenhouses

In greenhouses the conditions can be controlled to maximise the rate of photosynthesis.

- Sunlight heats up the inside of greenhouses.
- A carbon dioxide source can be used to increase the concentration of the gas.
- Paraffin heaters can be used in greenhouses to increase the temperature on cooler days and nights. As fuel burns it produces carbon dioxide.
- Watering systems deliver a regular supply of water.

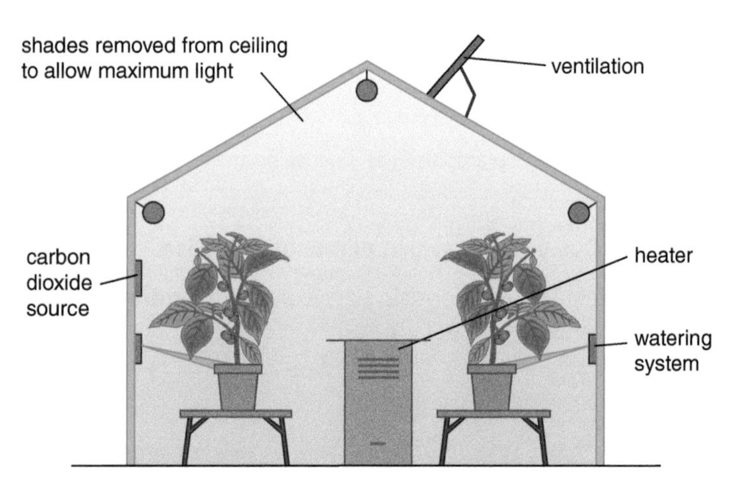

Figure 2.2.24 A greenhouse system

- Blinds can be used to control the amount of light.
- Humidifiers are used to add moisture to the air.

3. **Why do many greenhouses have vents in the roof?**
4. **Explain why paraffin heaters are used in greenhouses.**

HIGHER TIER ONLY

Limiting factors

Look at the graph of rate of photosynthesis against light intensity in Figure 2.2.25. Between A and B, the rate of photosynthesis increases as the light intensity increases. Because the rate depends on the light intensity, light intensity is called the **limiting factor.**

Between B and C, increasing the light intensity has no effect on the rate of photosynthesis. Another factor is now the limiting factor.

5. **What other factors might limit the rate of photosynthesis between B and C?**

Over time, light, temperature and carbon dioxide levels around a plant change.

- Carbon dioxide may be the limiting factor on a sunny day, especially when plants are crowded. (There is only 0.04% carbon dioxide in the atmosphere.)
- Temperature may be the limiting factor in cooler months.
- Light may be the limiting factor at dawn.

Increasing any one of the factors speeds up photosynthesis until the rate is limited by the factor that is in shortest supply.

It can be expensive to manipulate conditions in a greenhouse. Heating and artificial lighting must be paid for, which adds to costs. But it may be cost effective for farmers if they can significantly increase the value of the crops they can grow.

6. **Explain what the top graph in Figure 2.2.25 shows about the effect of temperature on the rate of photosynthesis.**
7. **Why is it important to monitor temperature, carbon dioxide concentration and light intensity levels in a greenhouse?**
8. **How does the use of a greenhouse increase the yield of a crop?**

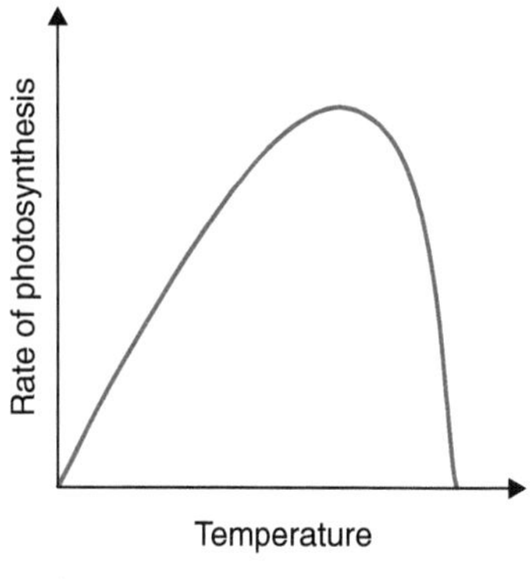

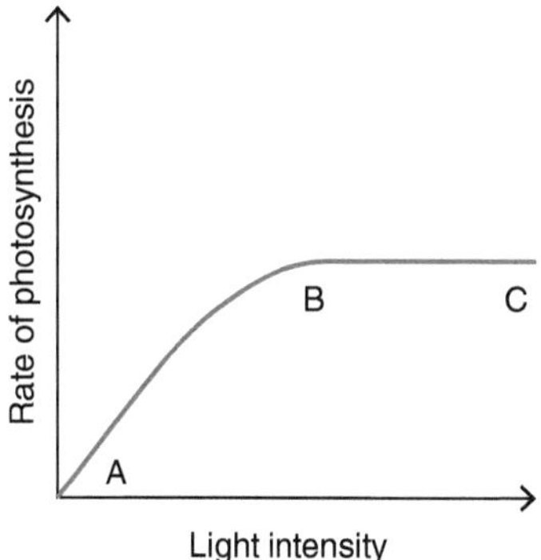

Figure 2.2.25 Graphs showing how the rate of photosynthesis depends on light intensity and temperature

Figure 2.2.26 Photosynthesis, and therefore food production, can be maximised in commercial greenhouses. Water and nutrient supplies are also controlled

MATHS SKILLS

2.2j Looking at tables and graphs

Learning objectives:

- draw and interpret graphs and tables
- understand and use the inverse square law in relation to light intensity and photosynthesis.

KEY WORDS

classify
compare
data
dependent variable
independent variable
line of best fit

Tables, charts and graphs give a lot of information in a small space. Extracting and interpreting data from graphics is a vital skill for all scientists.

Looking at tables

Tables allow us to **classify** and **compare** data. To read a table and extract information from it, you must first read the column headings carefully.

Light intensity (arbitrary units)	Rate of photosynthesis (mean number of bubbles per minute)
1	2
3	11
6	26
9	48
11	47
14	48

The heading of column 1 tells us that this table gives information about how something measured changes with light intensity. Column 2 tells us that the data collected was the rate of photosynthesis in a piece of pondweed, which was measured by counting the number of bubbles of oxygen released in one minute.

KEY INFORMATION

The variable that you change in an investigation is called the **independent variable**. In this example the light intensity is the independent variable. The number of bubbles per minute is the **dependent variable**. We want to find out how the rate of photosynthesis, measured by counting bubbles, depends on light intensity.

1 a **Look at the table. What was the light intensity when 26 bubbles were released in a minute?**

b **What happened to the rate between 9 light intensity units and 14 light intensity units?**

2 a **At what light intensity was the rate of photosynthesis highest?**

b **What trends do you notice in the data?**

MATHS

Some graphs will have a line or curve of best fit. A line of best fit goes roughly through the centre of all the plotted points. It does not go through all the points. The line of best fit could be a curve. If all the points fall close to the line (or curve), it suggests that the variables are closely linked. Look at the points carefully and see what would fit them well.

Plotting graphs

It is often easier to see trends in data when they are plotted on a graph. The values for the independent variable are plotted on the *x*-axis (horizontal) and the values for the dependent variable are plotted on the *y*-axis (vertical).

It is important to choose scales for the axes so you can include all the values without squeezing the plotted points too close together. Look at the largest value in the data, and choose a scale with 1, 2, 5 or 10 units per division of the grid that will allow the full range of values to fit on your graph paper.

Each pair of values in your table gives one plotted point. For each point, find the column 1 value on the *x*-axis and follow straight up from this until you are level with the corresponding column 2 value on the *y*-axis. Make a small cross at this point. Now repeat for all the pairs of values.

We often draw a **line of best fit**, to make the relationship clearer. The line (or curve) might not go through every single point, but it shows the overall relationship between the variables.

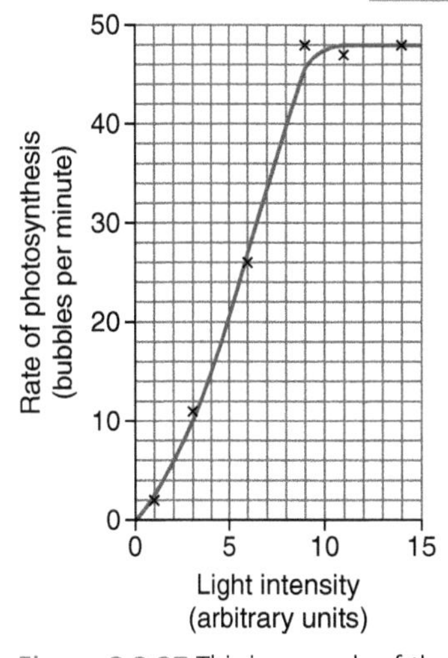

Figure 2.2.27 This is a graph of the data in the table above. On the horizontal axis the scale goes to 15 with 1 unit per division of the grid, and on the vertical axis it goes to 50 with 2 units per division

3 **Compare the table and the graph in Figure 2.2.27. Which do you think shows the data more clearly? Give your reasons.**

4 **Why might a line of best fit not go through every single point?**

Interpreting graphs

To interpret or read a line graph:

- Look at the labels, scale and units on each axis. What are they telling you?
- Look at the general **trend** on the graph. As *x*-axis values increase, what happens to the *y*-axis values?
- To read the *x*-value at a specific point of interest on the graph line, follow directly down from the point until you reach the *x*-axis and read off the value. To read the *y*-value, read across from the point on the *y*-axis.

5 **What does the graph in Figure 2.2.28 tell you? Explain the shape of the graph using the labels, A, B and C.**

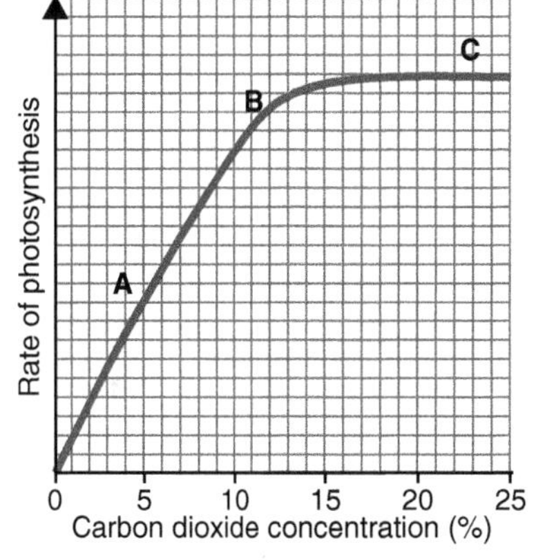

Figure 2.2.28 This line graph gives information about a factor that is relevant to plants

HIGHER TIER ONLY

Inverse square law

As the distance between the light source and a photosynthesising plant increases, the light intensity will decrease. This is described as being an inverse relationship. Light intensity obeys the **inverse square law**. The inverse square law states that if the light distance is doubled, the light intensity decreases by the square of the distance.

For example, if the distance between the light source and the plant is doubled, the light intensity is quartered (see Figure 2.2.29).

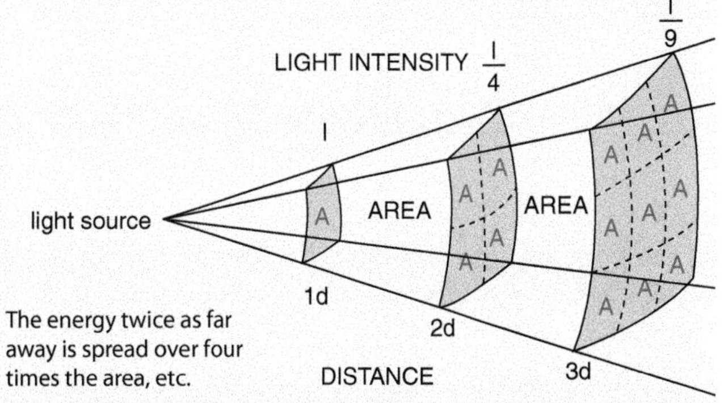

Figure 2.2.29 Light intensity obeys the inverse square law: if the distance between the light source and the plant is doubled, the light intensity is quartered.

REQUIRED PRACTICAL

2.2k How does light affect the rate of photosynthesis?

Learning objectives:

- use scientific ideas to develop a hypothesis
- use the correct sampling techniques to ensure that readings are representative
- present results in a graph.

KEY WORD

hypothesis

We can measure the rate of photosynthesis of a water plant like pondweed by measuring how much oxygen is given off each minute. If we change the intensity of light falling on the plant, we can see how this affects how fast oxygen is given off. Drawing a graph of rate of oxygen production against light intensity will help show the relationship between light intensity and photosynthesis.

These pages are designed to help you think about aspects of the investigation rather than to guide you through it step by step.

Developing a hypothesis

Laura and Amy are investigating how light intensity affects the rate of photosynthesis in pondweed. They will measure how much gas is released by the pondweed in 1 minute when they put a lamp at 10 cm, 15 cm, 20 cm, 25 cm and 30 cm away from the beaker containing the pondweed.

In this practical, take care when moving the lamp – depending on the type used, it may be hot.

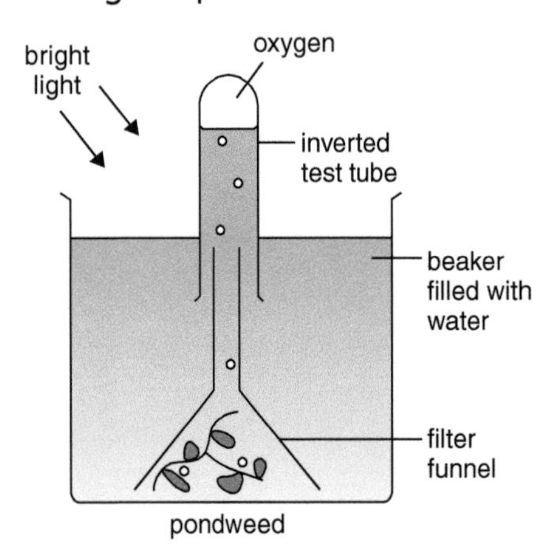

Figure 2.2.30 Pondweed gives off oxygen when exposed to light

Before they begin the investigation they are going to make a **hypothesis**. Hypotheses are developed using previous knowledge or observations. Laura and Amy know that:

- photosynthesis produces glucose and oxygen
- the oxygen will be released as bubbles of gas in the water

- light energy is absorbed by chlorophyll found in the chloroplasts in the leaf cells
- photosynthesis can be limited by different factors.

1. **When will the light intensity be greatest?**
2. **How will increasing the light intensity affect the rate of photosynthesis?**
3. **When the rate of photosynthesis increases what will happen to the amount of oxygen produced?**
4. **Could any other variable affect this investigation?**
5. **Suggest a hypothesis for the investigation that Laura and Amy are going to do.**

Improving accuracy

Amy and Laura decided that they would check that their measurements were repeatable by taking repeat readings at each light intensity. Look at their results.

Distance of lamp from pondweed (cm)	Number of bubbles per minute		
	Test 1	Test 2	Test 3
10	102	114	116
15	95	91	90
20	88	80	78
25	74	70	73
30	66	56	55

REMEMBER!

When you make **a hypothesis** you must try to explain your prediction or observations using what you already know about science.

6. **What do you notice about the results in Test 1 compared with the results in tests 2 and 3?**
7. **Suggest why this happened.**
8. **Suggest what Laura and Amy should do with the Test 1 results.**
9. **Why is it important to take repeat readings when carrying out investigations?**

Presenting results

Amy and Laura used a table to record their results because it was quick and the results were organised, but they found it hard to analyse them in this form. Their teacher said that they should think of a better way to present their results to help them. Laura thought that using a bar chart would be the best method but Amy disagreed. She presented her results as a line graph.

10. **Which would be the correct way to present these results? Explain why.**
11. **Plot the graph of Amy and Laura's results using only Test 2 and 3 data.**

MATHS SKILLS

2.2l Calculating rate of change from a line graph

Learning objectives:

- understand and use simple compound measures such as rate of change
- use the gradient of a graph to calculate the rate of change.

KEY WORDS

compound measure
gradient
linear
rate
tangent

It is often useful to see how one variable relates to another, such as how much mass an organism gains as it grows over time. A graph is a good way of displaying the relationship, and of calculating the rate at which a change happens.

Simple rate graphs

Rate tells us how much something changes with time. To find out how the temperature of some water changes as it is being heated, we record the temperature at set times. Then we can plot a graph of temperature against time, with time along the horizontal axis.

The **gradient**, or slope, tells us the rate of heating. In Figure 2.2.31 the line is straight. It is a **linear** graph, so the gradient is constant. The steeper the line, the faster the rate.

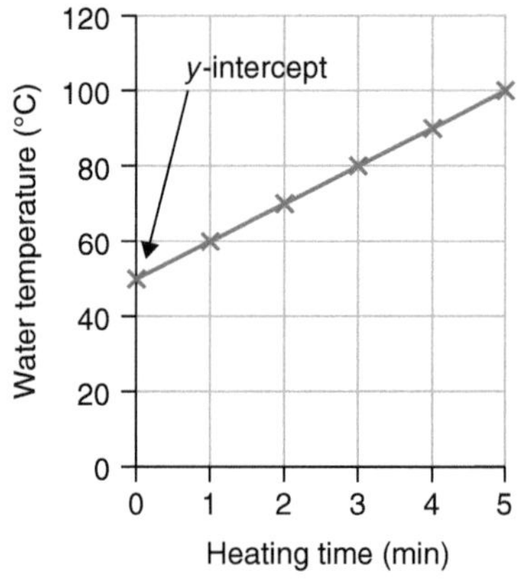

Figure 2.2.31 Linear graph showing how the temperature of water changes as it is heated

1. **In Figure 2.2.31, what was the water temperature at the start, before heating began?**

2. **On a copy of Figure 2.2.31 add a line starting at the same temperature that shows a slower rate of heating.**

Rate of change from a graph

Rate is a **compound measure**, which means it is made up of two or more other measurements, one of which is time. The rate of heating shown in Figure 2.2.31 is measured in degrees per minute (°C/min), while the rate of growth of a seedling might be measured in mm/day.

We calculate the gradient by dividing the change in the y-values by the change in the x-values.

$$\text{gradient} = \frac{\text{difference in } y}{\text{difference in } x}$$

DID YOU KNOW?

The independent variable always goes on the x-axis, and the dependent variable goes on the y-axis. Time goes on the x-axis, as we can't control how quickly it passes.

In Figure 2.2.31, from the first minute to the third minute the temperature went from 60 °C to 80 °C.

$$\text{gradient} = \frac{(80 - 60)\,°\text{C}}{(3 - 1)\,\text{min}} = \frac{20\,°\text{C}}{2\,\text{min}} = 10\,°\text{C/min}$$

So for every minute, the water temperature rises by 10 °C – that is, the rate of heating is 10 °C/min.

3 **Look at Figure 2.2.31, from the third minute to the fifth minute. What is the rate of heating in this period?**

4 **Why, in this case, would this trend not continue after the fifth minute?**

Finding a gradient on a curve

Figure 2.2.32 shows how puppies gain mass in the first two years of life. For all breeds, the rate of mass gain is higher when they are young, and then levels off as they become adults.

The gradient at any point is the rate of growth at that time. To find the gradient of a curve, draw the **tangent** to the curve. This is a straight line that touches the curve at the point of interest, making equal angles to the curve on either side.

In Figure 2.2.33, to find the growth rate of an English bulldog at 12 months, we draw a tangent to the curve at the point vertically up from 12 months on the *x*-axis. Then we find the gradient of the tangent, to tell us the growth rate at that point.

$$\text{gradient} = \frac{(30 - 15)\,\text{kg}}{(19 - 4)\,\text{months}} = \frac{15\,\text{kg}}{15\,\text{months}} = 1\frac{\text{kg}}{\text{month}}$$

So at 12 months, an English bulldog puppy grows at a rate of about 1 kg/month.

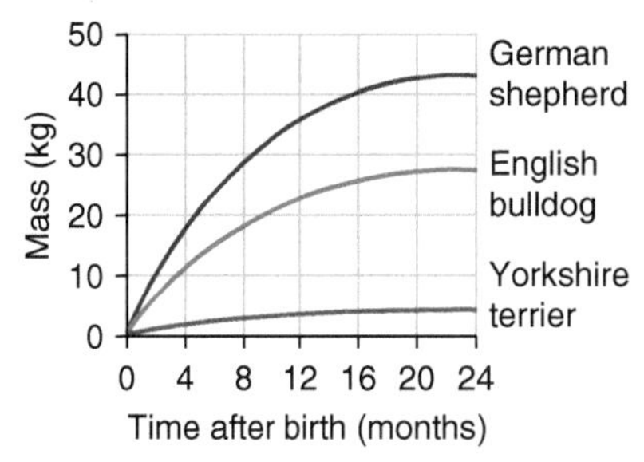

Figure 2.2.32 Rate of growth of different dog breeds

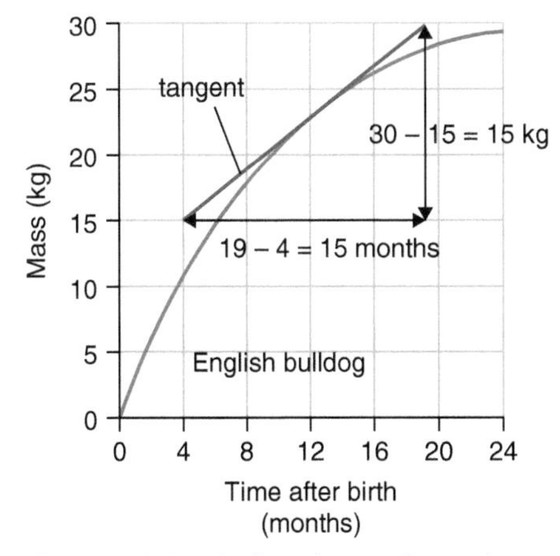

Figure 2.2.33 Finding the gradient of a curved line

5 **From Figure 2.2.32 find the growth rate of a German shepherd at:**

a 4 months

b 16 months.

6 **Find the *average* growth rate of a German shepherd in the first two years.**

2.2m Translocation

Learning objectives:

- describe the movement of sugar in a plant as translocation
- explain how the structure of phloem is adapted to its function in the plant
- explain the movement of sugars around the plant.

KEY WORDS

phloem
translocation

Sugars are needed by all cells for respiration. They are transported from the leaves, where they are produced by photosynthesis, to other parts of the plant, for immediate use or for storage.

Moving sugar

The movement of sugars in plants is called **translocation.** Cell sap containing sugars and amino acids is moved from where these substances are produced to where they are needed.

1. **Why are carbohydrates and proteins transported as sucrose and amino acids in the phloem vessels?**
2. **What is translocation?**

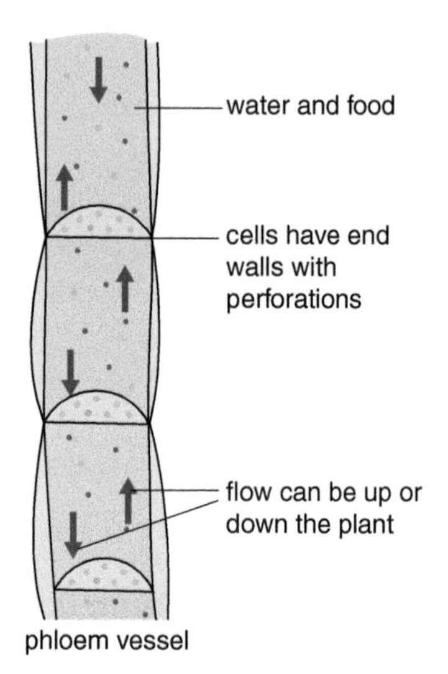

Figure 2.2.34 The structure of a phloem vessel

The structure of phloem vessels

Translocation happens in the **phloem**. Phloem cells are elongated, thin-walled living cells that form columns or tubes.

The end cell walls of phloem cells do not break down, but they do have many small pores in them. These perforated walls allow substances to move between the cells, along the phloem tubes.

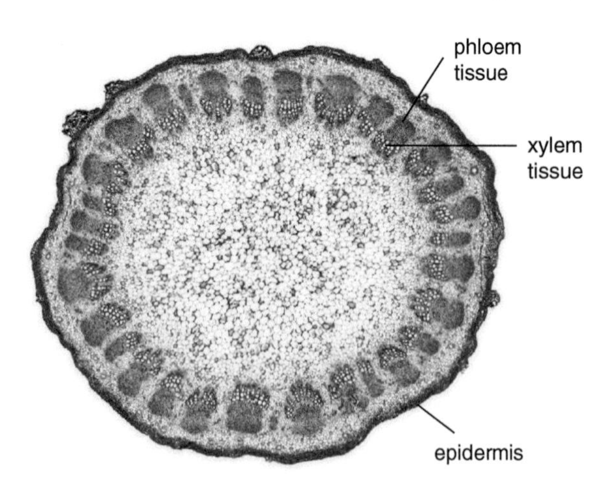

Figure 2.2.35 Light micrograph of a cross-section through a sunflower stem, showing where the phloem tissue is found (x 5.25). Note that the section has been stained with colour to show up the details

KEY INFORMATION

Phloem tissue is living. The cells have thin walls and perforated ends. Xylem tissue is made up of dead cells, with strengthened walls and no ends, forming hollow tubes.

Phloem cells contain cytoplasm but have no nucleus. They each have a companion cell next to them. The companion cell controls the activities of the phloem but does not help with translocation. Substances in phloem vessels are moved by a process that requires energy.

Substances are transported up or down the phloem tubes. Glucose made by photosynthesis is converted into a sugar called sucrose. Sucrose and amino acids are transported to all tissues in the plant. Different tissues use the substances in different ways. For example, while all cells require sugars for respiration, some convert it to starch for storage in organs such as potato tubers.

3. **Design a diagram to show how the plant uses the glucose made in photosynthesis.**
4. **Describe the structure of phloem cells.**
5. **Compare and contrast phloem and xylem cells.**

Adaptations of phloem

Phloem vessels are adapted for their function of transporting sugars because they have:

- companion cells with a nucleus and many mitochondria, which provide the energy needed to move substances in the phloem
- limited amounts of cytoplasm and no nucleus to allow efficient movement of substances
- perforated cell end walls to allow the movement of substances through the phloem
- flow both up and down the phloem tubes so that substances are transported all over the plant.

6. **Why do phloem vessels have companion cells?**
7. **Explain why phloem cell end walls are perforated.**

Figure 2.2.36 The strings that go up and down the length of bananas are phloem vessels

2.2n Plant diseases

Learning objectives:

- describe the causes, symptoms and identification of some plant diseases
- explain how communicable diseases are spread in plants
- explain applications of science to reduce or prevent the spread of communicable plant diseases.

KEY WORDS

communicable disease
pathogen
rose black spot
tobacco mosaic virus

Plant growth can be limited by diseases. Communicable diseases, transmitted from plant to plant by pathogens, can damage plants and prevent them from growing healthily.

Plant pathogens

Plants can be infected by diseases that affect their growth. Some plant diseases are caused by **pathogens**. Pathogens are microorganisms that cause diseases that can be passed from one plant to another. A disease that can be spread from one organism to another by a pathogen is called a **communicable disease**.

Some important plant pathogen are:

- viruses, such as the tobacco mosaic virus
- fungi, such as rose black spot.

Plants that are infected may have:

- stunted growth
- spots or discolouration on leaves
- areas of decay (rot)
- malformed stems or leaves.

Diseased plants grow poorly and may die.

1. **What is a pathogen?**
2. **What types of pathogen cause plant diseases?**

Symptoms of plant diseases

Tobacco mosaic virus (TMV) is a widespread plant pathogen. It affects many plant species, including tomatoes. It gives a distinctive 'mosaic' pattern of discolouration on the leaves. Other symptoms are curled leaves, stunted growth and yellow streaks or spots on leaves.

The disease affects the growth of the plant because photosynthesis is reduced.

Link: 2.2n ⟶ 3.1a

Figure 2.2.37 TMV causes a mottled 'mosaic' of discolouration on the plant's leaves

Rose black spot is a serious disease of roses caused by a fungus. Purple or black spots develop on the leaves, which often turn yellow and drop early. Untreated, it can quickly affect all the roses in a garden, resulting in plants with bare stems. Without leaves, the plants are unable to photosynthesise, so they cannot grow.

3 **Describe the symptoms of TMV.**

4 **What causes rose black spot?**

Controlling plant diseases

The tobacco mosaic virus can easily be transmitted from one plant to another by direct contact between the leaves or stem, and also by human handling of infected and healthy plants.

Some methods of controlling the spread of TMV in plant crops include:

- removal and destruction of infected plants
- careful washing of hands and equipment after contact with infected plants
- rotating crops to avoid planting in TMV-infected soil for at least two years.

Figure 2.2.38 Newly sprouted leaves are susceptible to black spot

Rose black spot is spread by spores of the fungus which are produced in the black spots. The spores are released in wet, humid conditions – for example, when it rains or the plants are watered. Wind also helps spores to disperse.

Treatments for rose black spot include:

- removing infected plants or dropped leaves to avoid spores spreading
- not planting roses too close together, so that air can flow freely around them
- avoiding getting the leaves wet when watering, as wet leaves encourage the growth of the fungus
- spraying with a fungicide to prevent or treat infection, especially in advance of warm, wet weather.

5 **Describe how TMV is transmitted from one plant to another.**

6 **Describe how rose black spot is transmitted.**

7 **Explain three treatments for rose black spot.**

Check your progress

You should be able to:

know that meristems are found at the growing tips of shoots and roots. →	describe the function of meristems in plants. →	explain the use of stem cells from meristems to produce clones of plants quickly and economically.
identify some parts of a leaf and their function. →	describe leaf adaptations for efficient exchange. →	explain how the leaf's structure is adapted for exchange.
know the definition of diffusion. →	explain diffusion using the idea of particles. →	explain how substances pass in and out of cells.
recall the word equation for photosynthesis. →	recall and use the word equation to describe photosynthesis. →	recall and use the balanced symbol equation for photosynthesis (HT only).
know that chloroplasts contain the plant pigment chlorophyll, which absorbs light and converts it to chemical energy. →	describe the use of light and chlorophyll in photosynthesis. →	explain that chlorophyll in chloroplasts absorbs energy to drive chemical reactions.
describe how to set up paper chromatography to separate plant pigments. →	explain how a chromatogram can be used to distinguish pure from impure substances. →	interpret chromatograms of plant pigments and determine R_f values.
recall that photosynthesis is an endothermic reaction. →	understand what is meant by an endothermic reaction. →	explain why photosynthesis is an endothermic reaction.
understand that plants respire and photosynthesise. →	explain why plants carry out respiration. →	describe the difference in gas exchange in plants during day and night.
name the factors that affect photosynthesis. →	describe how the rate of photosynthesis can be increased. →	explain the effects of limiting factors on photosynthesis.
describe experiments on how light intensity affects the rate of photosynthesis. →	develop and test hypotheses on how light intensity affects the rate of photosynthesis. →	understand and use the idea of inverse proportion in the context of light intensity and photosynthesis.
describe how water travels in plants. →	describe adaptations in xylem and phloem. →	explain adaptations of xylem and phloem.
describe experiments on the rate of transpiration. →	describe how different factors affect transpiration. →	explain how different factors affect transpiration.
recall that the movement of sugars is called translocation. →	describe how proteins and carbohydrates are transported in plants. →	explain how concentration gradients affect processes.
name some communicable plant diseases and describe their symptoms. →	describe how some communicable plant diseases are spread. →	explain how some communicable plant diseases can be controlled.

Worked example

Rekha is growing tomatoes in her greenhouse.

Rekha knows that to increase her crop of tomatoes, she must increase photosynthesis.

1 Write the word equation for photosynthesis, and the symbol equation (HT only).

carbon dioxide + water → glucose + oxygen

$CO_2 + H_2O \rightarrow C_6H_{12}O_6 + O_2$

The word equation is correct but the symbol equation has not been balanced. The correct answer is:

$6CO_2 + 6H_2O \rightarrow C_6H_{12}O_6 + 6O_2$

2 Rekha heats her greenhouse. She wants to know the best temperature to grow the tomatoes in. Look at the graph.

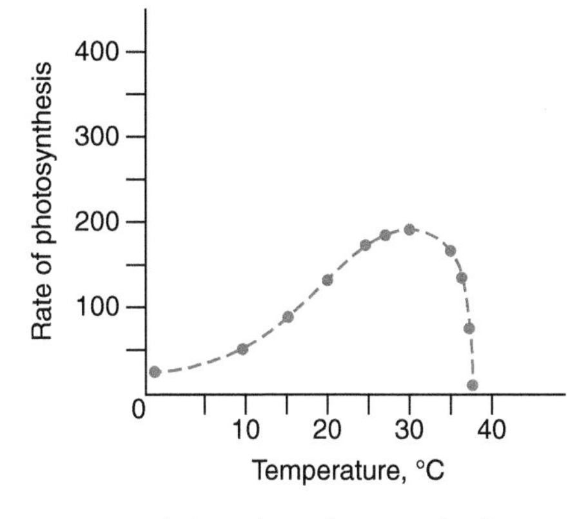

Explain what the graph shows.

The rate increases from 5 °C to 30 °C because the enzyme and reactant molecules gain more energy and collide more often, and then the rate decreases to 0 at 37 °C because the enzymes become damaged.

This answer gives an accurate description and explanation for what is happening in the graph.

3 What is the best temperature for Rekha to grow the tomatoes?

30 °C

Correct answer given.

How does water for photosynthesis get to the leaves?

Through the xylem by osmosis.

One correct answer is given (xylem).

Another answer would be '…by transpiration'.

4 Rekha notices that when she turns the heaters up, she needs to water the plants more often. Explain why.

The rate of transpiration increases with increasing temperature because more water evaporates through the stomata.

A good, full answer.

End of chapter questions

Getting started

1. a Why do plants photosynthesise? — 1 Mark

   b What structures inside plant cells absorb energy from light? — 1 Mark

   c In what form do plants store the glucose that they need for respiration? — 1 Mark

   d What type of cells carry dissolved sugars around the plant? — 1 Mark

   e What are the products of photosynthesis? — 2 Marks

2. The chromatogram shows the distances moved by four different colours, Q, R, S and T, and a mixed dye (labelled 'dye').

   a Copy the diagram and draw on it the spots that will appear on the chromatogram if the mixed dye contains colours Q, R and T. — 2 Marks

   b Calculate the R_f value for the red spot, S. — 2 Marks

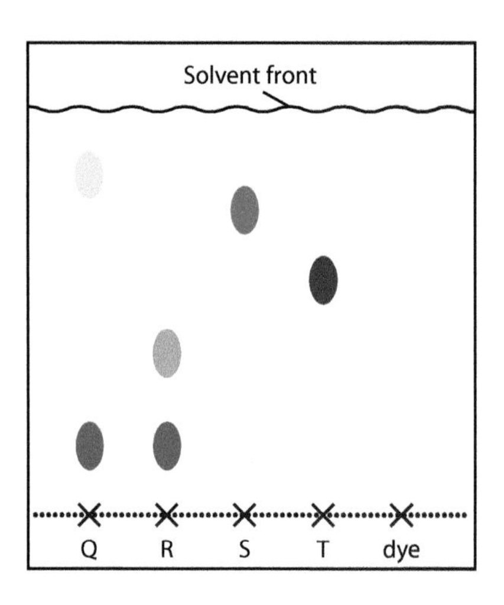

Going further

3. Give two uses of sugars in plants. — 2 Marks

4. Give a disadvantage of growing crops in greenhouses. — 1 Mark

5. Black spot is a fungal disease that affects roses. It causes purple or black spots to develop on the leaves.

   Describe four control measures that could be used to reduce or prevent the spread of rose black spot. — 4 Marks

6. A scientist measured water uptake and transpiration in a tree over 18 hours.

   The graph shows her data.

   a Transpiration is highest between 1 and 2 pm. When is water uptake highest? — 1 Mark

   b What conclusion can you make about water loss and uptake in this tree? — 2 Marks

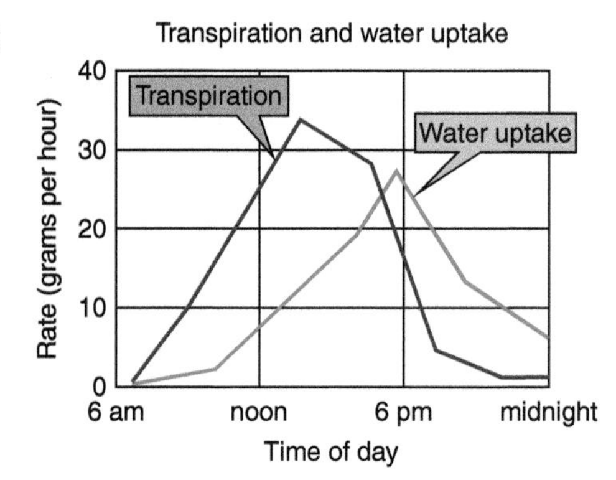

More challenging

7. Sally is growing tomatoes in her greenhouses. Describe and explain how Sally can maximise tomato production. 4 Marks

8. Water is transported from the roots to the rest of the plant by transpiration.

   An investigation was carried out over a 12-hour period to compare transpiration in the *Coleus* plant with transpiration in another type of plant called a *Begonia*.

   The mass of each plant was recorded before and after the 12-hour period to find out the effect of transpiration.

   The investigation was repeated five times with the same plants.

   The table shows the change in mass of each plant over each 12-hour period.

Trial	Decrease in mass of plant in grams over 12-hour period after being watered	
	Coleus	*Begonia*
1	3.7	1.1
2	4.5	1.3
3	2.8	0.8
4	1.6	0.6
5	3.2	1.0

   Compare the results from the *Coleus* plant with the *Begonia* plant. Suggest an explanation for the difference in the results. 6 Marks

Most demanding

9. Write a balanced symbol equation for photosynthesis. The formula for glucose is $C_6H_{12}O_6$. 2 Marks

10. Suggest two reasons why loss of water is important for a plant. 2 Marks

11. a If you are investigating the effect of light intensity on the rate of photosynthesis, what other variables must you control? 2 Marks

    b Draw a diagram to show how you could measure the effect of light intensity on photosynthesis using equipment you would find in a school laboratory. Show how you would set up the equipment and label each part. 2 Marks

    c The relationship between light intensity and distance from the light source can be explained by the inverse square law. Calculate the light intensity if your light source was placed at a distance of 10 cm. 2 Marks

Total: 40 Marks

How the ideas in this topic link together

Interactions between living cells and environmental factors have consequences for the health and wellbeing of both the organism and the environment. Lifestyle factors such as exercise, diet, alcohol consumption and smoking affect a person's chances of having certain non-communicable diseases, while exposure to ionising radiation carries a risk of damage to living cells that can lead to the development of malignant tumours.

The human body has defence systems to protect it from pathogens that cause communicable diseases. However, these defences can be breached. Vaccination helps to protect people from diseases that were once widespread. If the immune system fails, then antibiotics and other treatments can be used to treat diseases.

Scientific understanding of the ways our bodies interact with external factors, including medicines, helps people to control their fertility, treat diseases, protect themselves and avoid unnecessary exposure to risk. Research to develop new medicines and new technologies continues, to provide more effective treatments for diseases. The development and application of new technologies in medicine can raise ethical issues.

Working Scientifically Focus

- Explaining and evaluating personal, social, economic and environmental implications of applications of science
- Translating data from one form to another
- Considering ethical issues
- Evaluating risks

3 Interactions with the environment

Contents

LIFESTYLE AND HEALTH

IDEAS YOU HAVE MET BEFORE:

HEALTH CAN BE AFFECTED BY DRUGS AND DISEASE

- Smoking is damaging to health because cigarette smoke contains harmful chemicals.
- Drugs can be medicines and may help people suffering from pain or disease.
- Lack of exercise and a poor diet can lead to a higher BMI.
- Alcohol affects our response time and behaviour.
- Temperatures greater than 25 °C increase the likelihood that pathogens will grow.
- Bacteria multiply by simple cell division every 20 minutes, if conditions are favourable.

ORGANS WORK TOGETHER AS SYSTEMS

- Organs are aggregations of tissues.
- Organ systems work together to form organisms.
- Glucose is required by all cells for respiration; carbon dioxide and water are waste products.

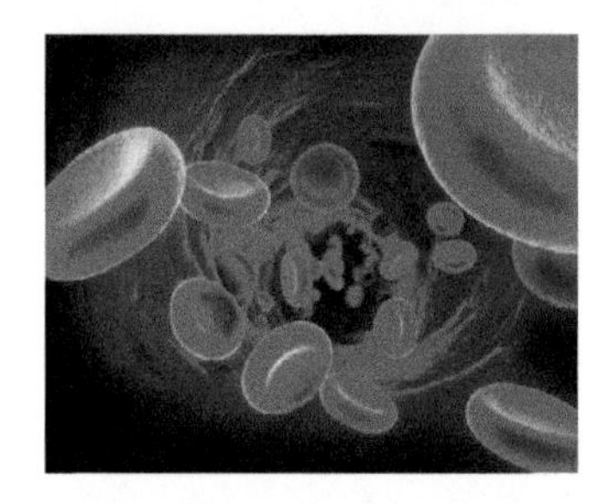

HOW HUMANS REPRODUCE

- The structure of the male and female reproductive systems.
- How gametes are produced and the processes of fertilisation, pregnancy and birth.
- Some diseases are transmitted sexually, and this can be prevented by some forms of contraception.

3.1

IN THIS CHAPTER YOU WILL FIND OUT ABOUT:

WHAT FACTORS AFFECT OUR CHANCES OF DEVELOPING DISEASE?

- Factors in our environment can increase our risk of disease.
- Our lifestyle can affect the chance of us developing a non-communicable disease.
- Sometimes a number of risk factors for developing a disease interact.
- Lifestyle factors can increase the risk of a person developing cancer.
- Pathogens are microorganisms that cause disease in plants and animals.
- Bacteria, viruses, fungi and protists can cause disease.

HOW ARE CONDITIONS IN THE BODY, PROCESSES AND ORGAN SYSTEMS COORDINATED AND CONTROLLED?

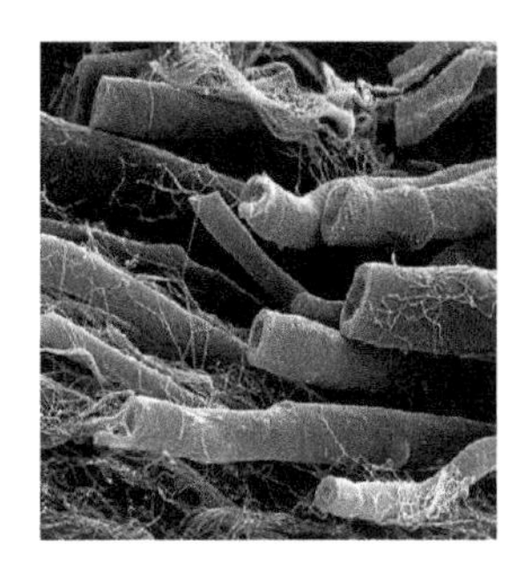

- Regulation of the internal conditions in the body is called homeostasis.
- The nervous and endocrine systems are involved in this coordination and control.
- The nervous system works using electrical impulses, transmitted using nerves; the endocrine system uses chemicals called hormones, which are secreted by endocrine glands.
- The concentrations of glucose, water and salts must be kept within strict limits.
- Glucose concentrations and water balance are controlled by hormones.
- Lack of insulin, or a loss of sensitivity to it, causes a condition called diabetes, which must be controlled.
- The control of hormone secretion by many glands is by negative feedback.
- Temperature regulation involves both the endocrine and nervous systems.

HOW ARE HUMAN SEXUAL DEVELOPMENT AND REPRODUCTION CONTROLLED?

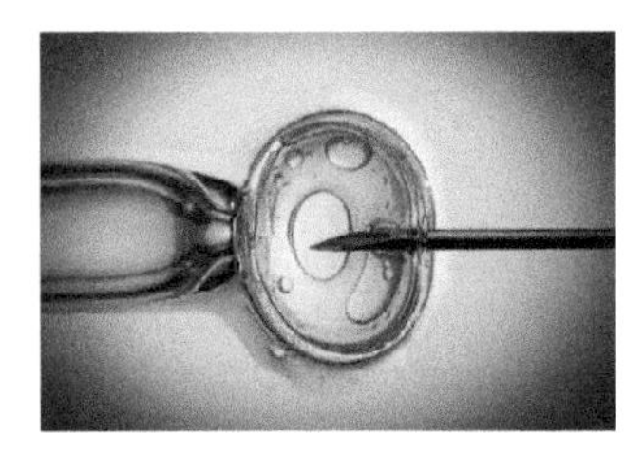

- Reproductive hormones cause secondary sexual characteristics to develop.
- Pituitary gland hormones regulate egg development and release and, along with reproductive hormones, prepare the body for a possible pregnancy.
- Different methods of contraception help to prevent unwanted births.
- Fertility drugs and *in-vitro* fertilisation are possible solutions to infertility.

3.1a Health and disease

KEY WORDS

communicable
mental health
non-communicable

Learning objectives:

- define what we mean by 'health' and describe the relationship between health and disease
- describe examples of communicable and non-communicable diseases
- discuss the costs of non-communicable diseases to people and communities.

Just like plants, animals including humans are affected by diseases. Diseases cause symptoms that stop the body from working as normal, and lead to poor health. Other factors such as diet and health can also affect our health.

Physical, mental and social wellbeing

Being healthy doesn't just mean 'not having a disease', it means being in a state of physical, mental and social wellbeing. **Mental health** is as important as physical health.

The major causes of physical and mental ill health include:

- disease
- diet
- life situations and stress.

Diseases are disorders that affect part or all of an organism. They can be **communicable**, like measles and HIV, or **non-communicable**, such as diabetes, cancer and cardiovascular disease.

Communicable or infectious diseases are caused by microorganisms called **pathogens**. Plants can also be affected by communicable diseases.

Non-communicable diseases can be caused by factors such as diet. For example, having high blood pressure damages arteries, leading to cardiovascular disease. High blood pressure and being overweight are strongly related.

Figure 3.1.1 Mental health is a very important part of health and wellbeing. Talk to someone if you have feelings of depression, anxiety or stress

1. **Describe what we mean by 'health'.**
2. **Name the major causes of ill health.**
3. **Name one communicable and one non-communicable disease.**

What are pathogens?

Pathogens are microorganisms that cause infectious diseases. They depend on their hosts to provide the conditions and nutrients they need to grow and reproduce. Pathogens can be

- viruses – for example, measles
- bacteria – such as *Salmonella*, which causes food poisoning

- protists – a protist causes malaria
- fungi – like rose black spot.

Pathogens are spread by direct contact, by water or through the air.

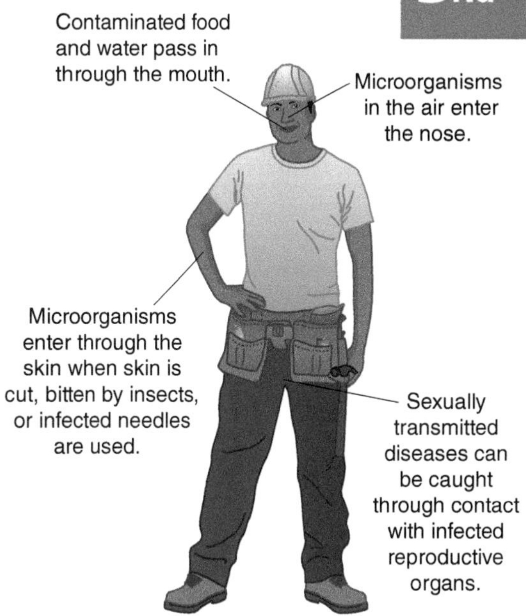

Figure 3.1.2 How are pathogens spread?

4 Give four examples of pathogens.

5 Describe how pathogens are spread.

Diet, life situations and stress

Non-communicable diseases are the leading cause of death in the world. Poor nutrition can contribute to the risk of developing some non-communicable diseases:

- High-fat or sugar-rich diets can cause high blood pressure, **depression**, heart disease and strokes, eating disorders and Type 2 diabetes.
- Low-calcium diets cause osteoporosis.
- Red meat and processed meat increase the risk of bowel cancer.

Stress develops when life situations occur, for example, moving home, death in the family or divorce.

When you are stressed, hormones are released, blood vessels constrict and blood pressure rises. Stress often causes depression. It increases the risk of obesity, heart disease, Alzheimer's disease and diabetes; it can also trigger asthma in people who already have it.

Figure 3.1.3 Name a cause of asthma

Non-communicable diseases can have devastating effects on individuals and families, as well as on local, national and global communities. For example, smoking increases the risk of many diseases including cardiovascular disease, lung disease, cancer, and having a low birthweight or premature baby. The human and financial costs of smoking include:

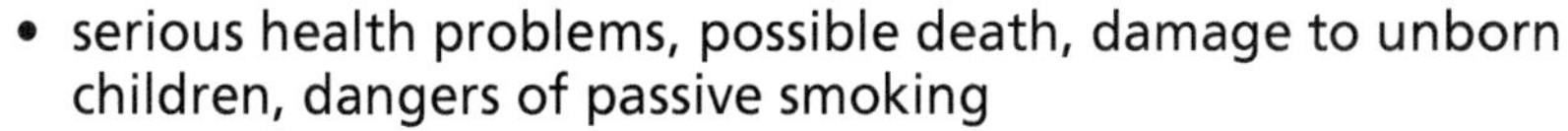

- serious health problems, possible death, damage to unborn children, dangers of passive smoking
- cost of cigarettes, loss of family income and a depleted workforce
- financial burden of healthcare costs on local, national and global economies.

6 Describe three links between diet and poor health.

7 Explain how lifestyle factors affect mental health.

8 Explain how non-communicable diseases affect national economies as well as the wellbeing of families.

3.1b Risk factors for non-communicable diseases

KEY WORDS

causal mechanism
risk factor

Learning objectives:

- understand what we mean by 'risk factors'
- explain lifestyle risk factors for non-communicable diseases
- recall that many non-communicable human diseases are caused by interactions of factors
- interpret and manipulate data about risk factors.

Non-communicable diseases are the largest cause of death in the world, so it's important to understand what factors may be causing these diseases.

Risk factors

Risk factors increase the chance of having a disease. Risk factors can be aspects of a person's lifestyle, and substances in the person's body or environment. They include diet, exercise, type of workplace, sexual habits, smoking, drinking and drug-taking.

Non-communicable diseases may be caused by the interaction of a number of factors:

- factors involved in cardiovascular disease may be diet, obesity, age, genetics and exercise
- lung disease factors are smoking and pollution of the environment
- alcohol, diet, obesity, genetics, drugs and viral infection may be involved in liver disease
- genetics, diet, obesity, and exercise may affect Type 2 diabetes
- carcinogens in cigarette smoke, foods and the environment, and ionising radiation such as X-rays, are all risk factors in cancer.

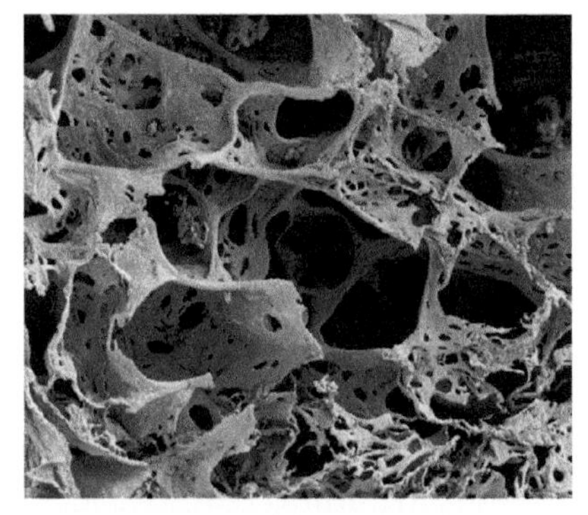

Figure 3.1.4 A scanning electron micrograph (SEM) with false colours added to show up the detail (x97.5). It shows human lung tissue that has been damaged by smoking

We cannot control some risk factors, such as genes and age, access to a good diet, and the cleanliness of the air in the region where we live, but we can control lifestyle factors, like drinking and smoking, to reduce the risk of contracting some diseases.

1. **What factors increase the risk of Type 2 diabetes?**
2. **How could the risk of liver disease be reduced?**

Causal mechanisms

Some risk factors are linked to an increased rate of a disease, but we don't know if that factor actually triggers the disease. A **causal mechanism** is a process that has been shown by research to be responsible for a factor causing a disease.

Disease/condition	Risk factor
cardiovascular disease	poor diet, smoking and lack of exercise
Type 2 diabetes	obesity
liver and brain damage	alcohol
lung disease, cancer	smoking
low birth weight and premature birth	smoking
abnormal foetal brain development	alcohol
cancer	carcinogens (including ionising radiation)

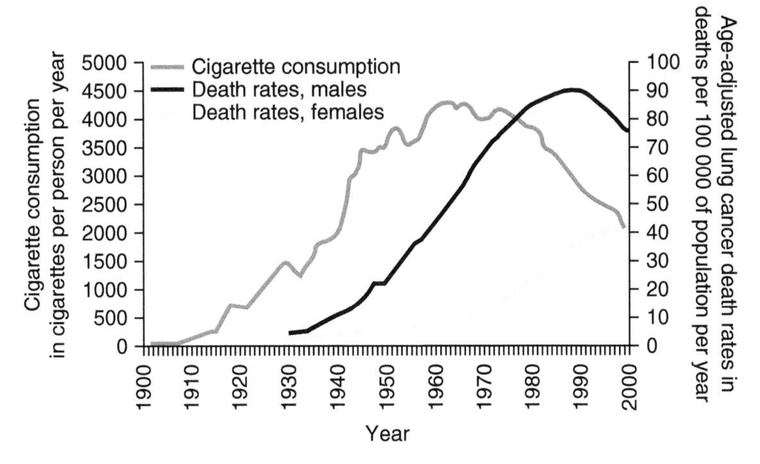

Figure 3.1.5 These data are for the US population. Is there evidence for a link between smoking and cancer?

The link between smoking and cancer was known for many years before being widely accepted, but we now have enough compelling evidence to prove smoking is a factor that causes lung cancer.

3 Why do you think it took a long time for people to accept that smoking was a causal mechanism for lung cancer?

Figure 3.1.6 Smoking is the biggest preventable cause of cancer worldwide. Lung cancer symptoms include coughing up blood, chest pains, shortness of breath, and weight loss

Lifestyle factors in populations

The effects of lifestyle factors vary at local, national and global levels, depending on trends and habits. For example, cigarette consumption is declining in countries like the USA and UK, so overall lung cancer rates should also fall here. But red meat consumption is increasing globally. Because eating red meat is a risk factor for some cancers, incidence of cancer may rise worldwide in decades to come.

DID YOU KNOW?

Each person with lung cancer in the UK costs the NHS over £9000 every year. The total workforce is reduced and the economy is weakened.

4 Describe and explain the graph in Figure 3.1.5.

5 Suggest why there is a time lapse between the rise in cigarette consumption and the rise in lung cancer deaths.

6 Suggest why the incidence of many non-communicable diseases is higher in countries where the average income is higher.

3.1c Treatments for cardiovascular disease

Learning objectives:

- identify the causes and symptoms of cardiovascular disease
- describe and evaluate the risks and benefits of treatments for cardiovascular disease.

KEY WORDS

artificial pacemaker
coronary heart disease
statin
stent

Cardiovascular disease is one of the main causes of death in the UK, so it's important to understand what causes it and how to treat it.

Heart problems

Cardiovascular disease includes **coronary heart disease**, which affects the coronary arteries – the blood vessels supplying the heart muscle (see topic 2.1f). Fatty material builds up inside, which reduces blood flow. Less glucose and oxygen reach the heart muscle for respiration, so it cannot contract efficiently. If cells are starved of nutrients, they can die and the person may have a heart attack. Factors that contribute to coronary heart disease include genetic factors, gender, age, diet and whether someone smokes or not.

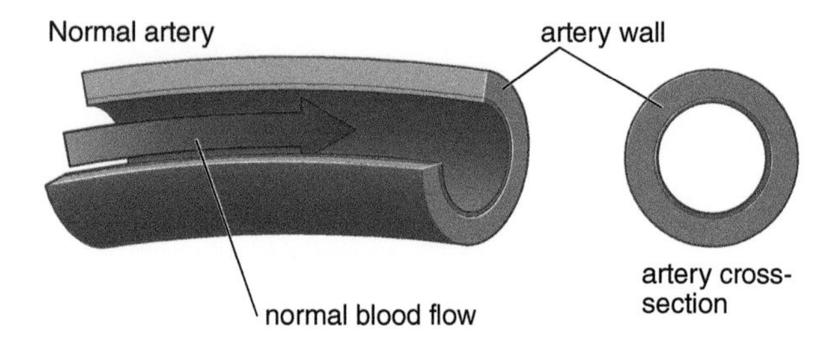

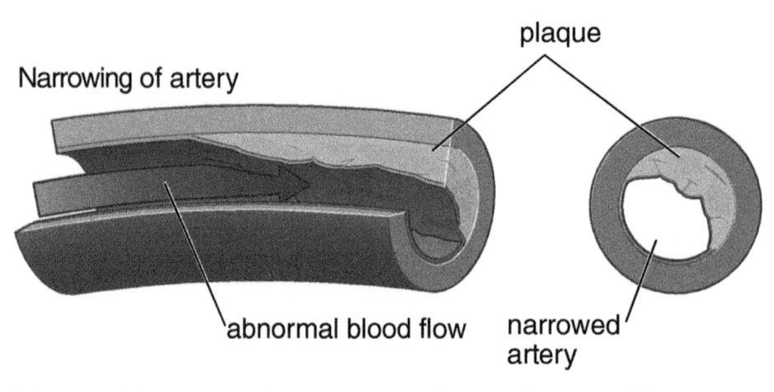

Figure 3.1.7 How does coronary heart disease affect blood flow?

Heart valves prevent the backflow of blood. They can become faulty due to heart attack, infection or old age. Faulty valves may not open fully or can leak, causing oxygenated and deoxygenated blood to mix. Symptoms of leaky valves include:

- tiredness and lack of energy
- breathlessness.

1 **Describe the symptoms of coronary heart disease.**

2 **What causes coronary heart disease?**

Treatments for heart problems

- Some people cannot control their heart rate. **Artificial pacemakers** can be fitted under the skin. A wire is passed from a vein to the right atrium. This sends electrical impulses to the heart to control heartbeat.
- Faulty heart valves can be replaced using biological (from humans or other mammals) or mechanical valves.

Stent pushed into position through catheter using X-rays.

Balloon inflated, expanding stent.

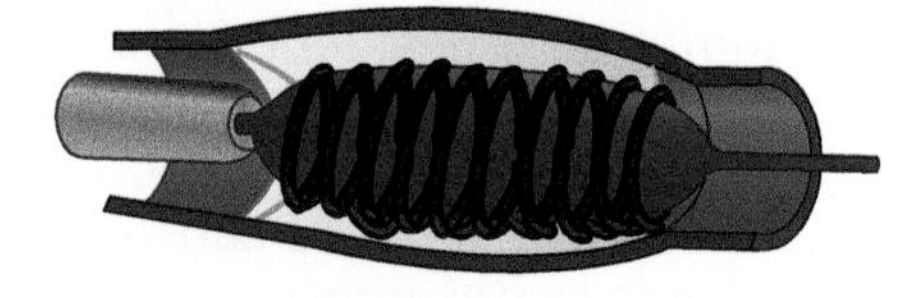

Catheter taken out, leaving stent to open up artery.

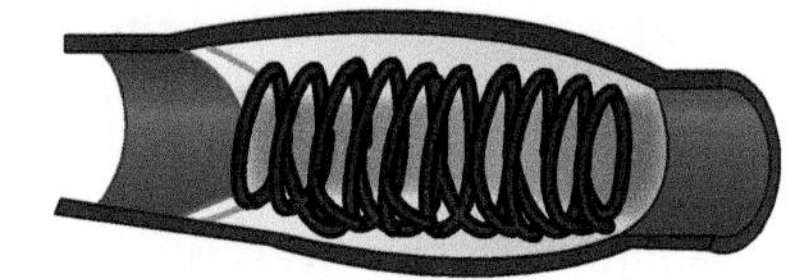

Figure 3.1.8 Inserting a stent

- High cholesterol levels are treated with drugs called **statins** which stop the liver producing as much cholesterol. Patients can also change their diet to help reduce cholesterol.
- **Stents** are used to treat narrow coronary arteries. If the coronary artery is too damaged, bypass surgery is used. A vein is transplanted from the leg to bypass the blockage.

When heart failure occurs, the only effective treatment is a donor heart transplant. An artificial heart can be used in the short term, while the patient is waiting for a heart transplant, or to allow the heart to rest to help recovery.

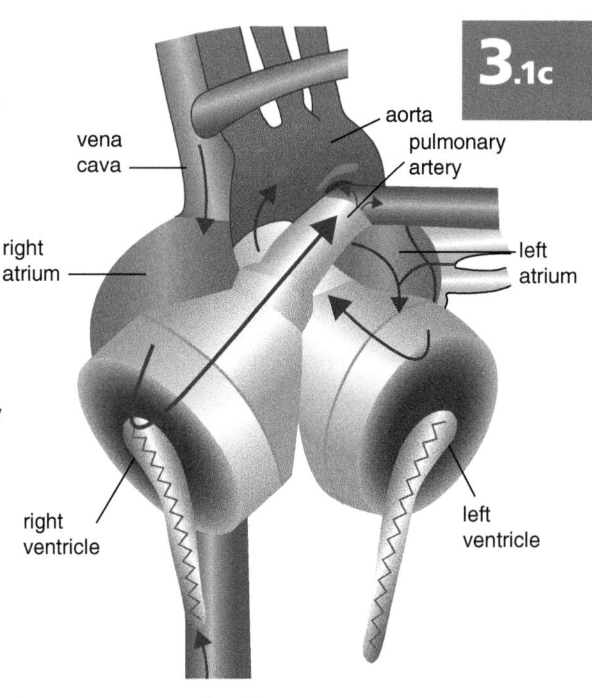

Figure 3.1.9 An artificial heart

3 How is coronary heart disease treated?

4 What are artificial valves and what do they do?

Evaluating heart treatments

All treatments help keep patients alive.

Treatment	Advantage	Disadvantage
artificial valves	no rejection	can damage red blood cells patient needs anti-clotting drugs
biological valves	red blood cells not damaged	valves can harden and need replacing
stents	little risk	fatty deposits can rebuild
bypass surgery	no rejection	major surgery
statins	reduce cholesterol	possible side effects, e.g. headaches
artificial pacemakers	major surgery not required	immune system can reject the pacemaker may need replacing
heart transplant	better quality of life	major surgery anti-rejection drugs needed (leading to greater infection risk) shortage of donors

5 Evaluate the use of artificial pacemakers or artificial heart valves over heart transplants.

6 Suggest why doctors might be increasingly prescribing statins for their patients.

KEY INFORMATION

The coronary artery takes oxygen and glucose to heart muscles for respiration.

DID YOU KNOW?

Over 300 000 people in the UK have a heart attack every year.

MATHS SKILLS

3.1d Analysing and interpreting data

Learning objectives:

- use information about risk factors from charts, graphs and tables
- use a scatter diagram to identify a correlation between a risk factor and incidence of a disease
- understand the principles of sampling data about risk factors
- interpret data about differences in the incidence of non-communicable diseases in different parts of the world.

KEY WORDS

correlation
trend

Analysing and evaluating data is a key skill for all scientists. Charts and graphs can help to show relationships between data.

Looking for patterns

Scientists try to identify links between variables. Sometimes there is no link. When interpreting graphs:

- identify patterns or **trends**
- use axis labels and units when describing the graph; for example, 'as vaccination uptake increases, cases of measles fall'
- look for particular features of the graph, such as maximum and minimum values, the range, outliers and patterns that do not fit trends
- quote numbers to clarify descriptions.

Useful words to describe graphs are: increased, decreased, faster, slower, constant, plateau, maximum, minimum.

1. **What do the data in Figure 3.1.10 show?**
2. **Describe the trends in the data shown in Figure 3.1.10. Are the two trends linked?**

Exploring links in data

A **correlation** is an association between two sets of data. A correlation does not prove that A affects B, or that B affects A. Both A and B could be caused by another variable. Scientists have to determine links between treatments and cures, or risk factors and disease.

(a)

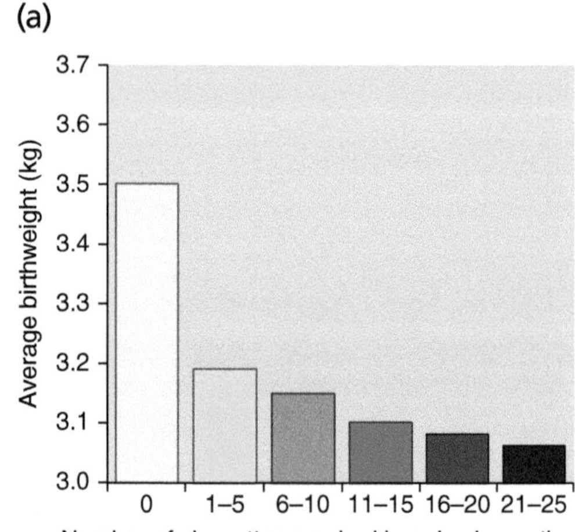

(b)

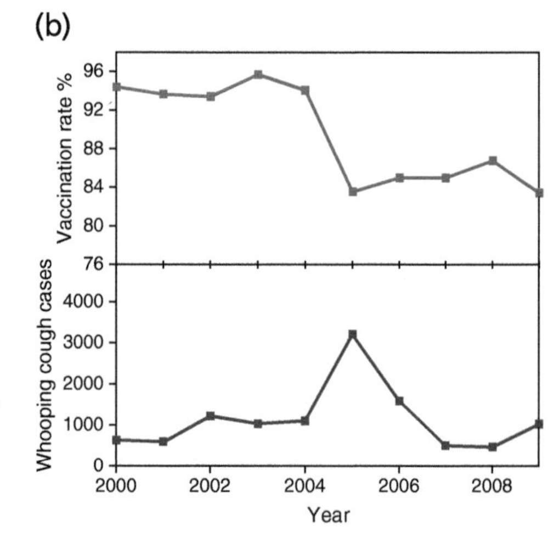

Figure 3.1.10 (a) Smoking in pregnancy and birthweight (b) Whooping cough incidence and vaccination uptake

Determining links can be difficult if:

- the proposed cause only sometimes results in the disease
- the disease has many possible causes
- there is a long delay between proposed cause and effect.

If one variable increases as the other increases, this is a positive correlation. If one variable increases as the other decreases, it is a negative correlation. To prove a causal mechanism, an investigation needs to be carried out.

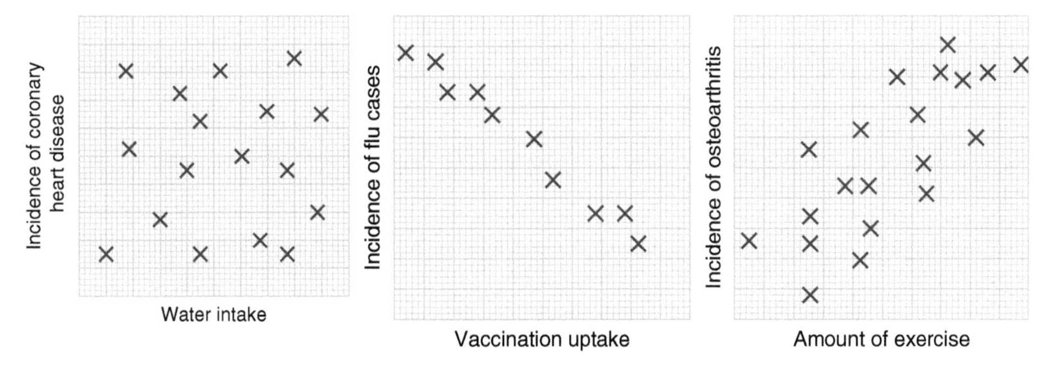

Figure 3.1.11 Scatter graphs comparing incidence of disease with different factors

3 **Describe the correlation in the graphs in Figure 3.1.11.**

4 **Why is it often difficult to determine links between variables?**

COMMON MISCONCEPTION

Remember: a correlation is not the same as a causal mechanism.

Evaluating data

When evaluating data think about these questions:

- Are the data repeatable and reproducible? Was there an appropriate control group?
- Are the data valid?
- Was the sampling good enough? Large data samples, collected over long periods, are needed to detect small effects. Sample groups must also be randomly selected, or matched for age and sex.
- How could more accurate data be collected? Do the data answer the question?
- Are there any anomalies in the data? Can these be explained?
- How confident are you that the evidence supports the conclusion? Has a causal relationship been proved? Could anyone use the same data to support a different conclusion?

The table shows data about the prevalence of diabetes and obesity among adults in 2010, in different countries. The data came from a report by the Organisation for Economic Co-operation and Development (OECD).

Country	Prevalence of diabetes (% of population)	Prevalence of obesity (% of population)
Iceland	1.6	20.1
UK	3.6	23.0
Italy	5.9	10.3
France	6.7	11.2
Germany	8.9	14.7
USA	10.3	33.8
Mexico	10.8	30.0
China	4.2	2.9
Australia	5.7	24.6
South Africa	4.5	18.1

5 **Plot a scatter graph of the data in the table. Is there any correlation between prevalence of diabetes and of obesity?**

3.1e Homeostasis

KEY WORD

homeostasis

Learning objectives:

- understand that homeostasis is maintaining a constant internal environment in the body
- explain why homeostasis is important
- explain how the body responds to internal and external change to keep conditions stable.

Diabetes is a disease in which the body cannot properly control blood glucose levels. It's vital to keep conditions like glucose level stable so that enzymes and all cell functions can work at their optimum rate.

Figure 3.1.12 Eve is running a marathon. While she is running, mechanisms in her body will try to keep conditions as constant as possible

Body changes

As Eve is running, the rate at which her body cells respire increases. Eve's body cells use glucose and oxygen in respiration to provide the energy she needs. Thermal energy is also released.

The thermal energy is carried around Eve's body by her blood. Normal body temperature is 37°C, the optimum temperature for enzyme action and other cell functions. The human brain, for instance, is very sensitive to changes in temperature.

Eve's body needs to control her temperature within strict limits.

One way it does this, as she runs, is to sweat. But sweating will also affect the water level in her body. She takes in extra water from sports drinks during the race. The drinks also contain some glucose. Her stores of energy (in the form of a carbohydrate called glycogen in the liver) decrease during the race and, as she suffers from fatigue at the end of the race, her blood glucose concentration falls.

1. **Why does body temperature need to be kept constant?**
2. **Name two other things that have to be controlled by the body.**

Homeostasis

Homeostasis is the regulation of the internal conditions in a cell or an organism. So when conditions change, either because of internal or external effects, processes in the organism work to counter the change and bring conditions back to the normal level.

In the human body, blood glucose concentration, body temperature and water levels are all controlled by homeostasis and kept within very narrow limits. It is vital to do this to maintain the optimum conditions for enzymes to work and for all body functions to take place efficiently.

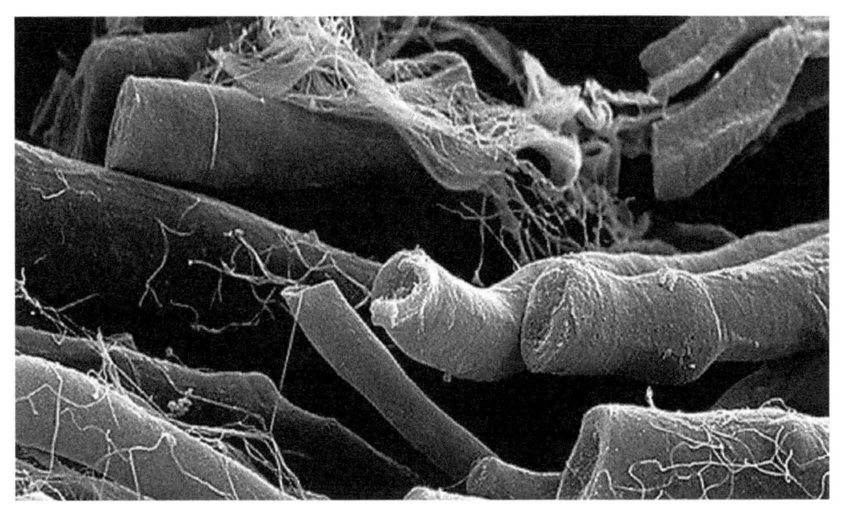

Figure 3.1.13 This scanning electron micrograph (SEM) has had false colours added to highlight the nerve cells (grey and pink) with some connective tissue (yellow), x845. A nerve is made up of a bundle of nerve cells, bound tightly together by connective tissue

3 **Write down a definition of homeostasis.**

4 **Why is it vital that water levels are controlled?**

Monitoring and control

For homeostasis to occur, the organism needs to monitor its internal conditions, and then respond appropriately when they change from their optimal state. In humans, special receptor cells in the body detect the levels of blood sugar, water and temperature, for example, and if they are too high or too low they trigger responses to bring them back to normal.

The body systems responsible for homeostasis are:

- the nervous system, which communicates rapidly using electrical impulses carried by nerve cells to specific effectors
- the endocrine system, which coordinates slower but longer-lasting responses using chemical messengers called hormones, carried around the body in the blood.

The control systems involved in homeostasis are 'automatic' – they happen when conditions change without us having to think about taking any action.

5 **How are changes detected by the body?**

6 **Compare and contrast the two control systems involved in homeostasis.**

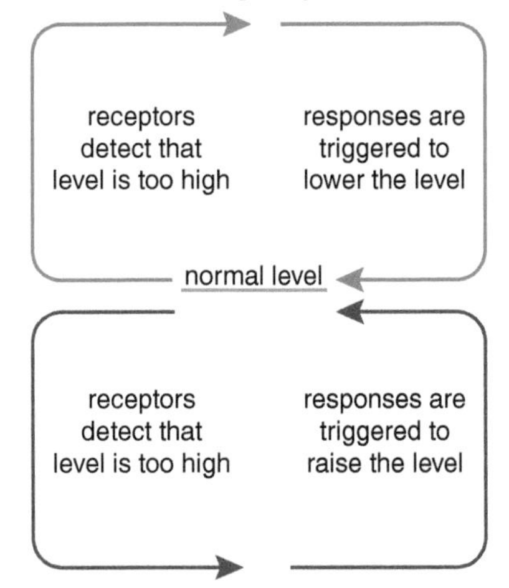

Figure 3.1.14 Homeostasis involves detecting a change and responding to counter the change (an example of negative feedback)

MAKING LINKS

Topic 2.1k The human nervous system
Topic 2.1n The endocrine system
Topic 2.1o Negative feedback

3.1f Controlling blood glucose

Learning objectives:

- recall that blood glucose is monitored and controlled by the pancreas
- understand how insulin controls the blood glucose level
- extract and interpret information about blood glucose control from graphs, charts and tables

- understand how insulin works with another hormone – glucagon – to control blood sugar level.

KEY WORDS

glucagon
insulin

Blood glucose levels are controlled within narrow limits by hormones secreted by the pancreas.

Controlling blood sugar

The pancreas secretes enzymes that digest carbohydrates, proteins and lipids.

The pancreas also has another function. It produces hormones that control the concentration of our blood glucose. Glucose is needed by all cells for respiration to release energy. It is carried to cells in our blood. But its concentration must be strictly controlled within certain limits.

After a meal, the concentration of glucose in our blood increases. The hormone **insulin** causes glucose in the blood to move into our body's cells. Here, this glucose can be used for respiration. In cells of the liver and muscle, the glucose is also converted into glycogen so that it can be stored.

1. **Name a hormone that controls blood glucose.**
2. **What effect does this hormone have on our body's cells?**

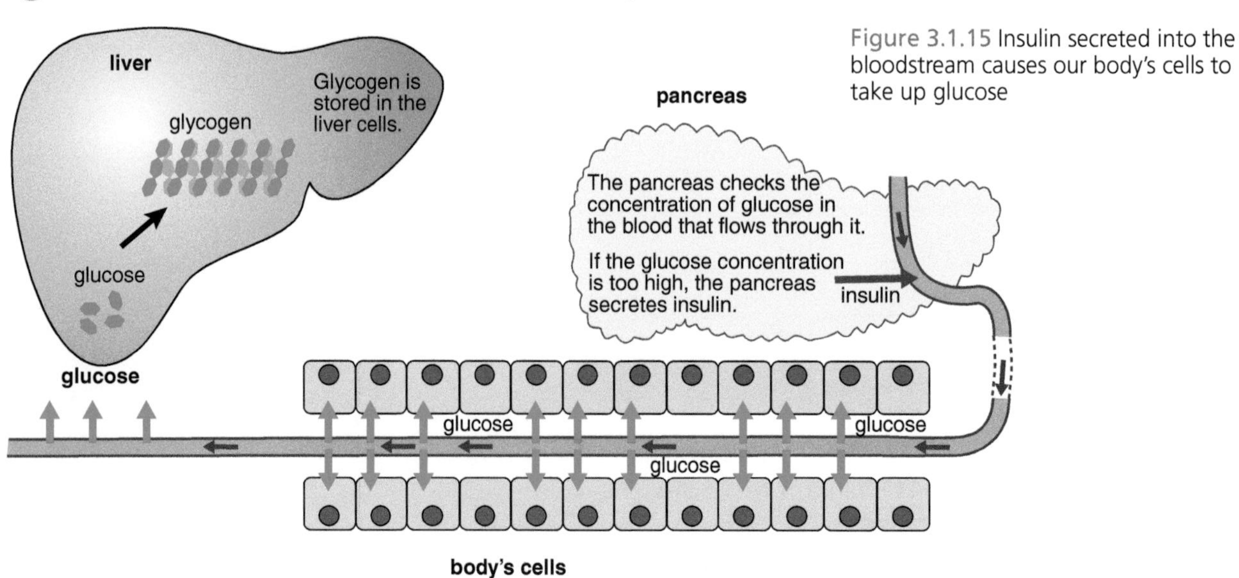

Figure 3.1.15 Insulin secreted into the bloodstream causes our body's cells to take up glucose

Blood glucose concentration

Insulin restores the blood glucose concentration to its normal level. A little insulin is produced as you first smell or chew food. As you eat food and it's digested, the blood glucose level rises, which causes a surge in insulin. The insulin level reaches a peak, then gradually falls.

Figure 3.1.16 shows what happens to blood glucose on eating a meal and then with the effect of insulin.

3. **Describe how a person's blood glucose concentration changes after a meal.**
4. **How long after having a meal did this person's blood glucose level start to fall?**

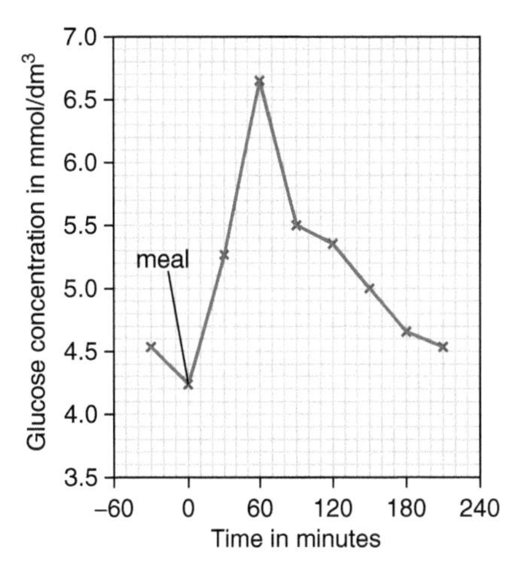

Figure 3.1.16 Blood glucose concentrations can rise to up to 8 mmol/dm^3 after a meal or snack, before falling as insulin is secreted

HIGHER TIER ONLY

Glucagon

Blood glucose is normally regulated at between 4 and 7 mmol per dm^3 of blood. To obtain this fine control, another hormone which is produced by the pancreas is also involved – **glucagon.**

Insulin and glucagon achieve this fine control by balancing glucose with carbohydrate stored as glycogen. This balance is maintained via a negative feedback cycle.

Insulin promotes the uptake of glucose by cells, and its conversion into glycogen in the liver and muscles.

Glucagon, which is secreted in response to low blood glucose concentration, promotes the conversion of stored glycogen into glucose, which is released into the bloodstream.

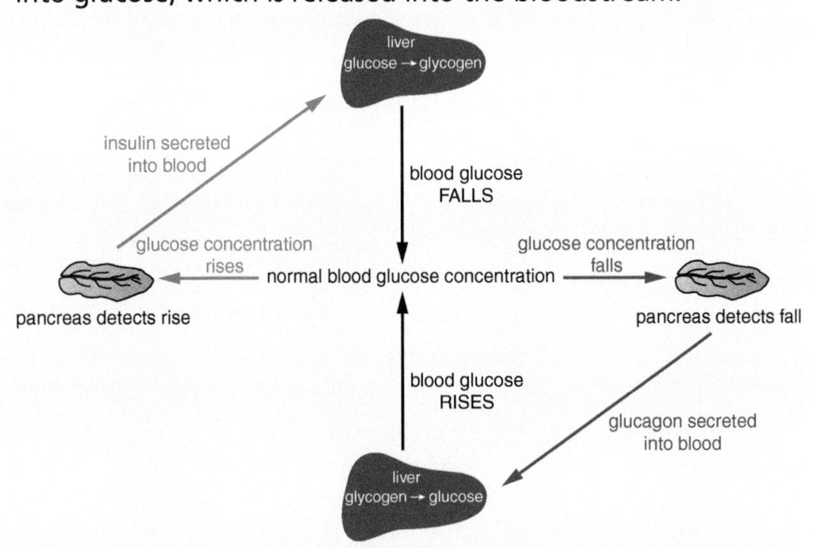

Figure 3.1.17 The control of blood glucose concentration by insulin and glucagon

5. **What is the normal concentration of blood glucose? Why is it important to maintain this level of blood glucose?**
6. **Explain how a constant level of blood glucose concentration is maintained.**

DID YOU KNOW?

Until the discovery of insulin in 1921 by Frederick Banting and Charles Best, diabetes wasn't treatable. Banting and Best carried out experiments on dogs and, later, themselves, before testing purified insulin on a teenager with diabetes in 1922.

REMEMBER!

Don't confuse the storage compound glycogen with the hormone glucagon.

3.1g Diabetes

Learning objectives:

- understand the causes of Type 1 and Type 2 diabetes
- compare Type 1 and Type 2 diabetes
- evaluate information on the relationship between obesity and diabetes.

KEY WORDS

body mass index
Type 1 diabetes
Type 2 diabetes

Diabetes is a non-communicable disease that means a person cannot control blood glucose levels. If uncontrolled, it can damage the circulation, nerves, eyes and kidneys.

Type 1 diabetes

In **Type 1 diabetes**, the pancreas is unable to produce enough, or any, insulin. Without it, the body's cells are unable to take up glucose. The blood glucose level becomes uncontrollably high, and glucose is excreted in the urine.

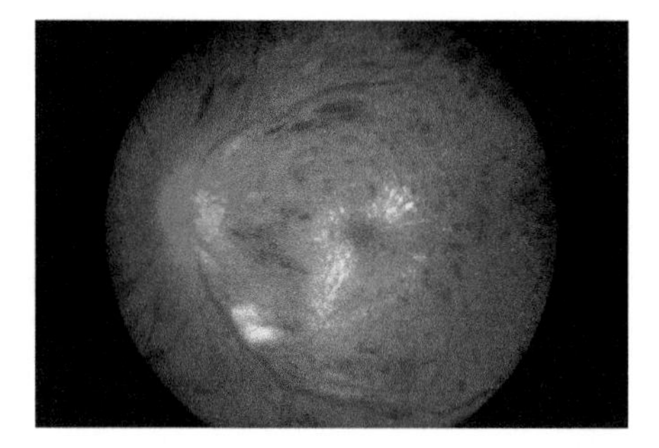

Figure 3.1.18 Diabetes can cause damage to the retina, at the back of the eye, if left untreated. This image shows blood leaking from the tihy blood vessels in the retina, as a result of unchecked high blood sugar

Without glucose, cells must use alternative energy sources. Fat and protein are used. The person will lose weight. If the condition is not controlled, kidney failure and death will result.

Patients with Type 1 diabetes control it with insulin injections, but there is currently no cure.

REMEMBER!

Remember the consequences of lack of insulin by working out the sequence of events if the body's cells are unable to take up glucose.

1. **What causes Type 1 diabetes? How is it treated?**
2. **Describe the sequence of events that will occur if a person has diabetes that isn't controlled.**

Type 2 diabetes

The main cause of **Type 2 diabetes** is that the body's cells lose their sensitivity – they no longer respond, or respond as effectively – to the insulin being produced.

Type 2 diabetes can be managed by modification of diet, and a controlled exercise regime.

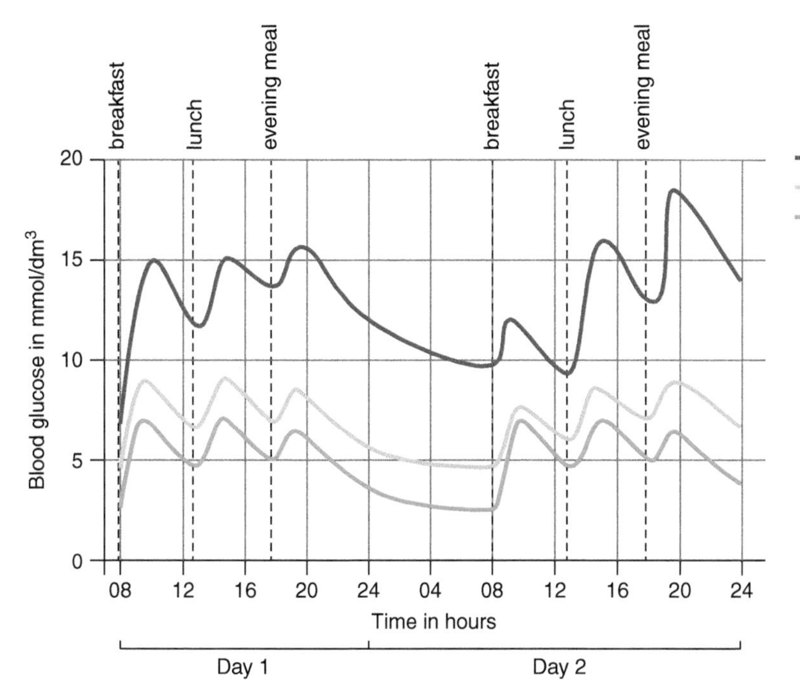

Figure 3.1.19 In a person with Type 2 diabetes, blood glucose concentration increases after a meal and is not brought back to normal

3 What is the cause of Type 2 diabetes?

4 Describe how Type 2 diabetes can be controlled.

5 Use Figure 3.1.19 to describe the effects of insulin secretion in the three groups of people shown on the graph.

What causes diabetes?

Only 10% of people with diabetes have Type 1. The cells that produce insulin have been destroyed. This can be an autoimmune condition – in which the immune system attacks the person's own body.

Type 2 diabetes tends to cluster in families. It's also associated with high-energy 'fast' food and an inactive lifestyle. Obesity accounts for 80–85% of the risk of developing Type 2 diabetes.

Type 2 diabetes is now emerging in young people. The first cases were in 2000, in children aged 9–16. Around three in ten children are now obese.

MAKING LINKS

Topic 4.3 Inheritance

6 Suggest reasons why Type 2 diabetes might 'cluster in families'.

7 Compare the correlation between obesity and Type 2 diabetes in men and women.

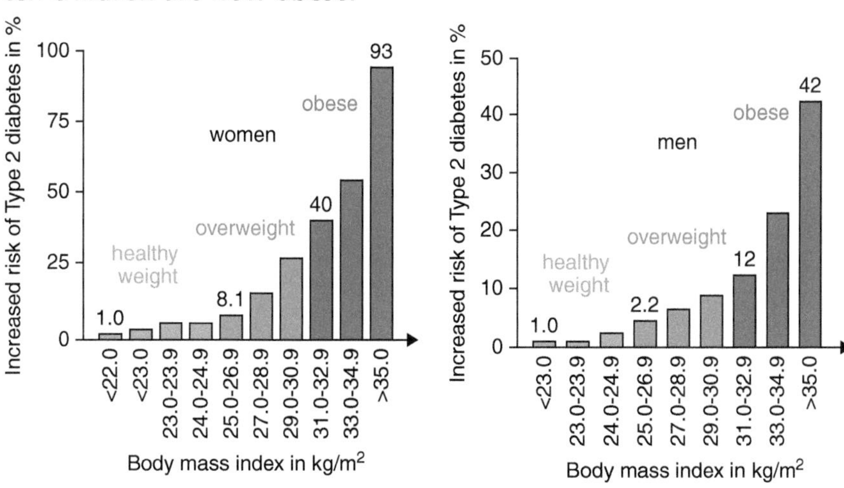

Figure 3.1.20 There is a correlation between increased risk of diabetes and obesity, as indicated by **body mass index** (BMI)

KEY INFORMATION

Body mass index is a measure of **body** fat based on height and weight.

3.1h Human reproductive hormones

Learning objectives:

- describe the roles of hormones in sexual reproduction
- extract and interpret data from graphs showing hormone levels during the menstrual cycle.

KEY WORDS

FSH
LH
oestrogen
progesterone
secondary sex characteristics
testosterone

Hormones control whether we develop as female or male. If male sex hormones are present in a fetus (which happens if there is a Y chromosome), it develops as male. If not, the foetus becomes female. Sex hormones also control puberty and reproduction.

MAKING LINKS

Topic 4.3 Inheritance

DID YOU KNOW?

Oestrogen, progesterone and testosterone are steroid hormones. Steroid hormones also help to control glucose and protein metabolism and water balance.

Reproductive hormones

Secondary sex characteristics develop as our bodies produce reproductive hormones at puberty.

Oestrogen is the main female reproductive hormone that is produced at this time. It is produced by the ovaries. Eggs start to mature in the ovaries and are released, at approximately one every 28 days.

In the male, **testosterone** is the main reproductive hormone. It is produced by the testes. It stimulates sperm production.

1. **Name the main reproductive hormones in males and females.**
2. **What happens in males and females under the influence of these hormones?**

The menstrual cycle

The menstrual cycle is the reproductive cycle in women, which – by convention – starts with a period (menstruation), if the woman is not pregnant.

Four hormones control the menstrual cycle:

- follicle stimulating hormone (**FSH**) causes eggs to mature in the ovaries
- luteinising hormone (**LH**) stimulates the release of an egg from an ovary
- oestrogen and **progesterone** maintain the lining of the uterus.

3. **Which four hormones control the menstrual cycle?**
4. **Which hormones maintain the lining of the uterus?**

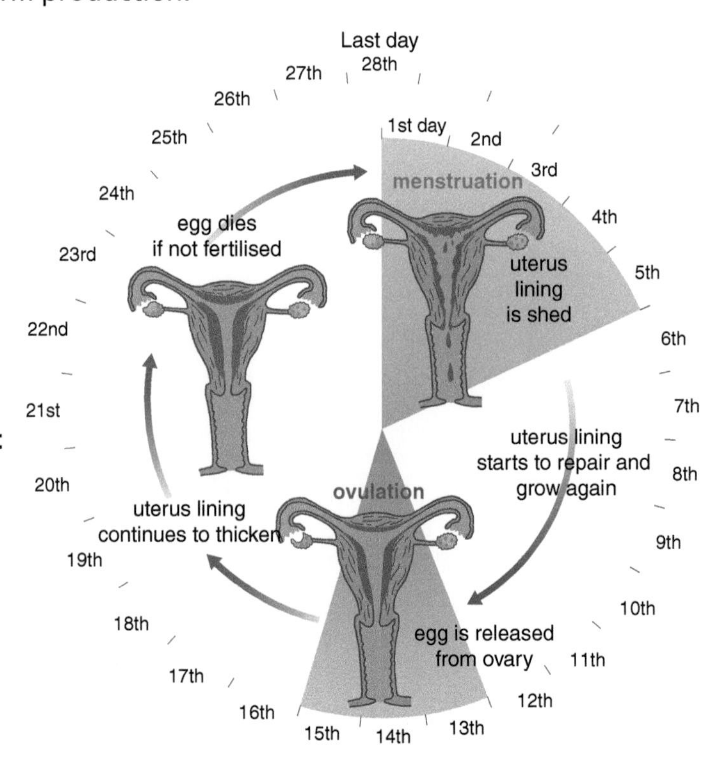

Figure 3.1.21 The menstrual cycle lasts approximately 28 days, but this is highly variable

Menstrual cycle hormones

The menstrual cycle is concerned with the maturation of an egg every month and preparing the uterus to receive that egg if it is fertilised, that is if the woman becomes pregnant.

These hormones interact with each other during the menstrual cycle.

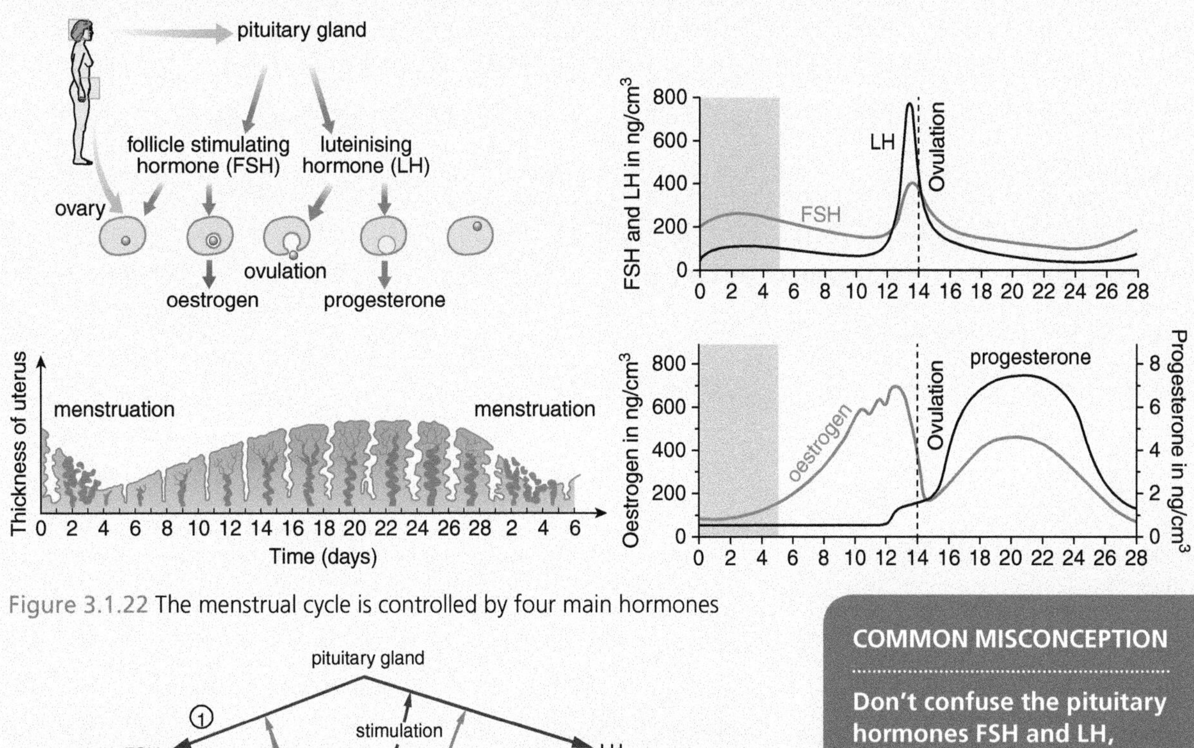

Figure 3.1.22 The menstrual cycle is controlled by four main hormones

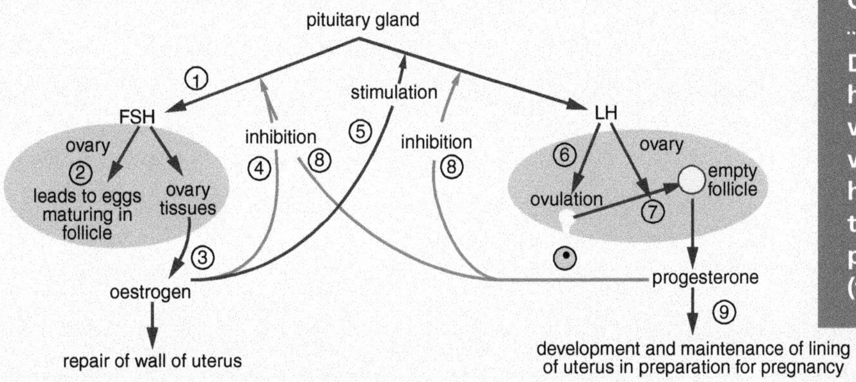

COMMON MISCONCEPTION

Don't confuse the pituitary hormones FSH and LH, which are regulatory, with the *reproductive* hormones produced by the ovaries (oestrogen and progesterone) and testes (testosterone).

Figure 3.1.23 The roles of the hormones as the cycle progresses are:

①FSH is secreted by the pituitary gland.

②FSH causes the eggs to mature in the ovaries.

③FSH stimulates the ovaries to produce oestrogen.

④ and ⑤Oestrogen inhibits further release of FSH and stimulates release of LH.

⑥LH triggers ovulation – the release of the mature egg from the ovary – and …

⑦… leads to the secretion of progesterone by the empty **follicle** that contained the egg.

⑧Progesterone inhibits the release of LH and FSH.

⑨Progesterone maintains the lining of the uterus during the second half of the menstrual cycle, in readiness for receiving a fertilised egg.

5 **Describe the changes in level of each hormone between days 12 and 14 of the cycle. What event happens as a result?**

6 **Suggest which hormones show:**

- **negative feedback**
- **positive feedback.**

3.1i Contraception

Learning objectives:

- explain how fertility can be controlled by different hormonal and non-hormonal methods of contraception
- evaluate the personal, social, economic and environmental implications of different methods of contraception.

KEY WORDS

cervix
condom
contraceptive pill
diaphragm
IUD
spermicidal cream

For personal or economic reasons, many people want to control their fertility, so that they can plan when to have children. Contraceptive methods allow people to prevent pregnancy, until they're ready to have a baby.

Contraceptive methods

Most contraceptive methods fall into one of two types:

- methods that use hormones
- non-hormonal methods, or barrier methods.

Some people choose *natural* planning methods. A woman's time of ovulation is linked with:

- her menstrual cycle, occurring at around 14 days
- a *slight* increase in body temperature
- thinning of mucus secreted from the **cervix**.

By estimating when ovulation occurs, it's possible to avoid having sexual intercourse when an egg might be in the oviduct. But eggs and sperm can live for several days, and women's cycles can be irregular.

Some people choose to have surgery. In the woman, the oviducts are cut, sealed or blocked by an operation. In the man, the sperm ducts are cut, sealed or tied. Surgical methods are designed to be permanent.

DID YOU KNOW?

Oral contraceptives were first approved for use in Britain in 1961. The hormone content was equivalent to seven of today's pills.

1. How can contraceptives be divided into broad categories?
2. List three indicators of ovulation.

Barrier methods

Barrier methods prevent sperm from reaching an egg.

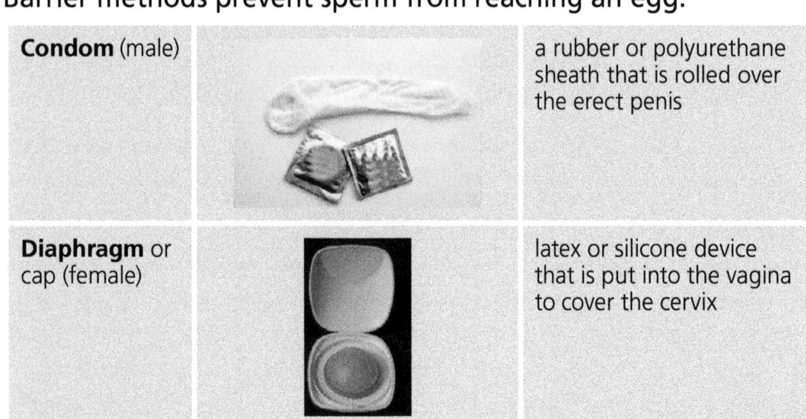

Condom (male)		a rubber or polyurethane sheath that is rolled over the erect penis
Diaphragm or cap (female)		latex or silicone device that is put into the vagina to cover the cervix

Figure 3.1.24 Two commonly used barrier methods

KEY SKILL

You may be provided with new information or data on contraceptives that you have to analyse and interpret.

A **spermicidal cream** is toxic to sperm. It can help with the effectiveness of other contraceptives, such as diaphragms. It should not be used on its own, or with condoms.

3 **Give two examples of barrier methods of contraception.**

4 **When should a spermicidal cream be used?**

Hormonal methods

Hormonal methods of contraception use reproductive hormones to prevent pregnancy. The combined **contraceptive pill** contains a synthetic oestrogen and progesterone.

For women for whom the combined pill isn't suitable, for instance, if they're older or have high blood pressure, there's also a pill called a progestogen-only pill (POP, mini pill).

These contraceptives inhibit the release of the pituitary hormones that control egg maturation and release. They also thicken cervical mucus which helps to prevent sperm reaching an egg. The combined pill is taken for 21 days, allowing periods to occur. The POP is taken every day.

Some other types of contraceptive can also include hormones. For example, some types of intrauterine device (**IUD**) release a synthetic progesterone. Others have fine copper wire wrapped round them. IUDs are placed in the uterus and work by inhibiting sperm from reaching an egg. They also prevent a fertilised egg from implanting in the uterus lining, if fertlisation does happen.

Ethical thinking helps us decide how we ought to act in a particular situation. Ethics provides a set of standards that help individuals work out reasons why something might be 'right' or 'wrong', within their own social and moral context. So, for example, some people object to IUDs on ethical grounds, because they can stop a fertilised egg from developing. They see this as terminating an existing life, rather than preventing conception.

5 **How do oral contraceptives work?**

6 **What is an IUD?**

7 **Outline a simple ethical argument about the 'rights' and 'wrongs' of using an IUD as a contraceptive method.**

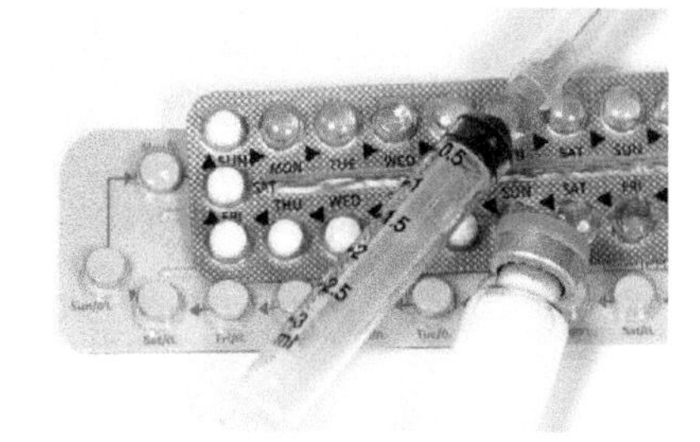

Injection: instead of a progestogen-only pill, an injection of a progestogen can be given.

Implant: a small flexible rod is implanted under the skin of the upper arm. Progestogen is released slowly.

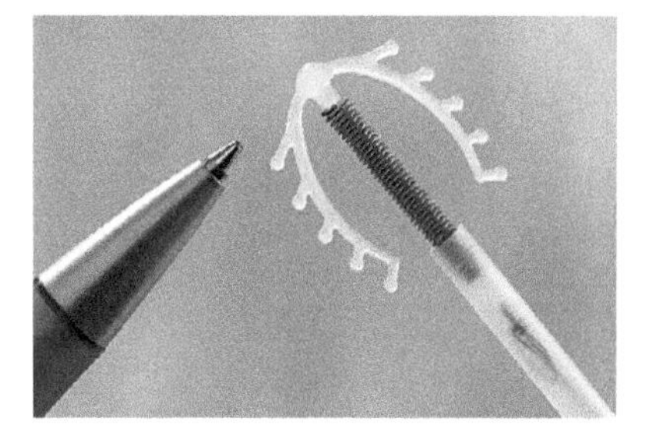

This IUD is wrapped in copper, which is toxic to sperm. The plastic applicator is used to position the IUD in the uterus. (The pen shows how small the IUD is.)

Patch: A sticky patch is put on the skin. It releases oestrogen and progestogen slowly.

Figure 3.1.25 As well as pills, hormonal contraceptives can be in the form of injections, implants or patches

3.1j Which contraceptive?

Learning objectives:

- explain how fertility can be controlled by different hormonal and non-hormonal methods of contraception
- evaluate the personal, social, economic and environmental implications of different methods of contraception.

Fifteen types of contraceptive are used in the UK – two designed specifically for men, and 13 for women. Each has different pros and cons that people must consider when deciding which suits them best.

Reliability

People who are considering using contraceptives should look at their success rate.

Successful surgery is 100% effective but irreversible. Natural family planning methods *can* be effective if used correctly, with guidance and teaching. Computerised devices that monitor hormone changes improve reliability of natural methods.

Type of contraception	Percentage of pregnancies prevented
condoms	98
diaphragm	92–96
implants	99
IUD	>99
oral contraceptives	>99

The effectiveness of contraceptives when used *perfectly*

1. **How can the reliability of natural family planning methods be improved?**
2. **Which method(s) is, or are, the most effective contraceptive?**

Advantages and disadvantages of different methods

With surgery, couples no longer need to think about contraception at all. But the decision should be carefully considered.

Hormonal contraceptives are convenient, although POPs do need to be taken at the same time every day to be effective. Implants last for 3 years, injections 12–13 weeks and patches 1 week. With injections, fertility may not return to normal straightaway.

With the combined pill, implant and patch, periods may be more regular, lighter and less painful. Oral contraceptives also reduce the risk of certain cancers, although those containing oestrogen slightly increase the risk of blood clots, and the risk of breast and cervical cancer.

DID YOU KNOW?

There is now a microchip implant that releases controlled doses of a contraceptive for up to 16 years. It can be controlled wirelessly, and turned off at any time to start a family.

Method	Advantages	Disadvantages
condoms	• widely available • can protect against transmitted infections, e.g. HIV	• may slip off • must withdraw after ejaculation and not spill semen
diaphragm	• put in just before sex • no health risks	• needs to be left in for several hours after sex • some people are sensitive to spermicide
IUD	• works immediately • can stay in place for 10 years (copper); 3–5 years (hormonal)	• insertion may be uncomfortable periods may be longer or more painful

Advantages and disadvantages of other methods

3 **Suggest why oral contraceptives are affected by vomiting or diarrhoea, but injections, implants and patches are not.**

4 **Give one advantage and one disadvantage of using an IUD.**

Wider issues

People must decide whether to use contraception or not. This could be influenced by religious factors.

If people do use contraception, they must choose a method.

Health factors are important in family planning, but for many people the decision to use contraception is *economic*, or can be a *lifestyle* choice:

- Can a couple afford to start a family?
- Is contraception likely to be needed every day?
- How soon would the woman like to become pregnant *after* using contraception?

A wider issue associated with contraception is one of world population and sustainability, although there are ethical issues with its use in population control. It can be seen as interfering with human rights and reproductive freedom.

5 **Name the most widely used contraceptive method in the UK.**

6 **Compare the personal, social, economic and environmental implications of using condoms and using a diaphragm for contraception.**

7 **Some governments have considered compulsory sterilisation to control population growth. Outline the ethical arguments for and against a compulsory sterilisation programme.**

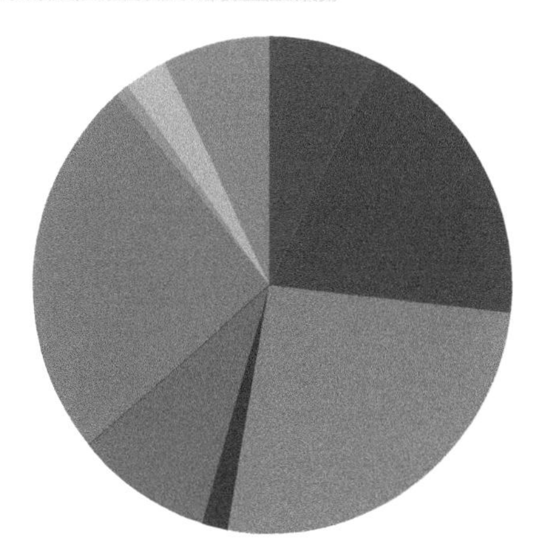

- sterilisation: female
- sterilisation: male
- contraceptive pill
- other methods, including emergency pill
- intrauterine devices (IUDs)
- male condom
- vaginal barrier, including spermicides
- hormonal injections and implants
- traditional methods, e.g. abstaining and withdrawal

Figure 3.1.26 Contraceptive use in the UK, 2008–9

KEY SKILL

Be prepared to evaluate data on the use of different types of contraception.

REMEMBER!

An ethical argument discusses the reasons why something might be 'right' or 'wrong'.

3.1k Treatment for infertility

Learning objectives:

- explain the use of hormones in technologies to treat infertility
- describe the technique of *in-vitro* fertilisation.

KEY WORDS

fertility drug
in-vitro fertilisation
IVF cycle

Louise Joy Brown was the first 'test tube baby', born on 25 July 1978 following *in-vitro* fertilisation (IVF) treatment. Her parents had been trying to conceive naturally for nine years.

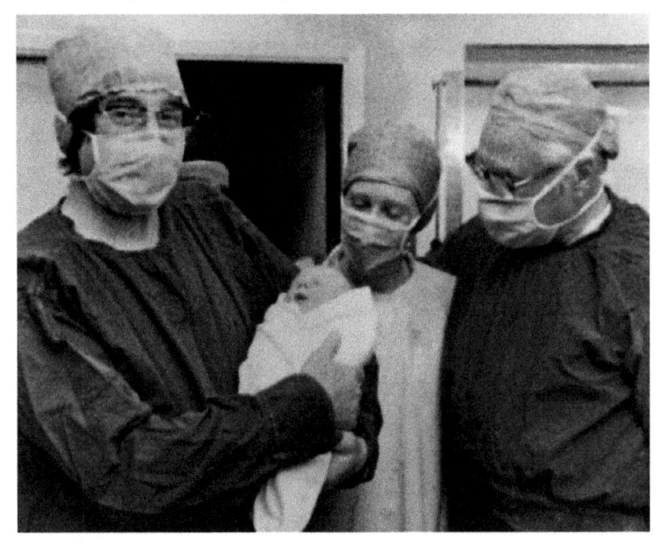

Figure 3.1.27 Louise was born following the pioneering technique of physiologist Dr Robert Edwards and gynaecologist Dr Patrick Steptoe

HIGHER TIER ONLY

Treating infertility

The NHS recommends that if a couple has been trying to conceive for a year – 6 months if the woman is over 35 – but with no success, it may be time to investigate.

For women whose levels of FSH are too low to conceive, combined hormones can be given as a **'fertility drug'**.

After treatment with a fertility drug, many women ovulate and become pregnant. They are warned of the possibility of multiple births.

If treatment is successful, it's usually within the first 3 months.

1. **Explain why FSH and LH are given as fertility treatments.**
2. **Why are the ovaries monitored after treatment?**

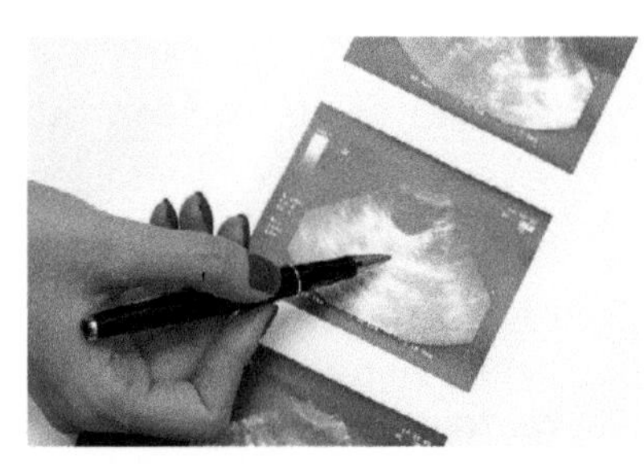

Figure 3.1.28 The women's ovaries are monitored with ultrasound to look at the number and size of developing follicles. The ultrasound image can then be printed out for a doctor to analyse

 Link: 3.1k ⟶ 3.1l

In-vitro *fertilisation*

***In-vitro* fertilisation** (IVF) may be an option to treat infertility. Here, eggs are fertilised outside the body. '*In vitro*' means, literally, 'in glass'. Eggs are removed from the mother and are fertilised, in the laboratory, with sperm collected from the father.

The technique is more successful if the woman:

- is younger
- has previously been pregnant
- has a BMI within the range of 19–30
- has low alcohol and caffeine intake, and does not smoke.

Counselling is important at this stage. The couple must be optimistic, yet prepared for failure.

3 What is *in-vitro* fertilisation?

4 Give two criteria that increase the possibility of successful IVF.

Stages of the process

The stages of IVF:

- The woman is given FSH and LH to stimulate the production of more eggs than normal in her ovaries.
- Eggs are then collected. The woman is sedated but conscious.
- Eggs are mixed with the father's sperm in the lab for 16–20 hours. They are monitored microscopically for fertilisation.
- Any embryos are allowed to develop for 5 days. They will contain around 100 cells.
- One or two embryos are selected and placed in the mother's uterus.

A number of **IVF cycles** – from stimulation of the ovaries to implantation – can be attempted. If a cycle is unsuccessful, a gap of 2 months is usually left as the treatment is emotionally and physically stressful.

5 If the father's sperm count is low, explain how the procedure is sometimes modified.

6 What is a cycle of IVF treatment?

7 Compare and contrast fertility treatment with IVF.

DID YOU KNOW?

Cryopreservation – or freezing – of eggs can be an option for someone who is undergoing a harmful cancer therapy and yet wants to be able to conceive later on.

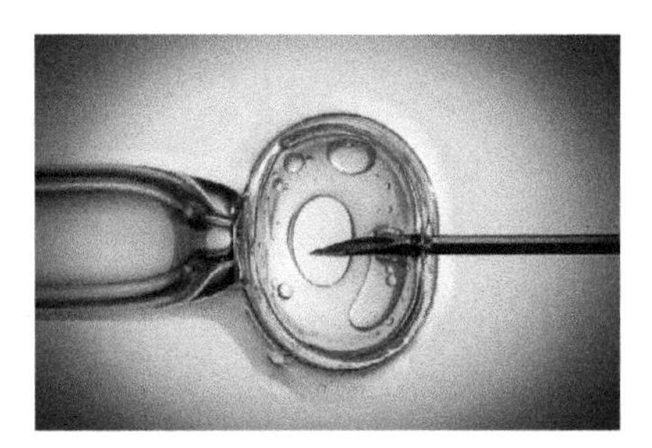

Figure 3.1.29 If the man's sperm count is low, a ***single*** sperm is sometimes selected and injected into the egg using a microscopic needle, as shown here

COMMON MISCONCEPTION

Don't confuse *fertility treatment* and *IVF*. IVF is just one type of fertility treatment.

3.1l IVF evaluation

Learning objectives:

- evaluate the scientific, emotional, social and ethical issues of *in-vitro* fertilisation.

KEY WORDS

eugenics
IVF

Fertility treatments like IVF give couples the chance to have a baby of their own, but they are emotionally and physically stressful, success rates are not high and they can lead to multiple births. Parents and doctors must balance many issues when deciding on the best course of action.

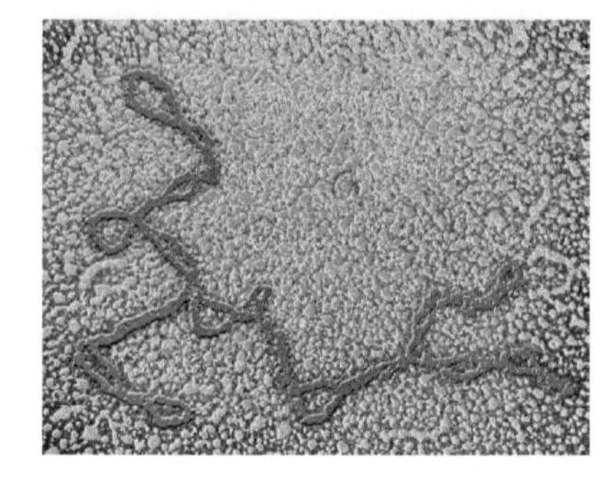

Figure 3.1.30 A transmission electron micrograph (TEM) of DNA taken from mitochondria, with false colours added to show up the detail (x300 000). The DNA is the long, red structure. Scientists in Oxford have devised a test based on mitochondrial DNA that could improve the implantation rate of IVF. A clinical trial resulted in a pregnancy rate of 80%

HIGHER TIER ONLY

The couple's perspective

IVF has been available on the NHS – provided certain criteria are fulfilled. The National Institute for Health and Care Excellence (NICE) produces guidelines, with the final decision is made by the couple's local NHS.

The process begins with counselling. This prepares the couple for the chances of success and failure. Emotional support is continuous throughout the process.

1. **Who determines the selection procedure for IVF on the NHS?**
2. **Why is counselling important for potential IVF couples?**

Medical and scientific evaluation

Around half the embryos produced by **IVF** have an incorrect number of chromosomes. One-third of the normal embryos actually selected will not implant in the uterus.

Age of woman in years	Proportion of live births in %
Under 35	32.2
35–37	27.7
38–39	20.8
40–42	13.6
43–44	5.0
Over 44	1.9

The woman's age is one factor affecting success

MAKING LINKS

To find out about chromosomes, see topic 4.3 Inheritance.

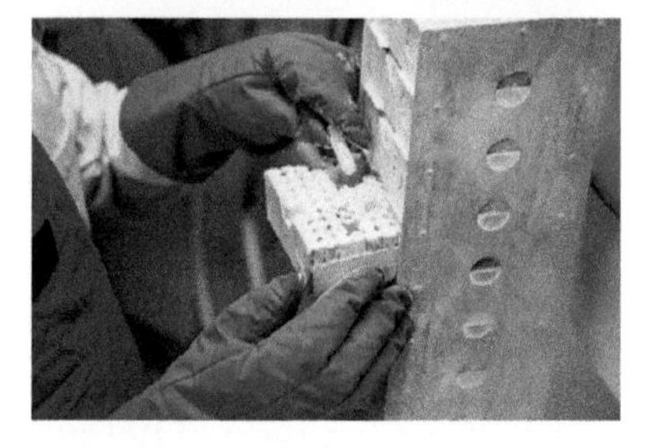

Figure 3.1.31 Eggs or embryos are frozen and stored, as shown here. That way, they can be used later on. But does freezing and then thawing affect them?

The success rate is not high, but it *is* rising – by 1% per year. There are reports of higher incidences, among IVF babies, of premature births, stillbirths, low birthweights and infant deaths. But rates are higher still in babies born to couples with infertility problems who eventually manage to conceive naturally.

One possible drawback of successful IVF is an increased possibility of multiple births, as more than one embryo is implanted. This increases the risk to the mother and babies.

3. **The data in the table are based on NHS 2010 statistics. Estimate the *current* success rate for a woman under 35.**
4. **Describe three medical issues of IVF.**

Ethical issues

An ethical issue is one involving questions about what is 'right' and 'wrong'. Discussion of ethical arguments may take into account people's values as well as the consequences of possible actions. Many people object to the technique of IVF and the treatment of embryos on ethical grounds.

One argument says that fertility treatments are just removing natural obstacles to fertility, but IVF is not removing a natural obstacle. It's replacing the physical and emotional relationship involved in conceiving with a laboratory technique.

Embryos that are not transplanted are eventually destroyed. Is a human embryo no more than a mass of cells to be used, selected and discarded? Should we accept these losses as the price for success? Or does an embryo demand the *unconditional* moral respect given to any human being?

Modern microscopic and genetic techniques have enabled embryos to be screened for abnormalities. One serious concern is that couples with no fertility problems may use IVF as a technique to select a child, possibly using dubious selection criteria – a form of **eugenics**.

5. **Discuss one ethical issue of IVF.**
6. **How could the screening of embryos be misused?**

DID YOU KNOW?

Nobel Prize winner Professor, later Sir, Robert Edwards was aware that his work was controversial because of ethical issues. After publication of a paper on producing embryos, he stopped research for 2 years while considering whether it was right to continue.

REMEMBER!

Use this spread to help you to evaluate the range of issues associated with IVF.

Check your progress

You should be able to:

describe the major causes of ill health. →	explain how diseases and lifestyle factors affect health. →	evaluate graphical data about lifestyle and health.
recall a number of interacting factors which cause different diseases. →	explain how risk factors are linked to an increased rate of some non-communicable diseases. →	evaluate evidence linking risk factors and increased rates of disease.
identify some treatments for cardiovascular disease. →	describe some treatments for cardiovascular disease. →	evaluate some treatments for cardiovascular disease.
recognise the need for homeostasis. →	recall that the nervous and endocrine systems are responsible for homeostasis. →	describe the components of body control systems and compare nervous and hormonal control.
recall that insulin helps control blood sugar levels in the body. →	compare Type 1 and Type 2 diabetes and explain how they can be treated. →	explain how glucagon interacts with insulin to control blood sugar levels in the body.
identify temperature, water and glucose concentrations as conditions requiring control. →	describe how these are controlled by the nervous and/or endocrine system. →	explain how monitoring and control is involved in homeostasis.
identify the reproductive hormones and their role in the development of secondary sexual characteristics. →	describe the role of hormones in the menstrual cycle. →	explain the roles and interactions of hormones in the menstrual cycle.
identify different methods of contraception. →	describe how different methods of contraception work. →	evaluate the use of different methods of contraception.
identify fertility drugs and *in-vitro* fertilisation as possible treatments for infertility. →	describe the techniques involved in fertility treatments. →	evaluate fertility treatments.

Worked example

The graphs show data about smoking and lung cancer in people in the UK.

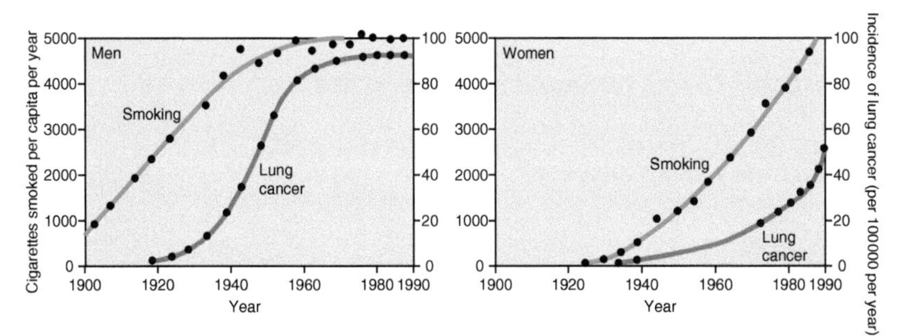

1 Describe the smoking habits of men from 1900 to 1990, according to these graphs.

Smoking increased in men between 1900 and 1990.

The answer is correct but not complete. It should say that, for men, the number of cigarettes smoked per person per year increased steeply until around 1950, when the rate of increase started to slow. From about 1970 onward there was no increase in the number of cigarettes smoked per person.

2 Describe the smoking habits of women from 1900 to 1990, compared to those of men.

For women, smoking also increased. It increased more steeply than for men.

The answer is slightly more complete, but it should also say that women smoked much less than men until around 1990. The rate of increase was still going up steeply in 1985, when it had stopped increasing in men.

3 Describe what happened to the incidence of lung cancer in men and in women during this time.

After a lag, the incidence of lung cancer increased in both men and women, closely following the pattern of the smoking data.

This is a good answer. The question asks you to 'describe' so you don't need to try to explain the pattern you see at this point.

4 What do the graphs suggest about a possible link between smoking and lung cancer?

The graphs suggest that there is a strong link. As the rate of smoking in the population goes up, so does the incidence of lung cancer. There is a correlation between the two factors, which suggests that smoking is a risk factor for lung cancer.

This answer correctly uses some key terms, such as 'risk factor' and 'correlation', which shows a good understanding of the ideas.

5 Do the graphs prove a causal mechanism?

No. They show a correlation, but there is no proof that one factor causes the other.

This is correct – correlation on its own does not prove causation. Further research is always required to understand the causes of an effect.

6 Lung cancer is a non-communicable disease. Explain what you understand by the term 'non-communicable disease'.

Non-communicable diseases are not infectious.

This is a good start but is incomplete. The answer should also say that they can be caused by factors such as diet, aspects of a person's lifestyle, and substances in the person's body or environment.

End of chapter questions

Getting started

1 a What is the name of the hormone responsible for reducing blood glucose after a meal, and in which gland is this hormone produced? 2 Marks

b If the body cannot produce or respond to this hormone properly, what disease results? 1 Mark

2 What body tissues are affected in coronary heart disease? 1 Mark

3 a What do we mean by a 'risk factor'? 1 Mark

b Name two risk factors for developing cardiovascular disease. 2 Marks

4 Name two barrier methods of contraception, and two hormonal methods. 2 Marks

5 Scientists have researched the relationship between the number of sugar-sweetened soft drinks consumed by a group of women and the relative risk of diabetes. Their results are shown here.

Describe the pattern shown by the graph and explain what this suggests about a possible cause of Type 2 diabetes. 3 Marks

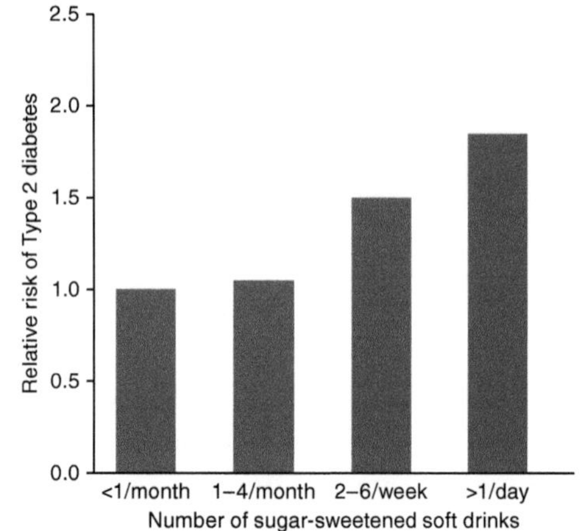

Going further

6 a Which hormone leads to oestrogen production by the ovaries? 1 Mark

b Which hormones triggers ovulation? 1 Mark

7 Why does body temperature need to be kept constant? 1 Mark

8 How is Type 2 diabetes controlled? 2 Marks

9 Explain how hormonal methods of contraception work. 3 Marks

10 The graph shows the effect of two risk factors on mortality. Describe the patterns in the data.

3 Marks

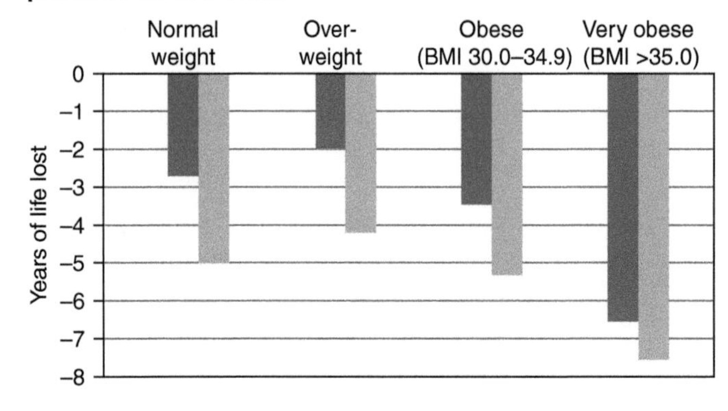

More challenging

11 a **Define 'homeostasis'.** 1 Mark

b **Compare and contrast the ways in which the nervous system and the endocrine system contribute to homeostasis.** 4 Marks

12 **Which hormones are involved in fertility treatments?** 2 Marks

13 **Endometriosis is a condition where the type of cells that normally line the uterus are found 'trapped' in the pelvic area and lower tummy. During the menstrual cycle the cells that have moved can react in the normal way to the hormones controlling menstruation. This can cause a number of problems including abdominal pain and painful periods.**

Using the information from the graph below, as well as your own knowledge, explain what happens to the cells that have moved and how doctors could treat the condition using sex hormones. 6 Marks

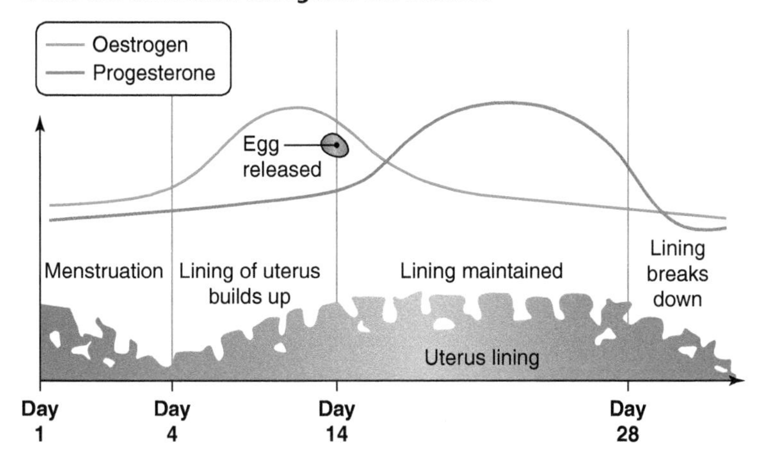

Most demanding

14 **The graph below shows smoking and lung cancer rates in the UK.**

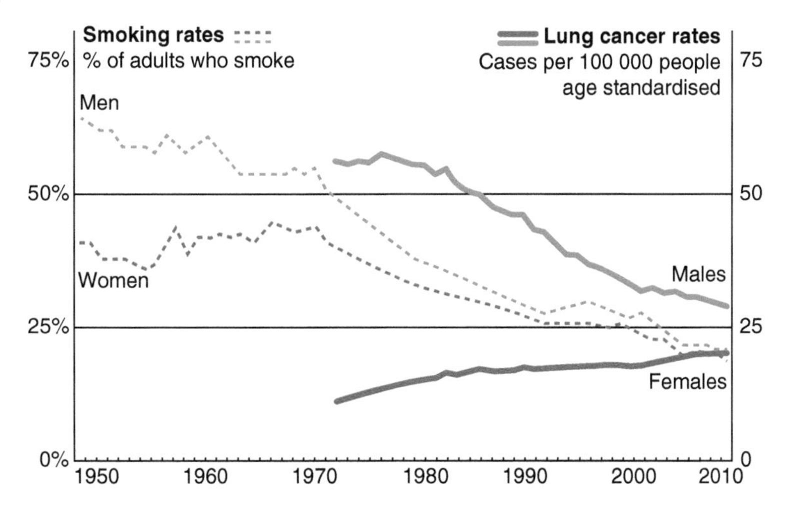

Describe the trends in the data and evaluate the evidence for a link between smoking and lung cancer. 4 Marks

40 Marks

RADIATION AND RISK

IDEAS YOU HAVE MET BEFORE:

AN ATOM CONTAINS PROTONS, NEUTRONS AND ELECTRONS.

- The energy level model of the atom describes electrons in 'shells' with particular energy values around the nucleus.
- The nucleus of an atom contains protons and neutrons. All the atoms of one element have the same number of protons.
- Isotopes are forms of the same element with different numbers of neutrons and different mass.

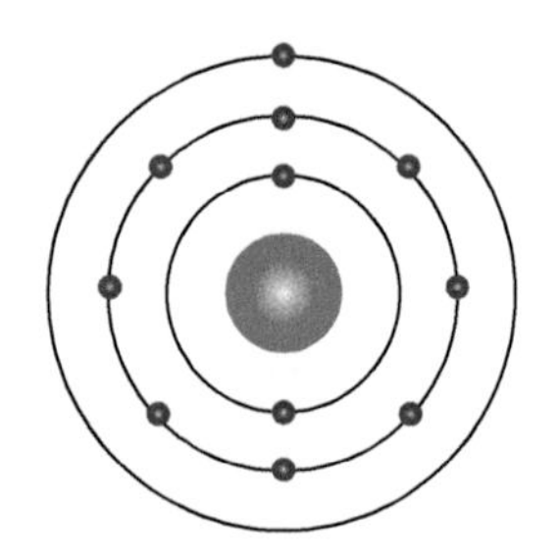

THE ELECTROMAGNETIC SPECTRUM INCLUDES RADIATION OF A RANGE OF WAVELENGTHS.

- The electromagnetic spectrum is a continuous range of wavelengths of radiation.
- Depending on their wavelength and frequency, different parts of the spectrum have different uses and dangers.

CHEMICAL EQUATIONS SUMMARISE WHAT HAPPENS IN A CHEMICAL REACTION.

- The atom is the smallest particle that can take part in a chemical reaction.
- There is always the same number of atoms before and after a chemical reaction.
- There is always the same number of each type of atom before and after a chemical reaction.

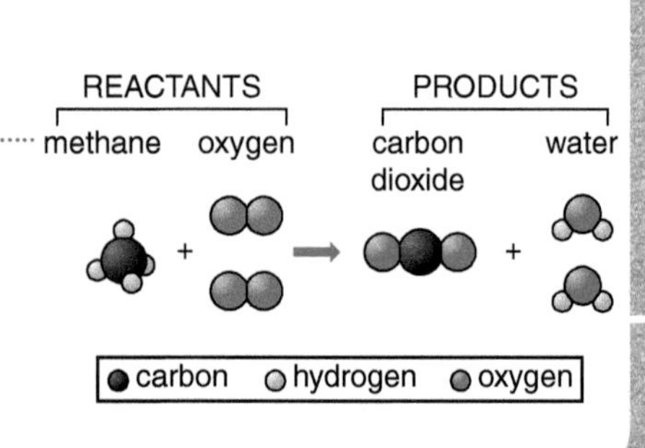

IONISING RADIATION CAN BE USEFUL, AND DANGEROUS.

- The ionising radiation emitted from nuclei has many uses in medicine and elsewhere.
- Ionising radiation can make atoms into ions, by loss of outer electrons, and it therefore damages substances, including living cells.

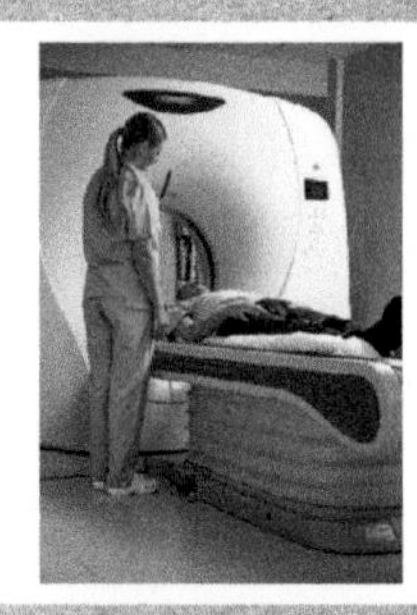

3.2

IN THIS CHAPTER YOU WILL FIND OUT ABOUT:

HOW CAN ATOMS CHANGE FROM ONE ELEMENT INTO ANOTHER?

- A radioisotope has nuclei that are unstable and randomly undergo radioactive decay, emitting ionising radiation.
- The emission of different types of ionising radiation may cause a change in mass and/or charge of the nucleus.
- The half-life of a radioisotope is the average length of time for half the nuclei in a sample to decay.

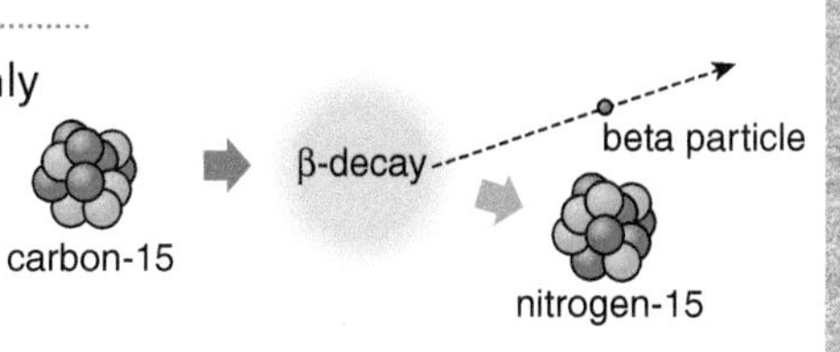

HOW ARE ATOMS AFFECTED BY ELECTROMAGNETIC RADIATION?

- The arrangements of electrons in atoms may change with absorption or emission of electromagnetic radiation.
- When atoms gain energy by absorbing electromagnetic radiation, some electrons jump to higher energy levels.
- Electromagnetic radiation is given out when the electrons drop back to lower levels. The wavelength and frequency depend on the size of the energy jump.

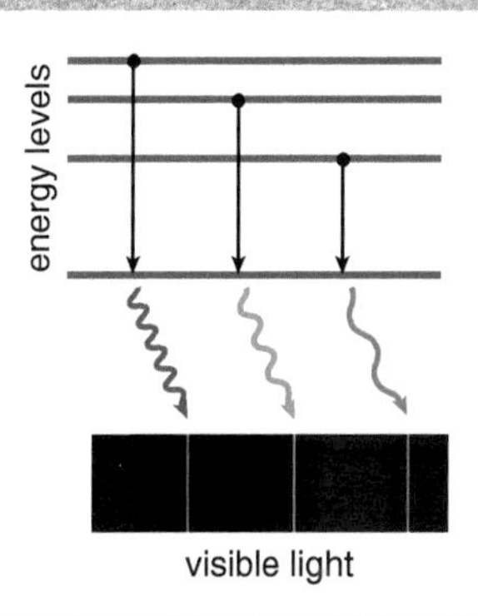

HOW CAN EQUATIONS BE USED TO REPRESENT NUCLEAR REACTIONS?

- An atom can be represented by its element symbol with its atomic number and mass number.
- A nuclear equation summarises what happens in a nuclear reaction.

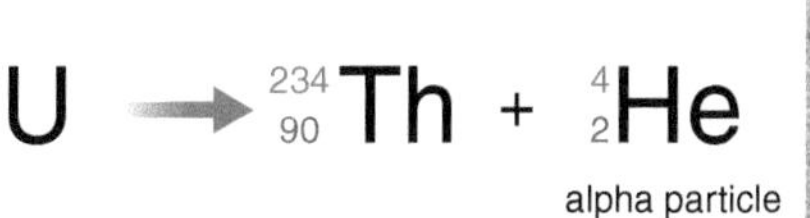

WHAT RISKS ARE ASSOCIATED WITH IONISING RADIATION?

- Contamination and irradiation by sources of ionising radiation carries risks of harm.
- Different types of ionising radiation have different penetration properties.
- Ultraviolet waves, X-rays and gamma rays can have hazardous effects on human body tissues.

3.2a Absorption and emission of radiation

Learning objectives:

- recall that the electron arrangement in atoms may change when electromagnetic radiation is absorbed or emitted
- be able to use the energy level model of the atom.

KEY WORDS

emission spectrum
energy level

The electrons in atoms are arranged in energy levels, but they can move between levels when energy is absorbed or emitted.

MAKING LINKS

Look back to topic 1.2g to remind yourself about electrons in atoms.

Model of the atom

According to our current model of atomic structure, the electrons in an atom are arranged in **energy levels** or 'shells' around the nucleus. The shells closer to the nucleus have a lower energy level, and each electron occupies the lowest available shell.

The electronic structure of an atom can be represented by a diagram (Figure 3.2.1) or by numbers – for example, the electronic structure of sodium is 2,8,1.

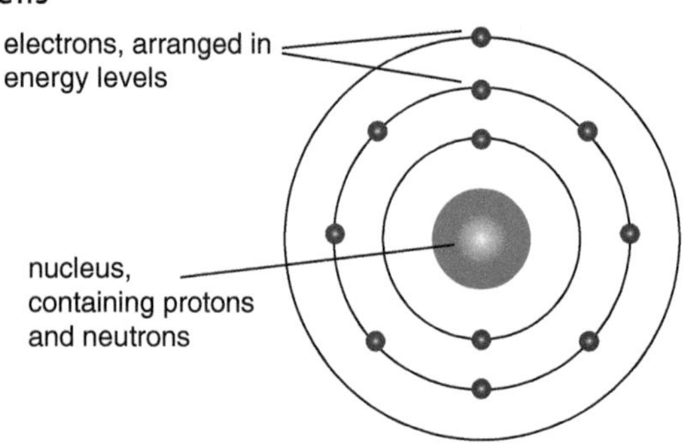

Figure 3.2.1 In a sodium atom, there are two electrons in the lowest energy level, eight in the second and one in the third

1. **How many electrons are in a sodium atom?**
2. **Draw a diagram of a chlorine atom, which has electronic structure 2,8,7.**

Gaining and losing energy

When atoms gain energy – by heating, from electricity, or by absorbing electromagnetic radiation – some electrons jump to higher energy levels. Because this 'excited state' is less stable, they often drop back to a lower level almost immediately, and as they do so the electrons give out the energy they had gained, as electromagnetic radiation.

The frequency and wavelength of the radiation emitted depends on the size of the energy jump.

Atoms of some elements, like neon and sodium, emit radiation in the visible part of the electromagnetic spectrum. Other atoms, such as mercury, give out ultraviolet radiation too.

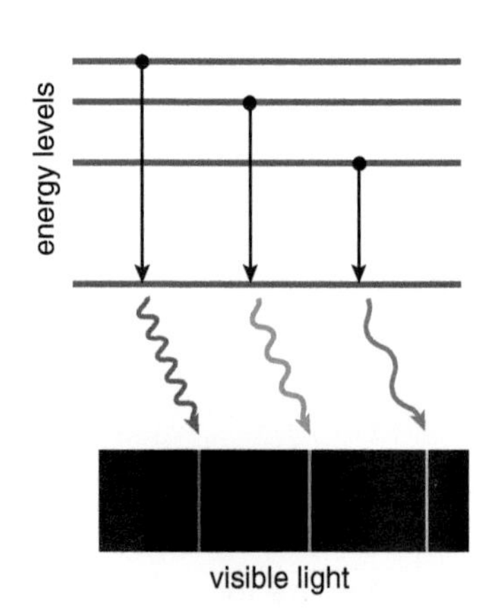

Figure 3.2.2 Electrons at higher energy levels fall back to 'ground state', emitting specific amounts of energy as radiation of characteristic wavelengths

3. **What do we mean when we say electrons are in an 'excited state'?**
4. **What do electrons need to get to an 'excited state'?**

Emission spectra

The arrangement of electrons in atoms is unique to each element. Because the energy levels have particular values, the energy jumps between them also have specific sizes. So, in atoms of a certain element, as excited electrons fall back to their ground state, they always give out particular amounts of energy, emitted as radiation of characteristic wavelengths for that element.

For radiation in the visible part of the electromagnetic spectrum, this means light of certain colours. We see this in flame tests, in which different elements give specific, identifying colours. An **emission spectrum** shows the wavelengths of radiation emitted by an element as sharp lines of different colours, which can be used as a 'fingerprint' to help identify elements more precisely.

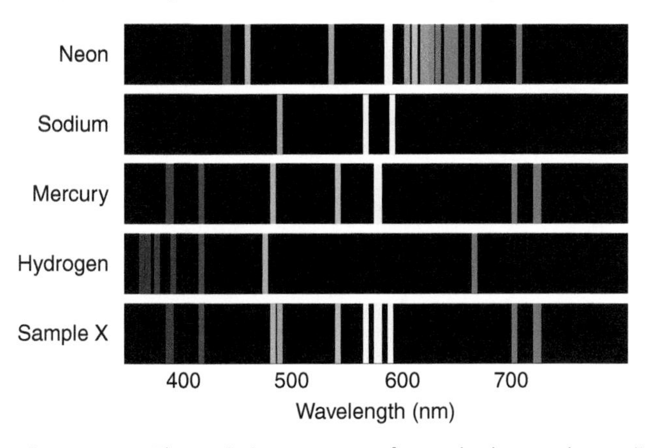

Figure 3.2.3 The emission spectrum for each element has a distinctive pattern. These can be compared with spectra of unknown samples, to identify the elements contained

5 **What elements does Sample X contain?**

6 **In a neon sign, electrical energy is applied to neon gas in a glass tube. The gas emits red light. Can you explain why this happens?**

REMEMBER !

The fact that the energy level model explains line spectra is good evidence for the model.

DID YOU KNOW?

Atoms absorb radiation of particular wavelengths, as well as emitting it. Absorption spectra can be used to analyse the atmospheres of distant planets, for example.

3.2b Radioactivity

Learning objectives:

- recall that some nuclei are unstable and may decay, emitting radiation
- recall that radioactive decay may change the mass or charge of the nucleus, or both
- write balanced nuclear equations.

KEY WORDS

alpha particle
beta particle
gamma ray
radioactive decay
radioisotope

MAKING LINKS

Look back to topic 1.2f to remind yourself about isotopes.

The nuclei of most atoms are stable but some are not. An unstable nucleus undergoes radioactive decay to become more stable, and emits radiation as it does so.

Radioactive decay

The largest stable nucleus is lead-208, which has a mass number of 208. (This is the total number of protons and neutrons in the nucleus.) Larger nuclei than this are unstable and break apart, emitting radiation. When they undergo **radioactive decay** like this, many of them eventually turn into lead.

An atom with an unstable nucleus is called a **radioisotope**. The nuclear radiation emitted when a radioisotope decays may be:

- an alpha particle (α)
- a beta particle (β)
- a gamma ray (γ)
- a neutron (n).

The emission of these different types of ionising radiation may change the mass or the charge of the nucleus, or both.

KEY INFORMATION

Isotopes of an element have the same proton number but different mass numbers, because their atoms contain a different number of neutrons.

Radioisotopes are isotopes whose atoms have unstable nuclei, which decay and emit radiation.

Not all isotopes of elements are radioactive.

1. **What is a radioisotope?**
2. **What is the mass number of an element?**

Alpha and beta decay

In alpha decay, an **alpha particle** is emitted from the nucleus. An alpha particle is identical to a helium nucleus. It has 2 protons and 2 neutrons.

We can write a balanced nuclear equation to represent radioactive decay. In the example in Figure 3.2.4:

$${}^{238}_{92}U \rightarrow {}^{4}_{2}He + {}^{234}_{90}Th$$

Because an alpha particle is identical to a helium nucleus, we can represent it as ${}^{4}_{2}He$.

When an alpha particle is emitted from a nucleus:

- The nucleus has two fewer protons (p), so the atomic number (Z, the proton number) decreases by two.

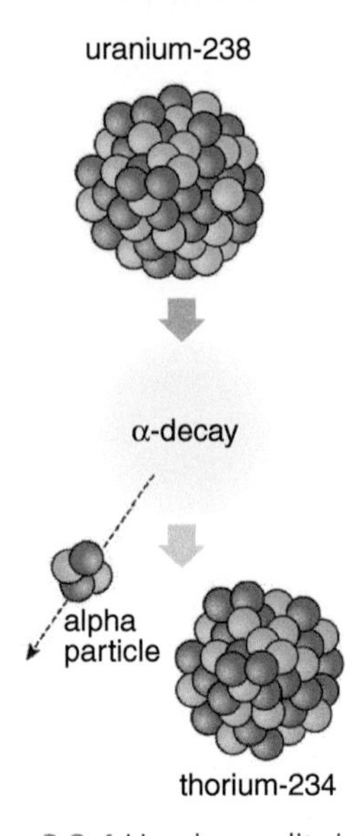

Figure 3.2.4 Uranium splits into thorium and an alpha particle during alpha decay

- The nucleus has two fewer neutrons, so the mass number (*A*) decreases by four.
- A new element is formed.

In beta decay one of the neutrons in the nucleus decomposes into a proton and an electron. This electron is emitted from the nucleus and is called a **beta particle**.

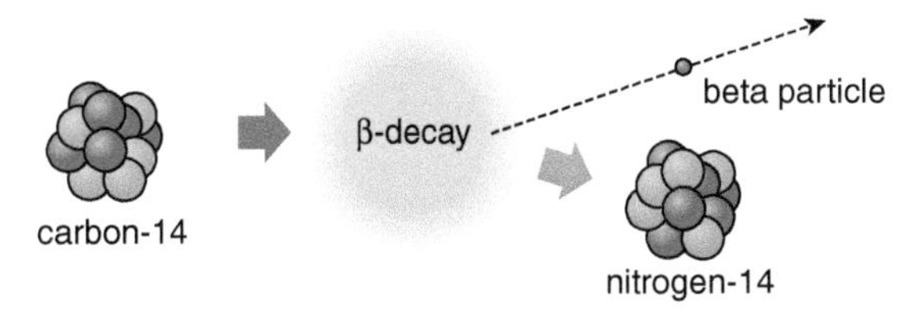

Figure 3.2.5 The carbon nucleus emits a beta particle and turns into nitrogen in beta decay

This is the balanced nuclear equation for the radioactive decay in Figure 3.2.5:

$${}^{14}_{6}\mathrm{C} \rightarrow {}^{0}_{-1}\mathrm{e}^{-} + {}^{14}_{7}\mathrm{N}$$

Because a beta particle is identical to an electron, which has practically no mass, we can represent it as ${}^{0}_{-1}\mathrm{e}^{-}$. (It has atomic number 0, but to conserve atomic number in nuclear equations we give it atomic number –1.)

When a beta particle is emitted from a nucleus:

- The nucleus has one more proton (p), so the atomic number (*Z*, the proton number) increases by one.
- The nucleus has one less neutron (n), but the mass number (*A*, the nucleon number) is unchanged.
- A new element is formed.

3 **In Figure 3.2.5, what are the differences between the nitrogen nucleus and the carbon nucleus?**

4 **What happens to a nucleus when a an alpha particle b a beta particle is emitted?**

Gamma decay

In gamma decay, **gamma rays** are emitted from a nucleus.

These are very high-energy electromagnetic waves. They have no charge and no mass.

The emission of a gamma ray does not cause the mass or the charge of the nucleus to change.

5 **When a nucleus decays, its mass number is unchanged but the atomic number increases by 1. What type of decay is it?**

6 **After a series of many radioactive decays, a uranium-238 atom eventually becomes a lead-206 atom. Explain how one element can turn into another.**

KEY INFORMATION

We can represent atoms using the notation where ${}^{A}_{Z}\mathrm{X}$ is the symbol for the element, *Z* is the atomic number (the number of protons) and *A* is the mass number.

Carbon has atomic number 6, so carbon-14 can be shown as ${}^{14}_{6}\mathrm{C}$.

A neutron has atomic number 0 and mass number 1. The symbol for a neutron is ${}^{1}_{0}\mathrm{n}$.

A proton has mass number 1 and atomic number 1. The symbol for a proton is ${}^{1}_{1}\mathrm{p}$.

An electron has practically no mass and atomic number 0, but to conserve atomic number in nuclear equations we give it atomic number –1. The symbol for an electron is ${}^{0}_{-1}\mathrm{e}$.

DID YOU KNOW?

Henri Becquerel discovered radioactivity by accident in 1896, when he left some uranium salts next to a wrapped photographic plate. The plate had become 'fogged', and he realised that some invisible radiation must be coming from the uranium.

3.2c Nuclear equations

Learning objectives:

- understand nuclear equations
- write balanced nuclear equations.

KEY WORDS

alpha decay
beta decay
nuclear equation

Uranium-238 undergoes a series of 14 radioactive decays to become lead-206 – some alpha and some beta decays. We could write a balanced nuclear equation for every decay to show how one element has become another.

Nuclear equations

Chemical equations show what happens in a chemical reaction. The number of atoms on each side has to be the same – the equation has to be balanced.

Nuclear equations show what happens when there are changes in the nucleus. They show the number of nucleons. As with chemical equations, they have to be balanced. The number of nucleons has to be the same on both sides of the equation.

1. **What do chemical and nuclear equations have in common?**

2. **What is a key difference between chemical and nuclear equations?**

Nuclear equations for alpha decay

Alpha decay and **beta decay** can be shown as nuclear equations. Figure 3.2.6 shows the **nuclear equation** for the alpha decay of uranium-238.

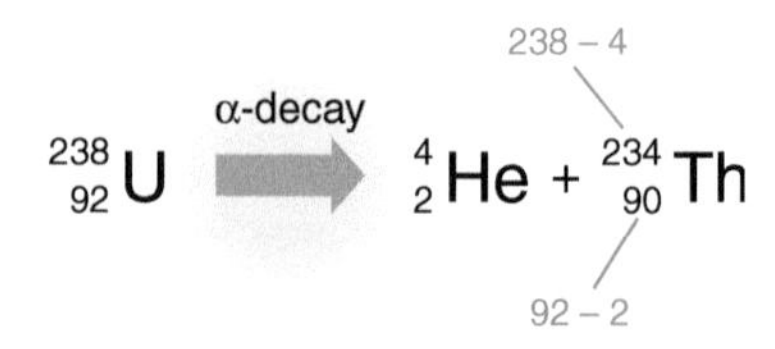

Figure 3.2.6 The equation for the alpha decay of uranium-238

The mass numbers add up to the same number on both sides of the equation (238 = 234 + 4). This means that mass is conserved.

The atomic numbers also add up to the same on both sides of the equation (92 = 90 + 2), so the number of protons is also conserved.

When you write nuclear equations, make sure that the mass numbers and atomic numbers balance on both sides of the equation.

DID YOU KNOW?

Smoke alarms contain a small amount of americium-241 which decays by emitting an alpha particle. It becomes neptunium-237.

3. **Copy and complete these nuclear equations for alpha decay.**

   a $^{226}_{\_}Ra \rightarrow ^{\_}_{86}Rn + ^{4}_{2}He$

   b $^{219}_{86}Rn \rightarrow ^{\_}_{\_}Po + ^{\_}_{\_}He$

$^{23}_{11}\text{Na}$ sodium	$^{24}_{12}\text{Mg}$ magnesium											$^{27}_{13}\text{Al}$ aluminium	$^{28}_{14}\text{Si}$ silicon	$^{31}_{15}\text{P}$ phosphorus	$^{32}_{16}\text{S}$ sulfur	$^{35.5}_{17}\text{Cl}$ chlorine	$^{40}_{18}\text{Ar}$ argon
$^{39}_{19}\text{K}$ potassium	$^{40}_{20}\text{Ca}$ calcium	$^{45}_{21}\text{Sc}$ scandium	$^{48}_{22}\text{Ti}$ titanium	$^{51}_{23}\text{V}$ vanadium	$^{52}_{24}\text{Cr}$ chromium	$^{55}_{25}\text{Mn}$ manganese	$^{56}_{26}\text{Fe}$ iron	$^{59}_{27}\text{Co}$ cobalt	$^{59}_{28}\text{Ni}$ nickel	$^{63}_{29}\text{Cu}$ copper	$^{65}_{30}\text{Zn}$ zinc	$^{70}_{31}\text{Ga}$ gallium	$^{73}_{32}\text{Ge}$ germanium	$^{75}_{33}\text{As}$ arsenic	$^{79}_{34}\text{Se}$ selenium	$^{80}_{35}\text{Br}$ bromine	$^{84}_{36}\text{Kr}$ krypton
$^{85}_{37}\text{Rb}$ rubidium	$^{88}_{38}\text{Sr}$ strontium	$^{89}_{39}\text{Y}$ yttrium	$^{91}_{40}\text{Zr}$ zirconium	$^{93}_{41}\text{Nb}$ niobium	$^{96}_{42}\text{Mo}$ molybdenum	$^{99}_{43}\text{Tc}$ technetium	$^{101}_{44}\text{Ru}$ ruthenium	$^{103}_{45}\text{Rh}$ rhodium	$^{106}_{46}\text{Pd}$ palladium	$^{108}_{47}\text{Ag}$ silver	$^{112}_{48}\text{Cd}$ cadmium	$^{115}_{49}\text{In}$ indium	$^{119}_{50}\text{Sn}$ tin	$^{122}_{51}\text{Sb}$ antimony	$^{128}_{52}\text{Te}$ tellurium	$^{127}_{53}\text{I}$ iodine	$^{131}_{54}\text{Xe}$ xenon
$^{133}_{55}\text{Cs}$ caesium	$^{137}_{56}\text{Ba}$ barium	$^{139}_{57}\text{La}$ lanthanum	$^{178}_{72}\text{Hf}$ hafnium	$^{181}_{73}\text{Ta}$ tantalum	$^{184}_{74}\text{W}$ tungsten	$^{186}_{75}\text{Re}$ rhenium	$^{190}_{76}\text{Os}$ osmium	$^{192}_{77}\text{Ir}$ iridium	$^{195}_{78}\text{Pt}$ platinum	$^{197}_{79}\text{Au}$ gold	$^{201}_{80}\text{Hg}$ mercury	$^{204}_{81}\text{Tl}$ thallium	$^{207}_{82}\text{Pb}$ lead	$^{209}_{83}\text{Bi}$ bismuth	$^{210}_{84}\text{Po}$ polonium	$^{210}_{85}\text{At}$ astatine	$^{222}_{86}\text{Rn}$ radon
$^{223}_{87}\text{Fr}$ francium	$^{226}_{88}\text{Ra}$ radium	$^{227}_{89}\text{Ac}$ actinium															

Figure 3.2.7 Part of the periodic table showing the relative atomic mass and atomic number for each element

Nuclear equations for beta decay

In beta decay, a neutron changes into a proton and an electron.

$^{1}_{0}\text{n} \rightarrow {}^{1}_{1}\text{p} + {}^{0}_{-1}\text{e}$

A beta particle actually has atomic number 0, because of course it contains no protons. To conserve atomic number in nuclear equations we give it atomic number –1, as shown in the symbol ${}^{0}_{-1}\text{e}$.

Figure 3.2.8 demonstrates how the mass number is conserved (15 = 0 + 15) during beta decay. It also shows that the atomic number is conserved (6 = 7 – 1) during beta decay.

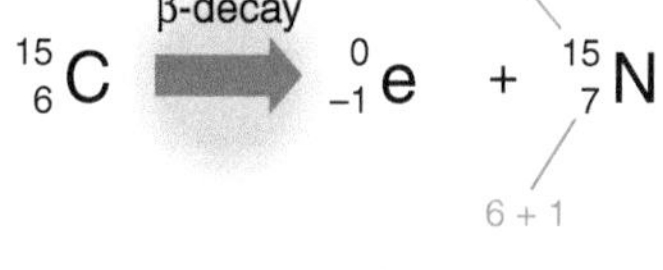

Figure 3.2.8 One of the radioisotopes of carbon is carbon-15. It decays by beta decay to nitrogen-15

4 Copy and complete these equations for beta decay.

a $^{90}_{\_}\text{Sr} \rightarrow {}^{\_}_{39}\text{Y} + {}^{0}_{-1}\text{e}$ **b** ${}^{\_}_{15}\text{P} \rightarrow {}^{32}_{\_}\text{S} + {}^{0}_{-1}\text{e}$

5 When radioactive sodium-24 decays, magnesium-24 is formed. One particle is emitted.

a Copy and complete the equation.

${}^{\_}_{\_}\text{Na} \rightarrow {}^{\_}_{\_}\text{Mg} + \_\_$

b What is the name of this particle?

6 This equation represents the decay of thorium-232.

$^{232}_{90}\text{Th} \rightarrow {}^{A}_{Z}\text{X} + {}^{4}_{2}\text{He}$

a What type of radiation is emitted?

b What are the values of A and Z?

7 Write a word equation and symbol equation for each radioactive decay:

a platinum-190, which emits an alpha particle

b rhenium-187, which emits an alpha particle

c copper-66, which emits a beta particle

d nickel-66, which emits a beta particle

e rhodium-105 which decays to palladium-105

f osmium-186 which decays to tungsten-182.

KEY INFORMATION

Radioactive decay by emitting a gamma ray causes no change in mass number or atomic number.

3.2d Half-life

Learning objectives:

- explain what is meant by radioactive half-life
- calculate half-life
- calculate the net decline in radioactive emission after a given number of half-lives.

KEY WORDS

activity
half-life
random
tracer

Not all of the nuclei in a radioisotope decay at the same time. Radioactive half-life is a way of measuring how quickly they decay. Different radioisotopes have different half-lives, ranging from fractions of a second to billions of years.

Radioactive half-life

Radioactive decay is a **random** process. We cannot predict when a particular nucleus in a radioisotope will decay – it could be seconds from now, or next week or not for a million years. But if we have a very large number of atoms, we can predict that some of them will decay each second.

The number of nuclear decays each second is called the **activity** of a radioisotope. The activity is measured in counts per second, using a detector such as a Geiger–Müller tube.

If we plot a graph of activity, or count rate, against time and draw a curve of best fit (Figure 3.2.9), we can see that the activity of a radioisotope gets less and less as time goes on. The line gets closer and closer to the time axis but never quite reaches it.

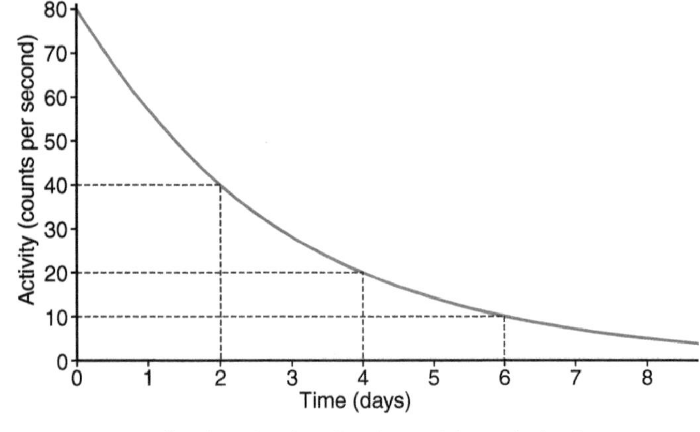

Figure 3.2.9 The time it takes for the activity to halve is constant

The **half-life** of a radioisotope is the average time it takes for the activity to fall to half its original value. This is also the average time for half the nuclei present to decay. Every radioisotope has its own characteristic half-life.

1 Radioactive decay is random. What does this mean?

2 Explain what is meant by 'half-life'.

3 What will the activity in Figure 3.2.9 be after: a 2 days? b 4 days? c 6 days? d 8 days?

KEY INFORMATION

When you draw a curve of best fit on a graph of activity against time for a radioisotope, not all of the points will be on the curve. Some will be above the curve and some will be below the curve because of the random nature of radioactive decay.

Calculating half-life

To calculate the half-life of a radioisotope, plot activity against time (Figure 3.2.10). Draw a smooth curve of best fit. Find several values for the time taken for the activity to halve (from 80 to 40, 40 to 20, and so on). The times may not be exactly the same, because radioactive decay is random, so calculate a mean value. This is the half-life of the radioisotope.

 Link: 3.2d ⟶ 3.2e

4 **Calculate the half-life of the radioisotope shown in Figure 3.2.10.**

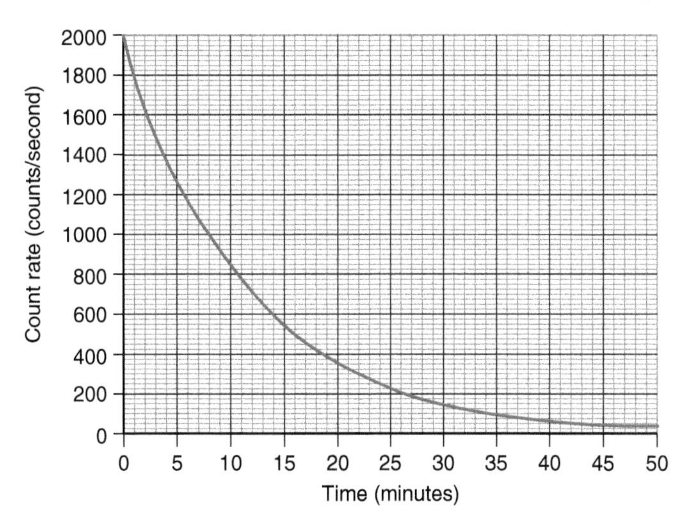

Figure 3.2.10

5 **The table shows the activity of a radioactive sample over time.**

a **Draw a graph of activity against time.**

b **Draw a best-fit curve through the points.**

c **Find the half-life at three different points on your curve. Calculate the average half-life.**

Time (minutes)	Count rate (counts/ second)
0	100
1	51
2	26
3	12

6 **The activity of a radioactive sample took 4 hours to decrease from 100 to 25 counts per second. Calculate its half-life.**

DID YOU KNOW?

We use half-life because we cannot predict the time it will take for all the atoms to decay – the activity halves each half-life, but it never quite reaches zero.

HIGHER TIER ONLY

Calculating decline in radioactivity

Radioisotopes have many everyday applications. For some, such as a smoke alarm, it's important that the activity stays fairly high over the lifetime of the product. For others, such as **tracers**, we want the activity to drop quickly. To select the best radioisotope for the job, we need to calculate what fraction of its activity remains after a certain time.

Technetium-99m, widely used as a tracer in medicine, has a half-life of 6 hours. What fraction of the original activity remains after one full day?

After 6 hours, ½ the activity remains.

After 12 hours, ¼ remains.

After 18 hours, 1/8 remains.

After 24 hours (1 full day), just 1/16 of the radioisotope's original activity remains.

7 **For the radioisotope in Figure 3.2.10, what fraction of the original activity will remain after 60 minutes?**

8 **Sulfur-35 has a half-life of about 90 days.**

a **What fraction of the original nuclei remain undecayed after 6 months?**

b **Would this be a good choice for use in a domestic smoke alarm? Explain your answer.**

DID YOU KNOW?

Radioisotopes are often used as medical tracers, to monitor biological processes in the body. For this, a short half-life is best to minimise the patient's exposure to ionising radiation.

MATHS SKILLS

3.2e Drawing and using lines of best fit

Learning objectives:

- draw a curve of best fit to calculate radioactive half-life
- calculate the net decline in radioactive emission after a given number of half-lives.

KEY WORDS

curve of best fit
net decline
scale
value

MATHS

Some graphs will have a line or curve of best fit. A line of best fit goes roughly through the centre of all the plotted points. It does not go through all the points. The **line of best fit** could be a curve. If all the points fall close to the line (or curve), it suggests that the variables are closely linked. Look at the points carefully and see what would fit them well.

When you plot a graph of activity against time for a radioisotope, the points do not always fit a smooth curve, because of the random nature of radioactive decay. To be able to make calculations from the graph, you have to draw a curve of best fit.

Using a graph to calculate half-life

Working out the half-life of a radioisotope usually involves drawing a graph and using the graph to calculate the half-life.

Figure 3.2.11 shows counts per second against time for a certain radioactive isotope. You can use the graph to work out the half-life of the isotope.

Look for numbers on the vertical scale that are easy to halve. For example, you can halve 40 easily to get 20.

From the 40 on the vertical axis, draw a line across to the graph. Then draw a line down to the time axis. So the count at 2 days is 40 counts/second.

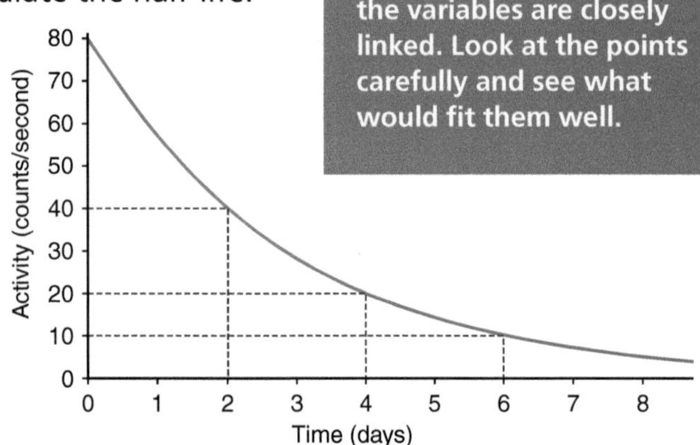

Figure 3.2.11

Now repeat this for 20 counts/second. The time is 4 days.

It takes 4 – 2 = 2 days for the activity to reduce by half from 40 to 20 counts/second.

So the half-life is 2 days.

You should always repeat for a second step, just to check. Choose another value that can be halved easily, e.g. 20. Repeating the process gives the same answer, 2 days.

Drawing a curve of best fit

When plotting experimental values, you need to draw a curve of best fit (Figure 3.2.12).

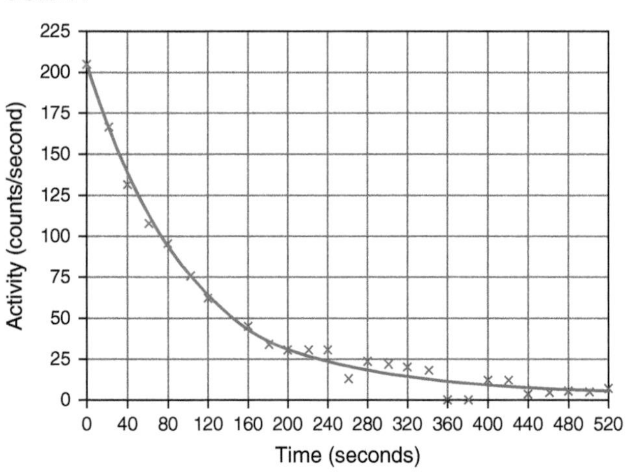

Figure 3.2.12

The curve should have approximately equal numbers of points above it and below it – it does not go through all the points. The curve should also be smooth, as shown in Figure 3.2.12.

1. **Calculate the half-life for the source shown in Figure 3.2.12.**

   Work out three values for the half-life.

   Calculate the average of the values.

2. **a From the data in the table plot a graph of activity against time.**

Time (minutes)	0	10	20	30	40	50	60	70	80	90
Activity (counts/second)	96	78	62	54	40	32	26	21	15	14

   b Calculate the half-life.

HIGHER TIER ONLY

Half-life calculations

A radioactive isotope has an activity of 160 counts/second and has a half-life of 2 hours. What is the count rate after 6 hours?

Every 2 hours the count will halve (i.e. one half-life). 6 hours means three half-lives. (2 + 2 + 2) days = 6 days.

Starting with 160 counts/second at time 0,

after 1 half-life, activity = 80 counts/second;

after 2 half-lives, activity = 40 counts/second;

after 3 half-lives, activity = 20 counts/second.

The count rate after 6 hours will be 20 counts/second.

This can also be expressed as a ratio: $\frac{20}{160} = \frac{1}{8}$

After 6 hours the count rate will have reduced to $\frac{1}{8}$ th of the initial activity.

This ratio is called the **net decline**.

3. **Sodium-24 has a half-life of 15 hours. A sample of sodium-24 has an activity of 640 counts/second.**

   a Calculate the activity after 60 hours.

   b Calculate the net decline.

4. **Iodine-131 has a half-life of 8 days. A sample of iodine-131 has an activity of 1800 counts per second.**

   a Calculate the count rate after 32 days.

   b Calculate the net decline in the activity of the sample.

5. **The activity of a sample decreases to $\frac{1}{16}$th of its original value over 24 hours.**

   Calculate the half-life of the sample.

MATHS

You can also work out the net decline using fractions and powers.

At the end of each half-life the activity has decreased by a factor of ½.

Suppose the initial activity of a sample is 160 counts/s.

After 1 half-life: activity = ½ × 160 = 80 counts/s

After 2 half-lives: activity = ½ × ½ × 160 = $(½)^2$ × 160 = 40 counts/s

After 3 half-lives: activity = ½ × ½ × ½ × 160 = $(½)^3$ × 160 = 20 counts/s

So after n half-lives, the activity is reduced to $(½)^n$ of the initial activity.

For 3 half-lives, we get a net decline of $(½)^3$ = 1/8. So the activity after 3 half-lives is 1/8 of the original activity.

3.2f Penetration properties of radiation

KEY WORD

ion

Learning objectives:

- recall that changes in atoms and nuclei can generate radiation
- recall that atoms can become ions by loss of outer electrons
- recall the differences in the penetration properties of alpha particles, beta particles and gamma rays
- compare the penetration of the different types of nuclear radiation and their ionising power.

Alpha particles, beta particles and gamma rays are all ionising radiations. They have different penetrating properties and ionising powers, and therefore have different potential effects.

Penetration properties

Gamma rays are the most penetrating of these forms of nuclear radiation. They can pass through most materials easily but are absorbed by a thick sheet of lead or by several metres of concrete.

Beta particles can pass through air and paper but are completely absorbed by a sheet of metal, such as aluminium, just a few millimetres thick.

Alpha particles are the least penetrating form of ionising radiation, and can be stopped by a few millimetres of air or by a thin sheet of paper.

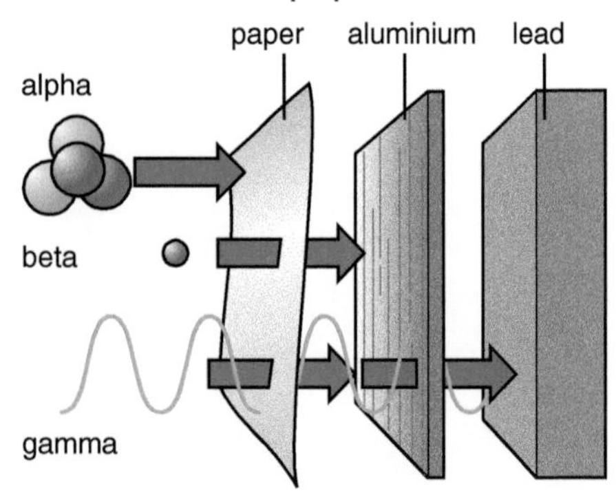

Figure 3.2.13 The three forms of nuclear radiation have very different penetrating powers

1. **Describe the penetrating powers of beta particles.**
2. **Which type of radiation can penetrate a few centimetres of aluminium?**

What is ionisation?

When unstable nuclei disintegrate, or decay, they give out radiation. Ionising radiation has so much energy that, when it interacts with another atom, it can knock out electrons from their energy levels around the nucleus. Because electrons are negatively charged, this leaves the atom with an overall positive charge. We say the atom has become an **ion**.

Ultraviolet waves, X-rays, alpha, beta and gamma rays are all examples of ionising radiation that can turn atoms into ions and break up molecules.

MAKING LINKS

See topic 5 for an introduction to ions.

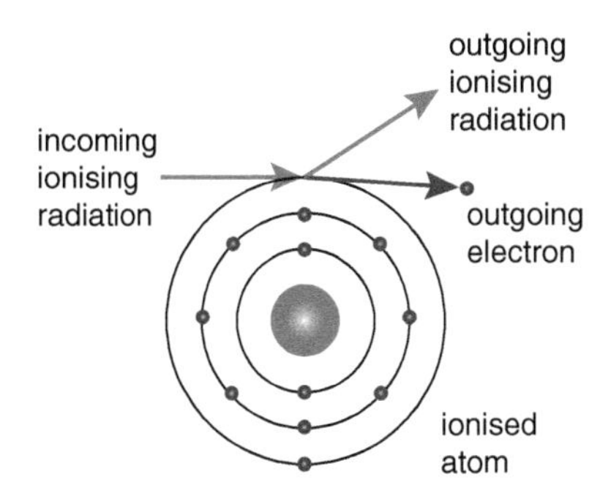

Figure 3.2.15 Ionising radiation has enough energy to knock electrons out of atoms, creating ions

Figure 3.2.14 It is considered safe to use alpha emitters in smoke detectors because they are easily stopped by the air

3. **What is an ion?**
4. **Why does an ion have a charge?**

Ionising powers

Alpha particles have the largest mass and the highest charge, so they have the highest ionising power. They are like the radioactive equivalent of cannonballs, crashing into molecules and splitting bits off. Beta particles are tiny in comparison, with a smaller charge, so they do less damage, and gamma rays are the least ionising.

Ionising radiation causes damage in living cells, because the ions and free electrons created by **ionisation** can go on to take part in chemical reactions that alter molecules vital to life processes.

5. **Explain why beta particles have much lower ionising power than alpha particles.**
6. **Why do you think alpha particles have the lowest penetrating power?**

3.2g Contamination and irradiation

Learning objectives:

- recall the differences between contamination and irradiation effects
- compare the hazards associated with contamination and irradiation.

KEY WORDS

contamination
irradiation

Radioactive materials in the environment, whether natural or artificial, can expose people to risks.

Radioactive contamination

Radioactive **contamination** is the unwanted presence of radioactive atoms. They can be on surfaces or within solids, liquids or gases, including in the human body and on the skin.

If radioactive material contaminates the body it can cause serious problems. Contamination occurs when people swallow or breathe in radioactive materials. They can also enter the body through an open wound or be absorbed through the skin.

Contamination usually results from an accident when radioisotopes are being made or used. Fallout from a nuclear explosion can also cause widespread radioactive contamination of the environment.

Figure 3.2.16 Radioactive sources are marked with this hazard symbol

1. **What do we mean by 'radioactive contamination'?**
2. **Describe how radioactive contamination can occur.**

DID YOU KNOW?

Radioisotopes occur naturally in the environment. Uranium and thorium are present in rocks and soil. Carbon-14, which is present in all living organisms, is continuously created by cosmic rays. Potassium-40 is present in our bodies.

Irradiation

When an object is exposed to radiation from an outside source, we say it is irradiated. **Irradiation** can originate from various sources, including radioactive contamination, nuclear industry wastes and natural sources.

Figure 3.2.17 Massive explosions at the Fukushima nuclear power plant in 2011 led to widespread radioactive contamination

Irradiation can be reduced by screening the source or moving the object away from it. The irradiated object does not become radioactive.

Irradiation with X-rays or gamma rays is used in radiotherapy to treat cancer. In hospitals, food for seriously ill patients is sometimes sterilised using gamma rays. Irradiation can also be used to kill bacteria on foods like soft fruits, so they stay fresh for longer.

KEY INFORMATION

An irradiated object does not become radioactive. A contaminated object could potentially become radioactive.

3. **Describe the process of irradiation.**
4. **Suggest where the low-level irradiation we are exposed to every day comes from.**

DID YOU KNOW?

The most unstable nuclei have the shortest half-lives, but they can give out a lot of radiation in a short time. Nuclei with longer half-lives may give out smaller amounts of radiation, but the dose builds up over a long period.

Hazards of contamination and irradiation

Contamination is a hazard because radioactive nuclei emit harmful ionising radiation as they decay. If a contaminant enters the body, the ionising radiation can severely damage living cells. It is often difficult to remove the contaminant, so it continues to add to the dose for as long as it emits radiation. The seriousness of the hazard depends on the radiation type, and the concentration and half-life of the contaminant.

For example, outside the body, gamma rays present the greatest hazard because they are the most penetrating. But inside the body gamma rays do least harm because they have the lowest ionising power. If you ingest an alpha emitter, however, its ionising power can be lethal.

Alexander Litvinenko was a former Russian Security Service officer, living in the UK. In November 2006, he suddenly became ill, and died three weeks later in hospital. At first, doctors could not work out what was wrong. They suspected gamma radiation poisoning, but tests with a Geiger counter on his skin were negative. Then, just before his death, Litvinenko was tested for alpha emitters. He had swallowed polonium-210, which emits alpha particles, and it had caused devastating damage inside his body. The poison was in Litvinenko's cup of tea.

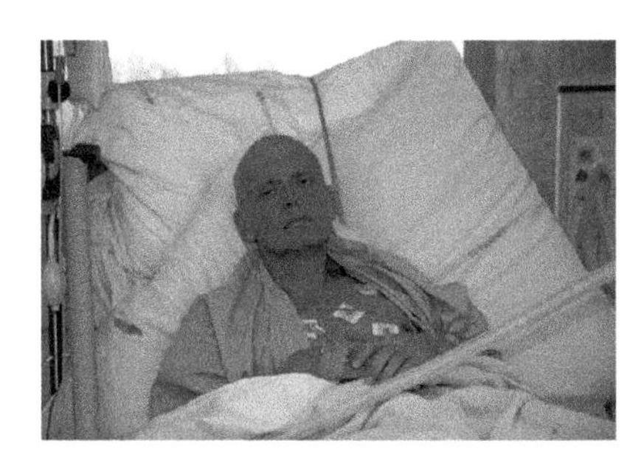

Figure 3.2.18 Alexander Litvinenko died following polonium-210 poisoning. The use of this alpha emitter as a poison had never been known before

Irradiation can also cause ionisation, which damages body cells. It is often easier to remove or screen against a potentially hazardous source of irradiation than a contaminant. For example, precautions are taken to screen health workers against the radioactive sources used in therapeutic irradiation.

5. **Explain the difference between irradiation and contamination, and the hazards they present.**
6. **Why couldn't the radiation be detected by Geiger counters outside Litvinenko's body?**
7. **Alpha particles could be thought of as a 'short-range weapon'. Do you agree? Explain your answer.**

3.2h Risks of ionising radiation

Learning objectives:

- describe how UV, X-rays and gamma rays can have hazardous effects on human tissues
- interpret simple measures of risk showing the probability of harm from radiation
- give examples to show that perceived risk can be very different from measured risk
- describe precautions to reduce the risks from radiation.

KEY WORDS

dose
hazard
mutation
risk

Perceived risk can be very different from real risk, especially if the cause is unfamiliar or invisible. It's important to understand the real hazards and risks from ionising radiation, and how to protect against them.

KEY INFORMATION

A **hazard** is a potential danger in a situation. **Risk** is the probability of that hazard actually causing harm.

Hazards of ionising radiation

When atoms in living cells become ionised, important molecules such as DNA may be changed, altering their function. Sometimes the cell can repair or replace damaged molecules, but if not the cell might die. Alternatively, changes in DNA might cause gene **mutations** that trigger uncontrollable cell division, leading to cancer.

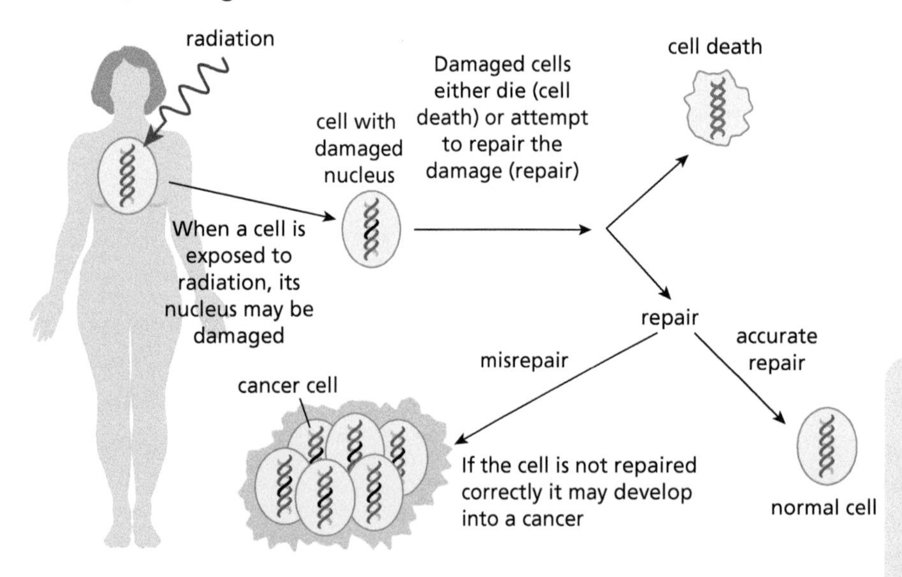

Figure 3.2.19 Possible effects of ionising radiation on body cells

DID YOU KNOW?

Ionising radiation does not affect all cells in the same way. Cells that are reproducing quickly are the most likely to be affected – for example, in a developing fetus.

So, while radiation can be useful, exposure also presents **hazards**. Gamma radiation and X-rays are used in medical diagnosis and treatment, but high doses can produce gene damage and cancer. We need ultraviolet radiation in sunlight to produce vitamin D in the skin, and UV is also used to treat eczema and jaundice. But excessive UV exposure causes sunburn, and can lead to premature skin aging, cataracts and skin cancer.

1. **What is a mutation?**
2. **Describe some hazards of excessive sunbathing.**

Risks and precautions

The hazardous effects of ionising radiation depend on the type of radiation – its penetrating and ionising powers – and the size of the dose. The bigger the effective **dose** from an exposure to radiation, the greater the **risk** of harm to a person. Effective dose takes into account that different radiation types have different effects on different tissues, and is measured in sieverts (Sv) and millisieverts (mSv).

Risk is measured as the probability of harm following exposure to a hazard. So, for example, the risk of getting cancer following a medical scan can be worked out from large amounts of past data on cancer rates after that type of scan.

The radiation dose that nuclear industry workers are exposed to is carefully monitored to ensure that their risk of harm is kept acceptably low. The 'safe' dose is based on the rate of cancer in workers exposed to radiation over many years. Each worker wears a badge containing a radiation-sensitive film. At the end of their shift, the film is used to calculate the radiation dose received.

KEY INFORMATION

The longer a person's exposure to radiation, and the higher the activity of the source, the greater the risk of harm. Reducing radiation dose, by avoiding or screening from radiation sources, reduces the risk of harm.

3 **Explain why the risk of harm from a highly active alpha source could be higher than for a low activity gamma source.**

4 **From Figure 3.2.20, what is the risk of someone getting cancer following a lung scan?**

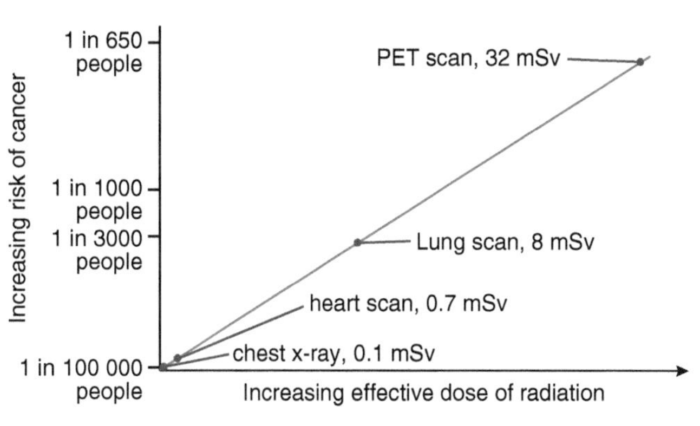

Figure 3.2.20 From a graph like this, if the dose a person has received is known, the risk of them getting a cancer can be found

Measured and perceived risks

People often perceive the risks from different sources of radiation very differently. For example, many people think nuclear power poses a very high risk, but see medical scans as much lower risk. Nuclear power stations might seem distant and threatening, which makes the risk seem worse than from something more familiar, like an x-ray machine in a local hospital, operated by friendly medical staff. The immediate benefits of medical care might also make people feel less concerned about the possible risks.

The real risk may be different from the perceived risk. Based on measured rates of cancer resulting from different sources, for instance, nuclear power is less risky than diagnostic scans.

REMEMBER!

Alpha particles have the most ionising power but the least penetrating power. Gamma rays are the most penetrating but have the weakest ionising power.

5 **Kevin uses a UV sunbed once a week. Gillian works at a nuclear power plant 5 days a week. Suggest who is at greater risk from ionising radiation. What data could help assess the risk?**

6 **Flight crew have approximately twice the measured risk of skin cancer compared with the general population. Suggest possible reasons for this.**

3.2i Cancer

Learning objectives:

- describe cancer as uncontrolled cell division and growth resulting from changes in cells
- identify risk factors for cancer
- describe the differences between types of tumours.

KEY WORDS

benign
carcinogen
malignant
tumour

In the UK, over 33% of people will develop a form of cancer during their lifetime. Certain risk factors increase the chance of a person getting cancer, because they can cause changes in cells that lead to uncontrolled growth.

Cancer facts

Sometimes cell division can become uncontrolled. This causes **tumours** to form. Tumours can be harmless and **benign**, or **malignant** and cancerous.

Doctors cannot explain why only some people develop cancer. Research shows that certain risk factors increase the chance of a person developing cancer. These include:

- smoking
- obesity
- some common viruses
- UV light and other ionising radiation
- age
- possible genetic causes.

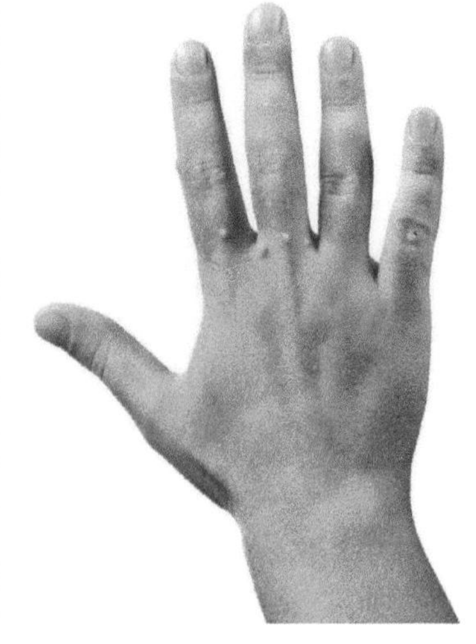

Figure 3.2.21 Warts are benign tumours

1. **How do tumours form?**
2. **Suggest which risk factor may increase the chance of:**

   a skin cancer **b lung cancer.**

Cancer in detail

Benign tumours grow slowly, do not spread and are harmless. Warts are benign tumours. They can be removed by simple surgery.

In malignant tumours, cells divide very quickly. They can invade tissues around them and spread via the blood to different parts of the body, where they form secondary tumours. Malignant tumour cells are cancers.

Substances or viruses that increase the risk of cancer are called **carcinogens**; for example:

- tar in tobacco smoke
- asbestos
- human papilloma viruses (HPV).

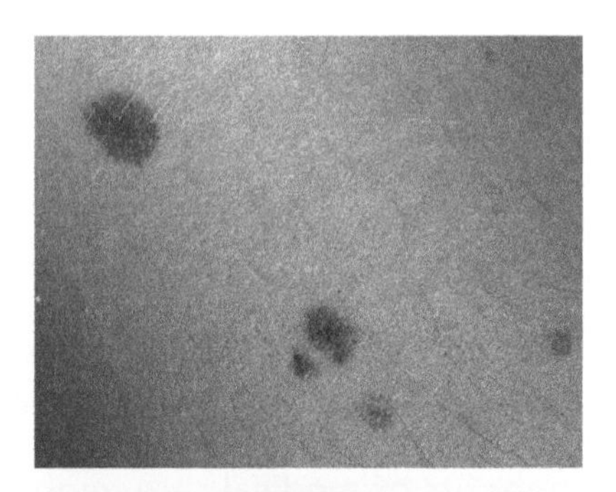

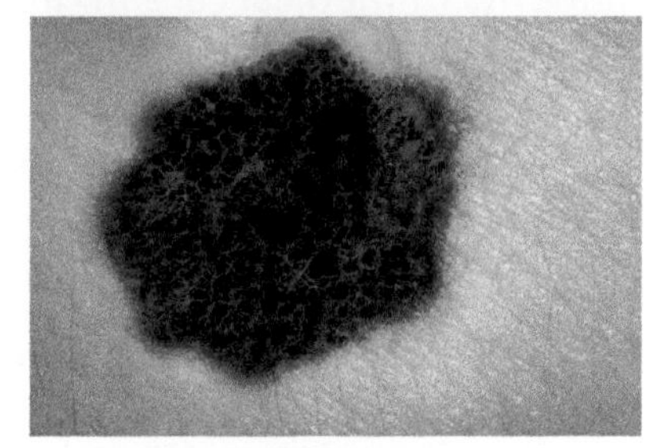

Figure 3.2.22 How are benign moles, as shown in the upper image, and malignant moles, as shown in the lower image, different?

 Link: 3.2i ⟶ 3.3a

Some risk factors that are associated with cancer cannot be changed, for example, aging and family history. Ways to reduce other risk factors include:

- not smoking
- staying out of the sun and using sunscreen
- drinking alcohol in moderation
- healthy diet and exercise.

3 Explain how benign and malignant tumours differ.

4 Describe some lifestyle changes to reduce the risk of cancer.

DID YOU KNOW?

Breast, lung, prostate and bowel cancers are the most common in the UK. They account for 50% of diagnosed cancers.

Causes and treatments

Usually, cells produce signals to control how much they grow and divide, to keep the body healthy. Cancer starts when a change called a mutation happens in the genes which means the cell no longer understands its instructions and starts to multiply out of control, forming a tumour.

The mutation could mean that too many proteins are made that trigger cell division, or that too few proteins are produced that normally stop it.

Mutations can happen by chance, or be triggered by carcinogens or ionising radiation. Genes get damaged every day and cells are good at repairing them, but over time, the damage may build up.

Cancer cells can be destroyed by exposure to extremely large amounts of radiation. This is called radiotherapy. Cobalt-60 emits high-energy gamma rays and is widely used to treat cancers. X-rays are often preferred to gamma rays, because their production is more easily controlled.

KEY INFORMATION

Remember: malignant tumours cause cancer, whereas benign ones do not.

KEY INFORMATION

Cancer starts with changes in one cell or a small group of cells. Where the cancer starts is called the primary tumour.

DID YOU KNOW?

About six separate mutations have to happen before a normal cell turns into a cancer cell.

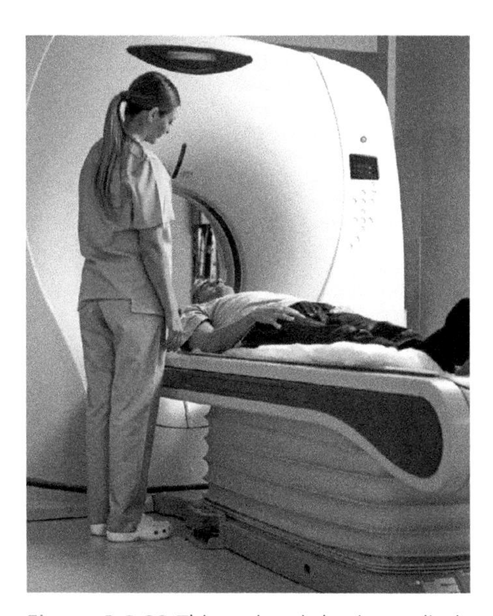

Figure 3.2.23 This patient is having radiotherapy to treat cancer

5 Describe how cancer begins.

6 Explain how ionising radiation can be both a cause and a treatment of cancer.

Check your progress

You should be able to:

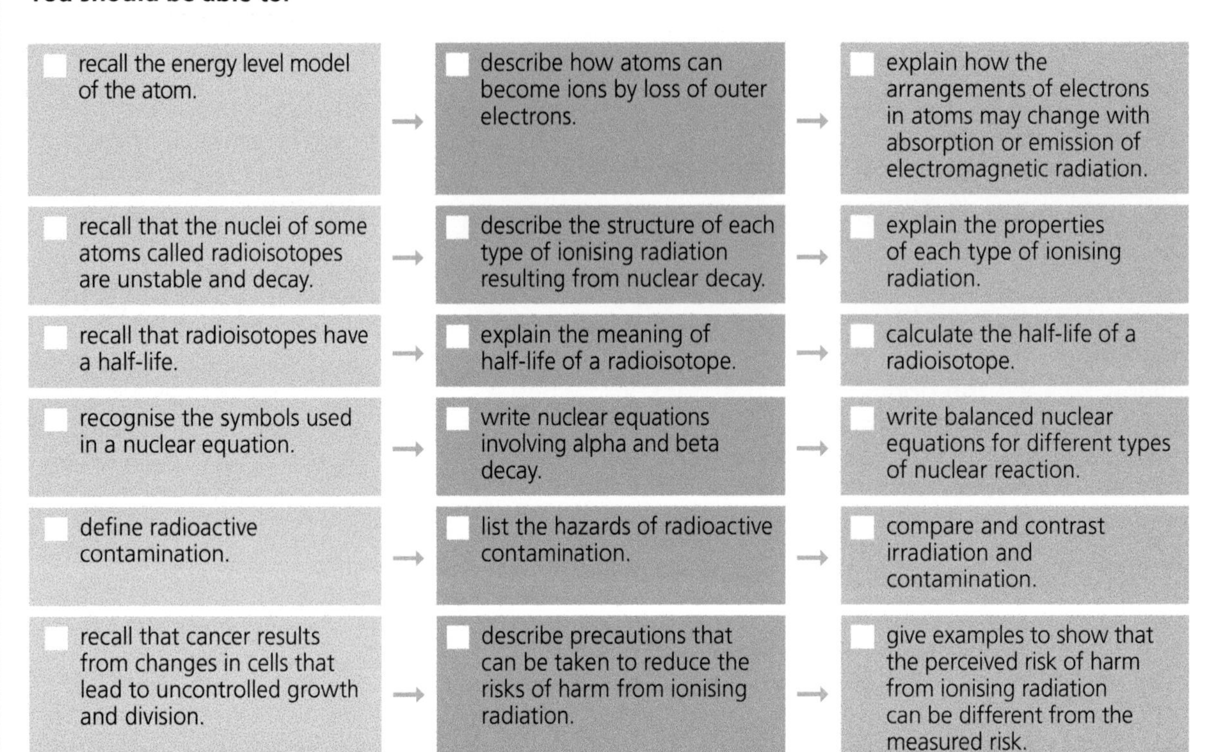

recall the energy level model of the atom.	→	describe how atoms can become ions by loss of outer electrons.	→	explain how the arrangements of electrons in atoms may change with absorption or emission of electromagnetic radiation.
recall that the nuclei of some atoms called radioisotopes are unstable and decay.	→	describe the structure of each type of ionising radiation resulting from nuclear decay.	→	explain the properties of each type of ionising radiation.
recall that radioisotopes have a half-life.	→	explain the meaning of half-life of a radioisotope.	→	calculate the half-life of a radioisotope.
recognise the symbols used in a nuclear equation.	→	write nuclear equations involving alpha and beta decay.	→	write balanced nuclear equations for different types of nuclear reaction.
define radioactive contamination.	→	list the hazards of radioactive contamination.	→	compare and contrast irradiation and contamination.
recall that cancer results from changes in cells that lead to uncontrolled growth and division.	→	describe precautions that can be taken to reduce the risks of harm from ionising radiation.	→	give examples to show that the perceived risk of harm from ionising radiation can be different from the measured risk.

Worked example

1 What is an alpha particle?

A helium atom.

Not strictly correct. It is a helium *nucleus*. It would also be helpful to give more detail of its characteristics like its charge, mass, ionising power and penetrating power.

2 Explain what you understand by the half-life of a radioisotope.

The half-life is the time it takes to lose half its size.

A good start, but it needs to be clearer and fuller in the explanation. The answer should refer to the activity of the radioisotope, not Its size. And don't forget to mention average time.

3 What is the difference between irradiation and contamination?

Contamination happens when a radioactive source gets on or inside a living organism or a material in the environment. Irradiation is where an object is exposed to nuclear radiation.

Good answer. You could also give examples of radioactive contamination and irradiation.

4 An experiment was carried out where the activity of a radioisotope was measured over time. The results are shown on the graph.

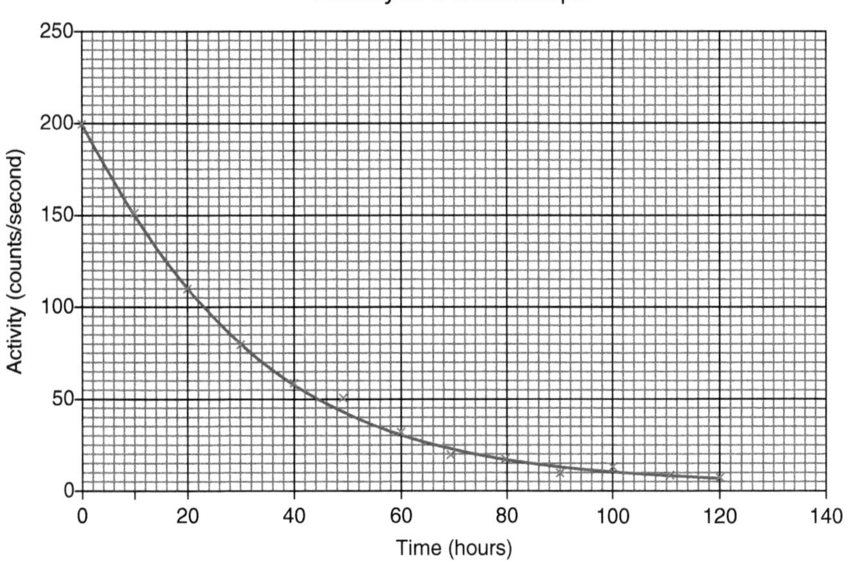

a Not all the plotted points fall exactly on the curve. Why is this?

Radioactive decay is random, so you can't say exactly how many nuclei will decay in each second. But if you have a large enough sample, you can say how many will decay on average. This is what the 'curve of best fit' shows.

A good, full answer.

b Explain how you would work out the half-life from the graph.

Pick a value, say 100 counts/per second and halve it, 50 counts/second.

Draw a line from 100 counts/second on the vertical axis across to the graph, then down to the time axis.

Repeat for 50 counts/second. Subtract the first time from the second time to get the half-life.

This gives a good answer for calculating one value of the half-life, but you should aim to calculate at least one more value from another part of the graph, e.g. 150 counts/second and 75 counts/second and calculate the average.

End of chapter questions

Getting started

1. Describe the structure of an atom using a diagram. — 1 Mark
2. State the three types of nuclear radiation. — 2 Marks
3. Define the half-life of a radioactive element. — 1 Mark
4. Radioactive decay is random. What does this mean? — 2 Marks
5. Describe the risks and advantages of using X-rays and gamma rays in medicine. — 2 Marks
6. A technician has a choice of three radioisotopes that can be used in a smoke alarm. The half-lives of the isotopes are 4 days, 4 years and 400 years. Which one should the technician use and why? — 2 Marks

Going further

7. What is the atomic number of an element? — 1 Mark
8. Name an instrument used to measure activity of a radioactive source. — 1 Mark
9. Explain the difference between the penetrating power and the ionising power of nuclear radiation. — 2 Marks
10. Sodium can be represented by the notation $^{23}_{11}Na$.
    a What are the numbers 23 and 11 and what do they stand for? — 1 Mark
    b Define what a radioisotope is. — 1 Mark

More challenging

11. Iodine-131 has a half-life of 8 days. Carbon-14 has a half-life of 5715 years.
    a Explain which isotope is better suited to use as a medical tracer. — 2 Marks
    b Evaluate the risks and advantages of using radioisotopes as medical tracers. — 2 Marks
12. What is a genetic mutation? — 1 Mark
13. State the difference between irradiation and contamination. — 1 Mark
14. $^{219}_{86}Rn$ decays to $^{x}_{y}Po$ by emitting an alpha particle.
    Write a balanced nuclear equation for the decay. — 2 Marks
15. The activity of a radioactive sample took 60 minutes to decrease from 400 to 50 counts/second. Work out its half-life. — 2 Marks

16 **Figure 3.2.24 shows an idealised graph of activity against time for a radioisotope.**

a Describe how you would expect a graph to differ from this if you measured the activity of the same radioisotope over time. 2 Marks

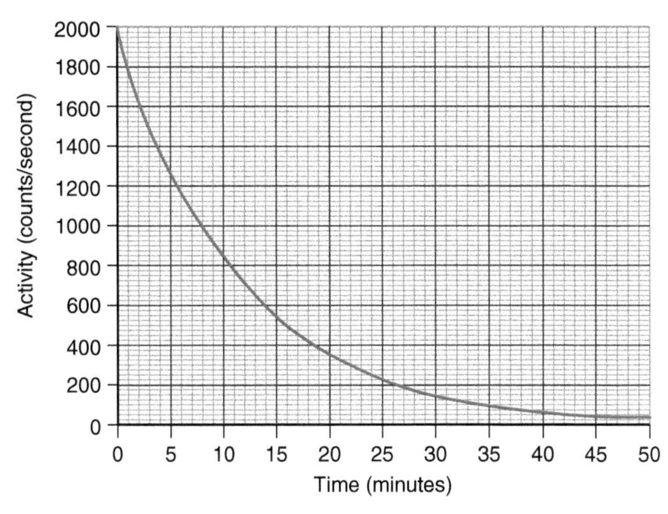

Figure 3.2.24

b Calculate the half-life for this sample. 2 Marks

17 **Draw up a table to compare the penetration and ionising properties of alpha and beta particles and gamma rays.** 3 Marks

Most demanding

18 **Explain how an emission spectrum can be used as a 'fingerprint' to help identify elements in an unknown sample.** 3 Marks

19 **Explain why irradiation from an alpha source carries a far smaller risk of harm than contamination of food by the same source.** 4 Marks

40 Marks

PREVENTING, TREATING AND CURING DISEASES

IDEAS YOU HAVE MET BEFORE:

BACTERIA ARE SINGLE-CELLED LIVING ORGANISMS

- Diseases have symptoms that stop the body from working as normal.
- Communicable (infectious) diseases are caused by microorganisms called pathogens, which include bacteria.
- Bacterial (prokaryotic) cells are smaller than eukaryotic cells. They have cytoplasm and a cell membrane surrounded by a cell wall.
- Pathogens may infect plants as well as animals and are spread by direct contact, water or air.

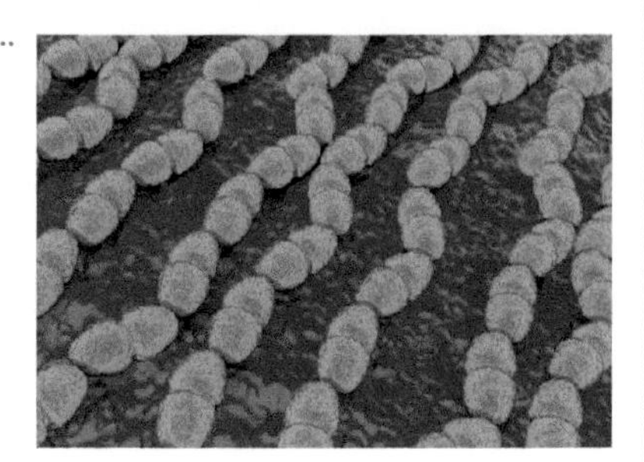

ORGANS WORK TOGETHER IN SYSTEMS TO PERFORM CERTAIN FUNCTIONS

- Air enters the respiratory system through the nose.
- The trachea and bronchi are organs in the respiratory system.
- The stomach produces hydrochloric acid.
- The skin covers the outside of the body.
- Blood is the transport system in the body; red blood cells transport oxygen from the lungs to all cells.

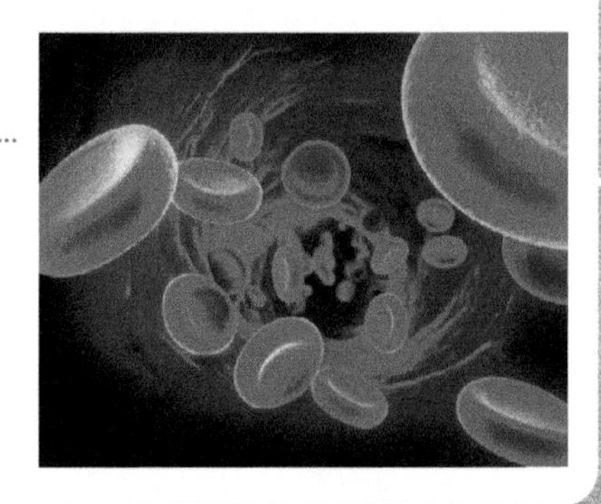

3.3

IN THIS CHAPTER YOU WILL FIND OUT ABOUT:

HOW ARE COMMUNICABLE DISEASES SPREAD?

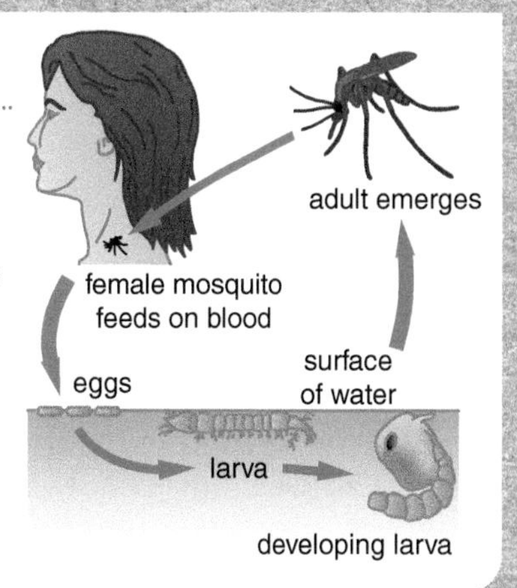

- Pathogens that cause disease in plants and animals include bacteria, viruses, fungi and protists.
- Toxins and cell damage caused by pathogens make us feel ill.
- Understanding the life cycles of some pathogens allows us to control the spread of disease.

HOW DO WE CONTROL AND TREAT DISEASE?

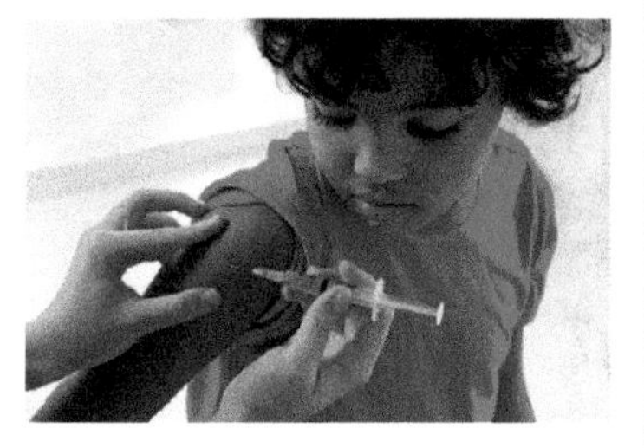

- The skin, nose, respiratory system and stomach all protect us from pathogens.
- The immune system is our major defence system against disease.
- White blood cells protect us from bacterial infections in a number of ways.
- Vaccination protects us from viral and bacterial pathogens.
- Medicines such as antibiotics and painkillers help us to control the spread of disease and relieve symptoms.
- Organisms can be genetically modified to produce medically useful substances.
- Stem cell treatments offer enormous potential in medicine, but their use is controversial.

3.3a Spread of communicable diseases

KEY WORDS

pathogen
protist
vector

Learning objectives:

- explain how communicable diseases are spread in animals
- know how the common cold, flu, cholera, athlete's foot and malaria are spread.

Communicable diseases are caused by pathogens, which can spread from one organism to another by a variety of means.

What is a communicable disease?

A communicable disease is an illness caused by harmful microorganisms entering an organism's body and damaging tissues. Disease-causing microorganisms are called **pathogens**, and include viruses, bacteria, protists and fungi. As pathogens pass from one organism to another, the disease spreads through the population.

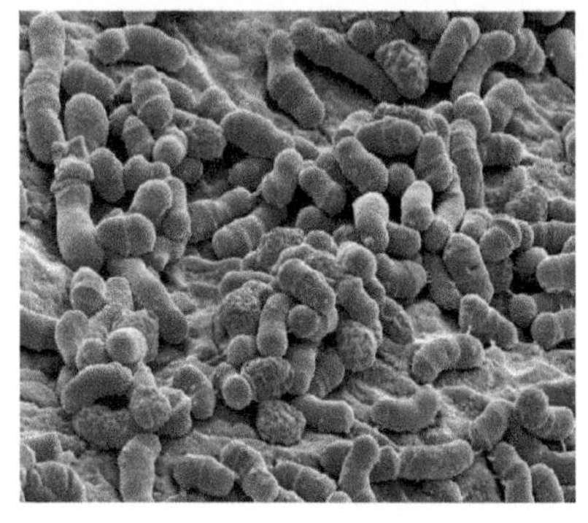

Figure 3.3.1 Athlete's foot is a condition caused by a type of fungus called *Malassezia* yeast. In this scanning electron micrograph (SEM), false colours have been added to make the yeast cells stand out against the skin of a human foot (x6000)

Pathogens can spread:

- through the air – for example, the common cold virus is spread when people cough or sneeze
- through food that is contaminated with bacteria
- through drinking water – for example, cholera is spread in drinking water contaminated with bacteria
- through contact with other people, or surfaces that infected people have touched – many diseases can be spread like this
- by animals that scratch, bite or draw blood – for example, malaria is spread when mosquitoes bite, passing a single-celled organism called a **protist** into the bloodstream.

1. **Define the term 'pathogen'.**
2. **Explain how touching a contaminated surface might spread a disease.**

Figure 3.3.2 Malaria is spread by mosquitoes carrying the plasmodium protist. In 2015, there were 214 million cases of malaria worldwide and 438 000 people died

Pathogens in detail

Bacteria and viruses can reproduce rapidly in the body.

- Bacteria produce toxins that damage tissues and make us feel ill.
- Viruses live and reproduce inside cells, damaging them.

The toxins released by pathogens cause the symptoms of infection. These include high temperatures, nausea, headaches and rashes. People get a communicable disease because they receive the pathogen from someone who has it. Measures to prevent or reduce the spread of diseases include:

- simple hygiene, such as covering your mouth when coughing, using a handkerchief when sneezing and washing your hands after using the toilet
- isolation of infected individuals
- destroying **vectors** – for example, mosquitoes are the malaria vector
- vaccination.

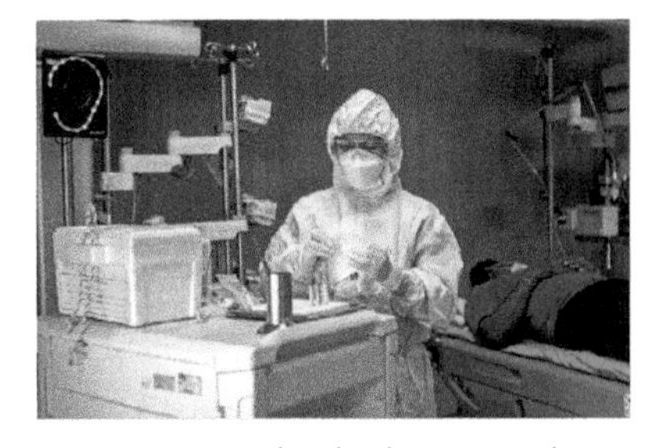

Figure 3.3.3 Why do doctors and nurses wear protective clothing?

3 Why do pathogens make us feel ill?

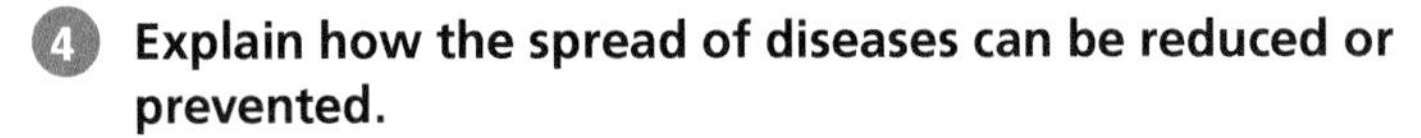

4 Explain how the spread of diseases can be reduced or prevented.

A case study

Ebola haemorrhagic fever (EHF) is a deadly virus: 50–90% of people infected with EHF die. The last outbreak of EHF began in December 2013, in Guinea, and spread to Nigeria, Sierra Leone and Liberia.

Fruit bats carry EHF and may infect animals and humans. The disease is caught by direct contact with body fluids (blood and saliva) of an infected individual. Symptoms include fever, headache, diarrhoea, nausea and rashes. Currently, there is no cure for EHF but two vaccines are under development.

Figure 3.3.4 Fruit bat meat is a delicacy in West Africa

Prevention measures include:

- reducing the risk of contact with infected animals
- wearing protective clothing
- washing hands frequently
- isolation of infected people and safe burials of the dead
- travel restrictions.

5 Evaluate the control measures for EHF.

DID YOU KNOW?

In 1918, Spanish influenza killed 40–50 million people.

REMEMBER!

Pathogens come in many different forms, and each one causes a different disease.

3.3b Viral diseases

Learning objectives:

- describe the symptoms and transmission of some viral diseases
- explain applications of science to prevent the spread of some viral diseases.

KEY WORDS

antiretroviral drug
vaccination

Viral diseases are caused by viruses, which invade body cells.

MAKING LINKS

See topic 4.3 to find out more about genetics.

Viral diseases

Viruses are very small pathogens. They are not living cells. They have a strand of genetic material inside a protein coat. When the virus gets inside a host cell, it uses the cell's own 'machinery' to make copies of its genetic material and proteins, forming many new viruses. Then the cell bursts and dies, and the new viruses are released.

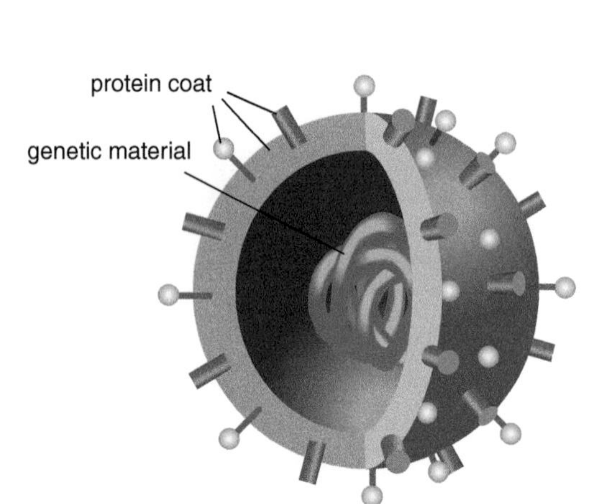

Figure 3.3.5 This diagram shows the basic structure of a virus, such as a flu virus. It is only about 100 nm across – about 1000 times smaller than a human egg cell

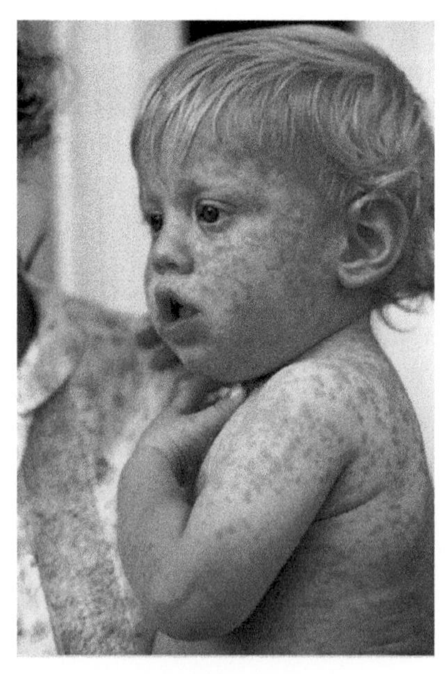

Figure 3.3.6 Measles is a serious disease caused by a virus

Examples of viral diseases are:

- the common cold. The cold virus frequently mutates, so you can catch a cold many times. The symptoms are always similar, and usually include a runny nose, sore throat and cough.
- influenza, or flu. Again, new strains of the flu virus arise often. Symptoms include aching muscles, fever, headache, fatigue, sore throat, runny nose and sometimes nausea. Flu can be fatal, especially in people who are already frail.
- measles, a serious illness. Symptoms include fever and a red rash over the skin. It can be fatal if there are complications.
- HIV, which initially causes a flu-like illness.

1. **How do viruses reproduce?**
2. **Describe the symptoms of two viral diseases.**

Spread and control

Viruses spread and are controlled in various ways.

Virus	Mechanism for spread	Control
Measles	**Droplet infection** Talking, coughing and sneezing all cause the expulsion of tiny droplets into the air. Inhaling droplets that carry viruses causes measles to spread.	Most young children have **vaccinations** to protect them against this possibly fatal disease.
HIV	**Direct contact** HIV is spread by sexual contact or exchange of body fluids such as blood. This can occur if drug users share needles.	**Antiretroviral drugs** are prescribed to stop the virus entering the lymph nodes.
Flu	**Droplet infection** As for measles, the flu virus is mainly spread through tiny droplets made when infected people cough, sneeze, or talk. These droplets can be inhaled by people nearby, or people may touch surfaces where they have landed and then take in the virus when they touch their eyes or mouth.	Vaccination is becoming more common, and is recommended for certain people at higher risk. New vaccinations are developed regularly to combat the latest new strain of flu virus.

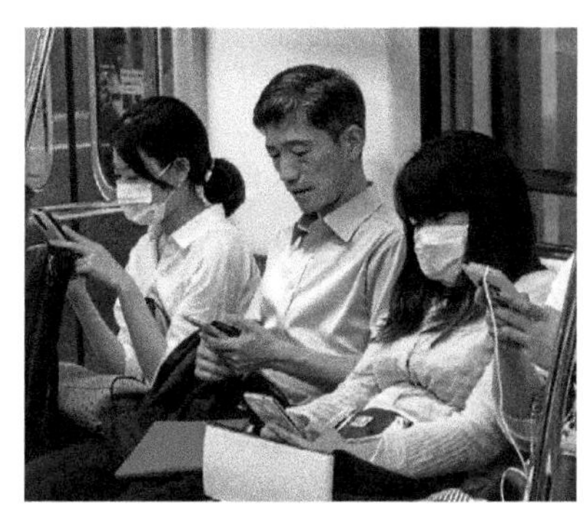

Figure 3.3.7 Suggest why these commuters are wearing masks

3. **Describe how HIV is spread and suggest control measures.**
4. **Describe how measles is spread and controlled.**

Counting the cost

Viral diseases are harder to treat than bacterial diseases. Bacteria are living cells that survive outside body cells. They can be treated with prescribed drugs. Viruses are only found inside host cells, where they are protected from drugs. This is why we are vaccinated against some viral diseases.

HIV is very difficult to control. If it is not successfully controlled by antiretroviral drugs, the virus enters the lymph nodes and attacks the body's immune cells. Late-stage HIV, or AIDS, occurs when the body's immune system is no longer able to deal with other infections or cancers.

5. **Explain how HIV leads to AIDS.**
6. **Why is it harder to treat viral diseases than bacterial diseases?**
7. **Compare the mechanisms for spreading measles and HIV. Explain which disease is most infectious.**

DID YOU KNOW?

Each type of virus attacks a specific cell. For example, the measles virus attacks skin and sensory nerve cells.

KEY INFORMATION

Remember: viruses are not living things. They have many different shapes (just one of which is shown in Figure 3.3.5).

3.3c Bacterial diseases

Learning objectives:

- describe the symptoms and transmission of some bacterial diseases
- explain applications of science to prevent the spread of some bacterial diseases.

KEY WORDS

diarrhoea
gonorrhoea
Salmonella

Some pathogens are bacterial. Bacteria can reproduce very quickly in our bodies.

Bacterial diseases

Many bacteria are not harmful and some are actually very useful. We use them to make cheese and yoghurt, to break down our waste and make medicines. Some bacteria, however, are pathogens and cause diseases.

Examples of bacterial diseases are:

- ***Salmonella***, which causes food poisoning. A build-up of toxic bacterial waste products causes symptoms that include:
  - fever
  - abdominal cramps
  - vomiting
  - **diarrhoea.**
- **Gonorrhoea**, a sexually transmitted disease. The toxic bacterial products cause symptoms that include:
  - a thick yellow or green discharge from the vagina or penis
  - pain when urinating.

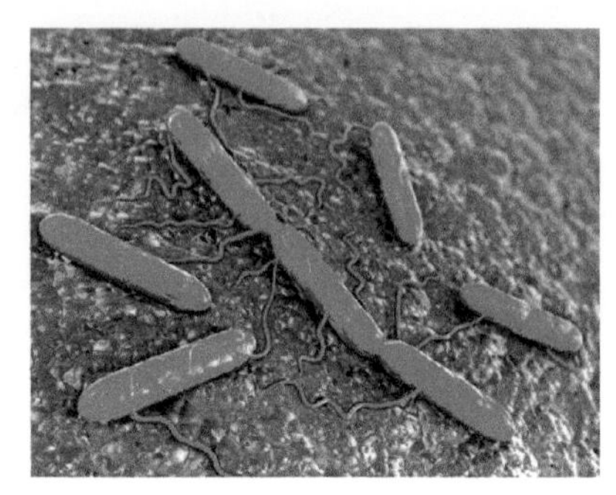

Figure 3.3.8 Scanning electron micrograph (SEM) of *Salmonella* bacteria, which cause food poisoning and are 2–5 μm long and 0.7–1.5 μm in diameter

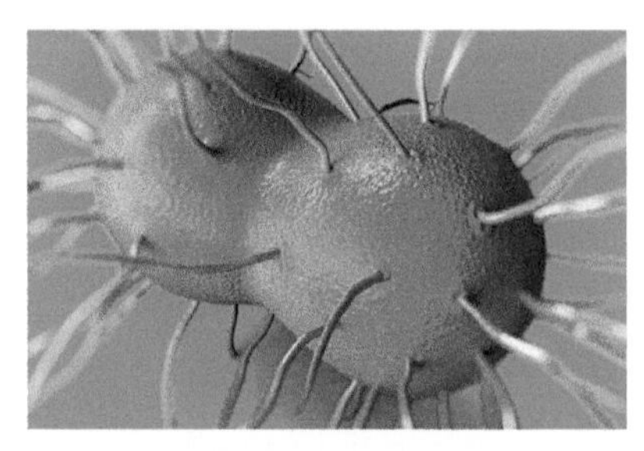

Figure 3.3.9 SEM of a pair of gonorrhoea bacteria, each of which is between 0.6 and 1.0 μm in diameter. Like the *Salmonella* bacteria in Figure 3.38, these have had false colours added to make the details show up better

1. **Name some bacterial diseases.**
2. **Describe the symptoms of *Salmonella* and gonorrhoea.**

Spread and control

Bacteria are spread and controlled in various ways.

Salmonella is spread by:

- ingesting (eating) food that is contaminated with *Salmonella* bacteria
- preparing food in unhygienic conditions; for example, using contaminated knives or chopping boards.

In the UK, the spread of *Salmonella* is controlled by vaccinating poultry against the bacterium. Food should be prepared in hygienic conditions and cooked thoroughly. Washing hands before and after food preparation and after using the bathroom also help.

Figure 3.3.10 Why must food be cooked thoroughly?

Gonorrhoea is spread by sexual contact. The spread can be controlled by:

- treatment with antibiotics. In the past, the disease was easily treated with the antibiotic penicillin, but in recent years many resistant strains of the bacteria have appeared
- use of a barrier method of contraception, such as a condom, to prevent contact.

3 **Describe how *Salmonella* is spread.**

4 **Explain how gonorrhoea can be controlled.**

MAKING LINKS

See topic 4.4d for more information about antibiotic-resistant bacteria.

Symptom delay

People with communicable infections do not develop symptoms as soon as they are infected with the pathogen. There are distinct stages of infection:

- The pathogen enters an organism.
- The pathogen reproduces rapidly in ideal conditions to increase numbers. This is the incubation period.
- Pathogens make harmful toxins, which build up. The more bacteria that are present, the quicker the toxins build up.
- Symptoms develop, for example, fever and a headache.

The graph shows *Salmonella* cases over a period of 12 months.

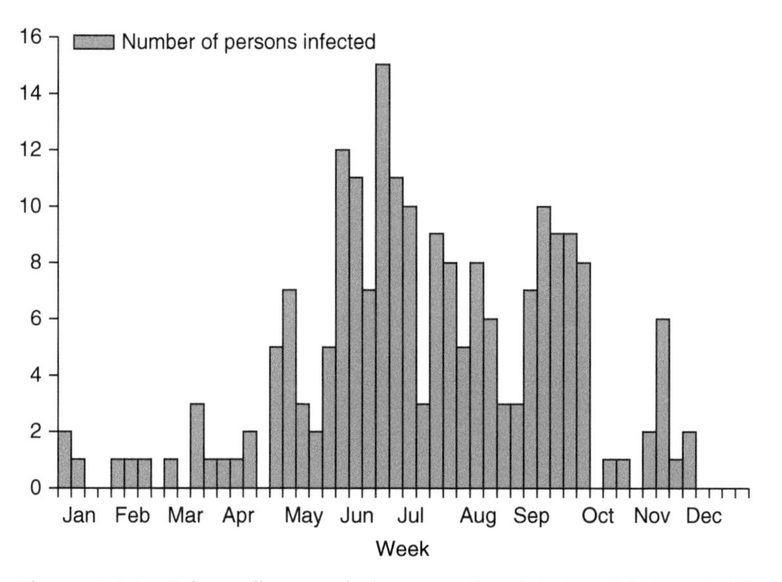

Figure 3.3.11 *Salmonella* cases during an outbreak in two US states in 2011. This outbreak was caused by a batch of infected chicken livers, which were sold for home consumption and also to make processed foods that were sold and eaten later. Salmonella symptoms arise within 1 to 4 days of infection

5 **Describe the pattern of *Salmonella* cases shown in the bar chart and suggest explanations for it.**

6 **Explain how knowledge of the spread of bacterial diseases can lead to their control.**

7 **Compare and contrast bacterial and viral diseases.**

DID YOU KNOW?

Milk has been routinely pasteurised to eliminate bacteria such as *Salmonella* for many decades. Now you can also buy eggs pasteurised in the shell, to use in uncooked or only lightly cooked dishes.

DID YOU KNOW?

In 2015 there was an outbreak of a highly drug-resistant strain of gonorrhoea bacteria in the UK.

KEY INFORMATION

Remember: bacteria are living things, unlike viruses. Bacterial diseases can be treated with antibiotics, unlike viral diseases.

3.3d Defence against pathogens

Learning objectives:

- describe and explain how the human body defends itself against pathogens.

KEY WORDS

cilia
goblet cells

The body has defence mechanisms for each way that pathogens are transmitted.

How does the body defend itself?

Millions of pathogens are around us each day, but your body protects you from being infected:

- Your skin acts as a barrier and produces antimicrobial secretions via glands in the skin.

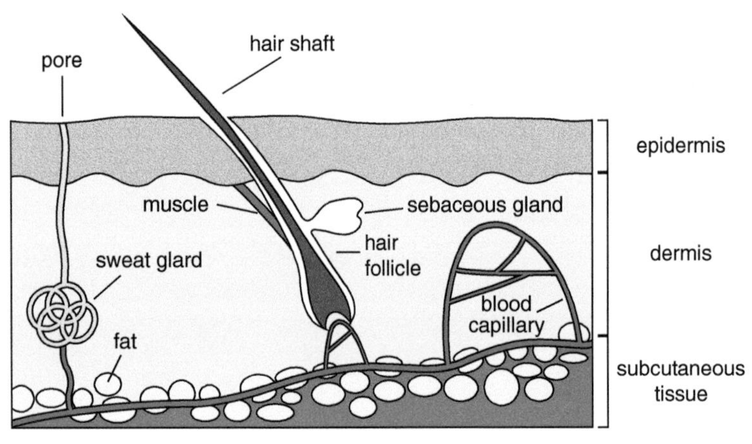

Figure 3.3.12 A schematic diagram, not to scale, drawn to show the key features of the skin

- The nose traps particles that may contain pathogens.
- Your trachea and bronchi secrete mucus, which traps pathogens.
- The stomach produces acid, which kills the majority of pathogens that enter via the mouth.
- Platelets (cell fragments in your blood) start the clotting process at wound sites. Clots dry to form scabs which seal the wound.

1. **How does the body protect itself from pathogens?**
2. **Which of these defence mechanisms will protect you against pathogens in water and in food?**

Hygiene and wounds

When you cough or sneeze, thousands of tiny drops of liquid are sprayed into the air. If you have a disease, the droplets may contain pathogens. This is why you should use a paper tissue

Link: 3.3d ⟶ 3.3e

and then put it in the bin. Coughing up phlegm and then spitting it out can also spread infections to those around you. Children playing may fall on it and any pathogens can enter the body through wounds in the skin.

When the skin is cut, platelets in the blood are exposed to the air at the wound site. They make protein fibres (fibrin) that form a mesh over the wound. The platelets and red blood cells get caught in the fibres to form a clot (Figure 3.3.13).

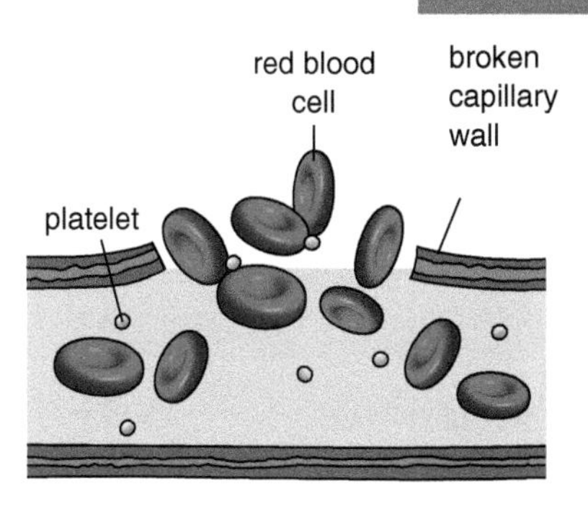

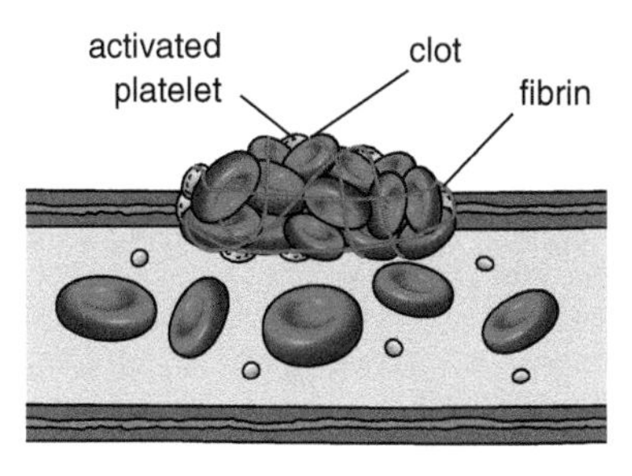

Figure 3.3.13 A schematic diagram to show how wounds are sealed

3 Explain how coughing and sneezing spread pathogens.

Defence mechanisms in detail

Skin protects the body from physical damage, infection and dehydration. The outer layer of skin cells is dry and dead. Pathogens cannot easily penetrate these dead cells. Sebaceous glands in the skin produce antimicrobial oils.

Every time you breathe, you take in many microbes. Hairs in the nose trap larger microbes and dust particles. The trachea and bronchi have a thin lining of tissue with lots of ciliated cells. **Cilia** are tiny hair-like structures. **Goblet cells** in the lining produce mucus; mucus traps smaller dust particles and microbes. Cilia beat together to waft mucus to the back of the throat, where it is swallowed. Ciliated cells have many mitochondria to supply the energy needed to do this.

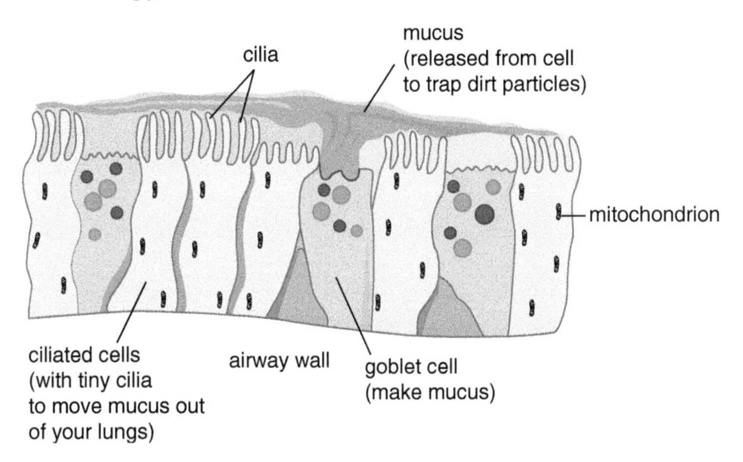

Figure 3.3.14 A schematic diagram, with colours added for clarity, to show the ciliated cells and goblet cells lining the inside of the trachea and bronchi

DID YOU KNOW?

It has been estimated that a sneeze expels mucus at 160 km/h.

REMEMBER!

Cilia are like hairs; they are not cells and cannot be killed.

4 How does skin act as a defence against pathogens?

5 Explain the adaptations of the respiratory system to protect us against pathogens.

6 Smoking can damage and paralyse the cilia. Explain why smokers are more susceptible to respiratory infections.

3.3e The human immune system

Learning objectives:

- explain how the human immune system defends against disease using phagocytosis, antibodies and antitoxins.

KEY WORDS

antibody
antitoxin
immunity
lymphocyte
phagocyte

Although the body's defence mechanisms prevent many pathogens from entering the body, some will succeed. The second line of defence is the immune system.

White blood cells

The immune system recognises and destroys pathogens that enter the body. White blood cells are an important part of the immune system.

They attack invading pathogens. If a pathogen enters the body, white blood cells defend it by:

- ingesting pathogens
- producing antibodies
- producing antitoxins.

There are two main groups of white blood cell: **phagocytes** and **lymphocytes**. Phagocytes can leave the blood by squeezing through capillaries to enter tissues that are being attacked. They move towards pathogens or toxins and ingest them. This is called phagocytosis.

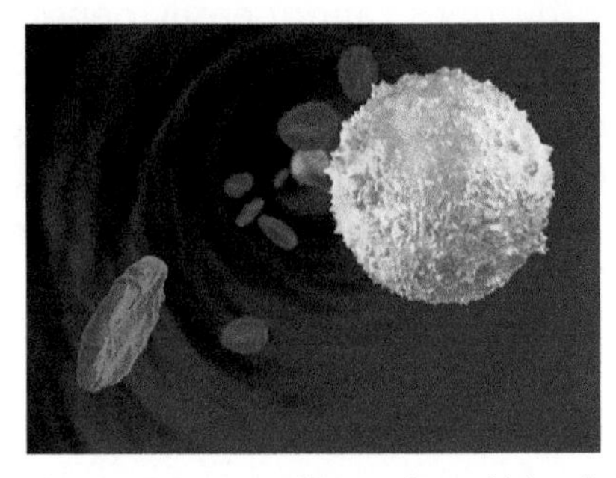

Figure 3.3.15 An illustration of blood cells, with a red blood cell in the foreground on the left and a white blood cell on the right to show their relative size. White blood cells are the largest blood cells – the one pictured here is likely to be 15–30 μm in diameter

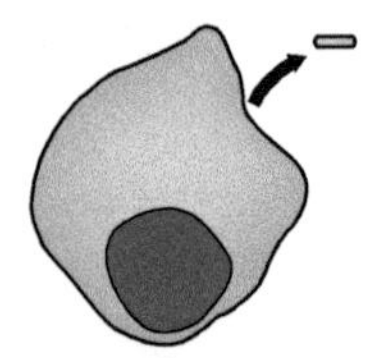

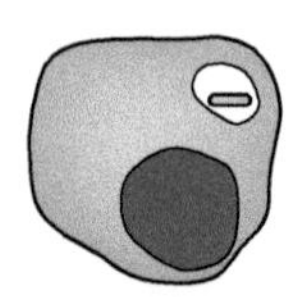

Figure 3.3.16 Phagocytes ingest pathogens during phagocytosis. In this schematic diagram, the phagocytes are shown in blue

1. **How do white blood cells protect the body?**
2. **What happens to a pathogen during phagocytosis?**

Antitoxins and antibodies

Pathogens make us feel ill because they release toxins into our body. White blood cells, called lymphocytes, produce **antitoxins** to neutralise toxins made by the pathogen. Antitoxins combine with the toxin to make a safe chemical. Antitoxins are specific to a particular toxin.

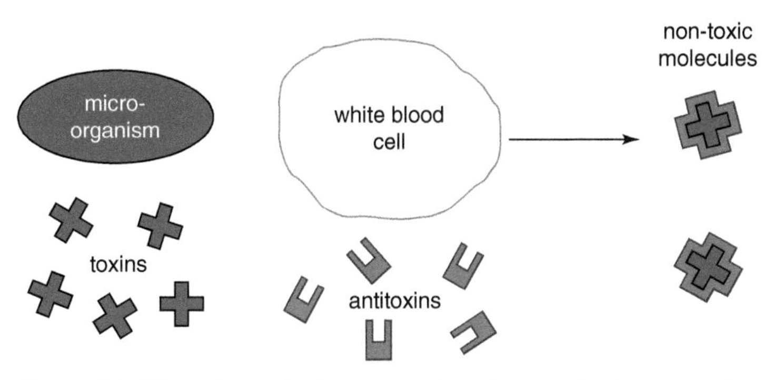

Figure 3.3.17 A schematic diagram to show how antitoxins neutralise toxins made by pathogens. Note that antitoxins and antibodies are large molecules such as protein molecules and so are actually very much smaller than the bacterial cells

Lymphocytes also produce chemicals called **antibodies** that destroy pathogens. Lymphocytes recognise when pathogens are present. They quickly reproduce to make lots of antibodies that:

- cause cell lysis (the pathogens burst)
- bind to the pathogens and destroy them
- cover the pathogens, sticking them together. Phagocytes then ingest them.

If the same type of pathogen enters the body again, lymphocytes recognise it and immediately make lots of antibodies. This is **immunity**: the person is immune to that disease.

3 **How do antitoxins work?**

4 **Explain how antibody production leads to immunity.**

5 **Explain how phagocytosis and antitoxin production protect the body.**

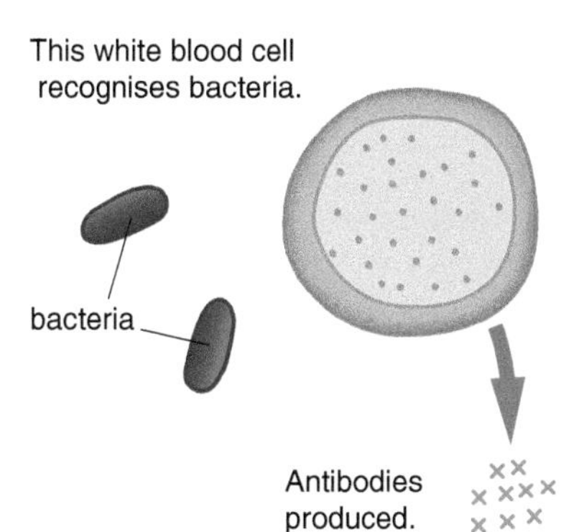

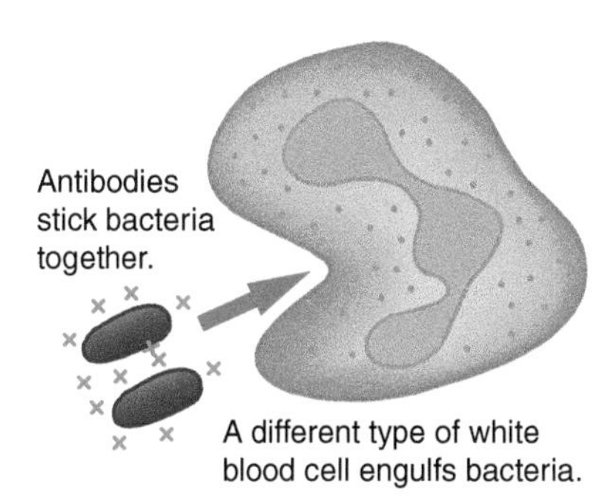

Figure 3.3.18 A schematic diagram to show the principle of how white blood cells protect the body against pathogens. Here the white blood cells are shown in yellow

Producing antibodies

Each lymphocyte has a specific antibody to attack a specific pathogen. Pathogens carry chemicals called antigens on their surface (look back to Figure 3.3.5). The appropriate antibodies lock onto the matching antigens, sticking the pathogens together to destroy them.

HIV attacks the white cells in the blood that help to control the response of the immune system to pathogens. Then the infected person has AIDS. Damage to the immune system by HIV exposes infected people to a variety of diseases that can kill, such as tuberculosis, pneumonia and skin cancer.

6 **Why can a person with AIDS die from a simple infection?**

DID YOU KNOW?

Only 1% of all the cells in blood are white blood cells. There are 5000–10 000 of them in a microlitre of blood.

COMMON MISCONCEPTION

Remember: white blood cells do not *eat* pathogens; they *ingest* them.

3.3f Vaccination

Learning objectives:

- recall how vaccinations prevent infection
- explain how mass vaccination programmes reduce the spread of a disease.

KEY WORDS

immunity
vaccination
vaccine

By understanding the immune system, scientists have been able to develop vaccines to help prevent the spread of disease.

What is vaccination?

A **vaccination** introduces a small quantity of an inactive or dead form of a pathogen into the body to protect us from disease. Lymphocytes produce antibodies to fight the 'infection' but we don't actually become ill. When live pathogens of the same type infect you, your immune system starts to protect you immediately.

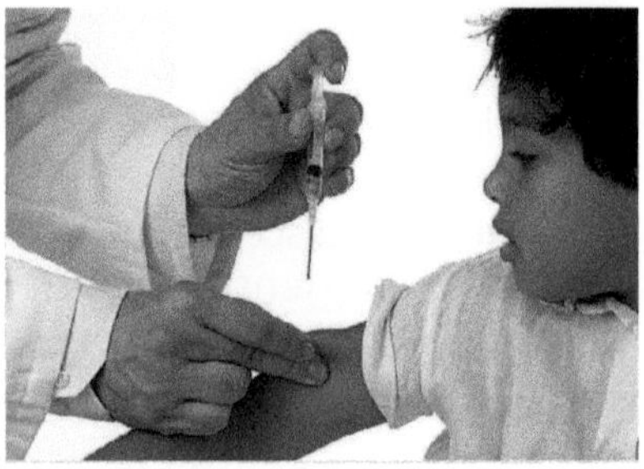

Figure 3.3.19 Suggest why vaccinations are given to children

Different **vaccines** are needed for specific pathogens. For example, polio, whooping cough, flu and HPV (human papilloma virus) all have a different vaccine. MMR vaccinations contain three vaccines (for measles, mumps and rubella).

Vaccinations are usually given to children and people who are going to travel to countries where there is a risk of serious disease. Vaccines are given by injection, orally or via nasal sprays.

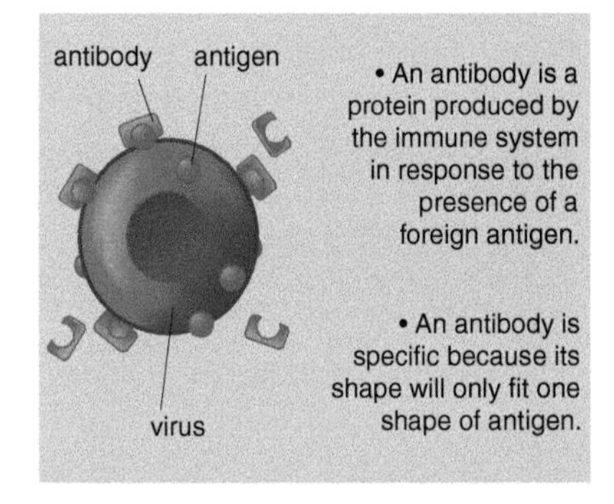

Figure 3.3.20 Antibodies are specific to a particular antigen on a pathogen

1. **What is a vaccination?**
2. **What do vaccinations do?**

How do vaccines work?

After the vaccination:

- lymphocytes detect antigens on the dead or inactive pathogen; they produce a specific antibody for these antigens
- antibodies lock onto the antigens
- lymphocytes 'remember' the shape of the antigens
- when there is a real infection due to a live pathogen entering the body, lymphocytes instantly recognise the pathogen because it has the same antigens as the vaccine
- lymphocytes quickly make many specific antibodies
- antibodies lock onto the pathogens and kill them before they have a chance to make you feel ill. This is **immunity**.

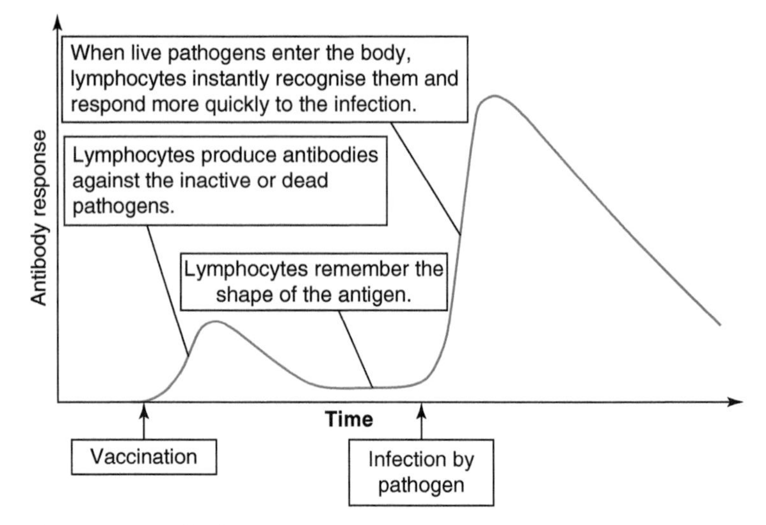

Figure 3.3.21 The immune response

Mass vaccination programmes increase the number of people who are immune to a pathogen, making it difficult for the pathogen to pass to people who are not immunised. If a large proportion of the population is immune to a pathogen, its spread is very much reduced. Global vaccination programmes have eradicated polio, except in Pakistan and Afghanistan. Nigeria did not have any new polio infections in 2015.

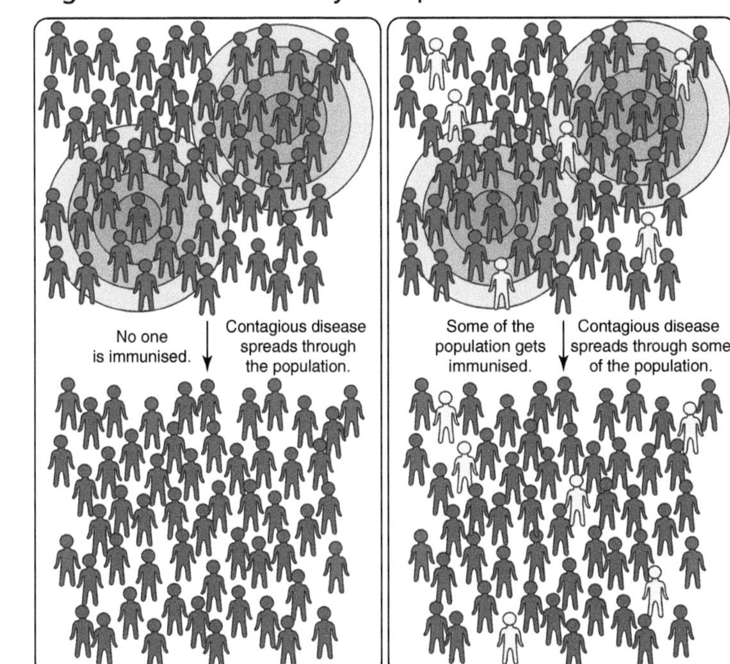

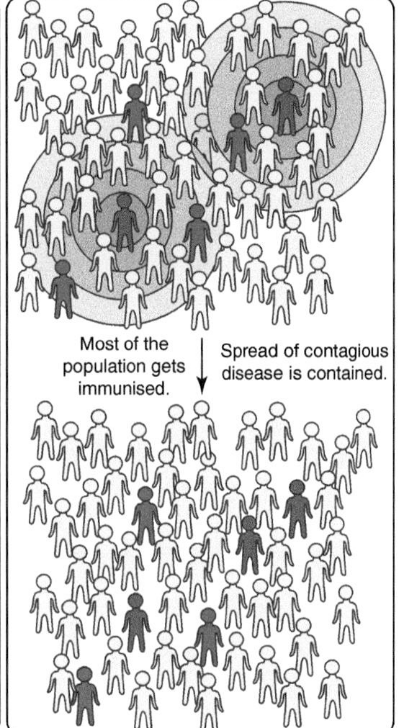

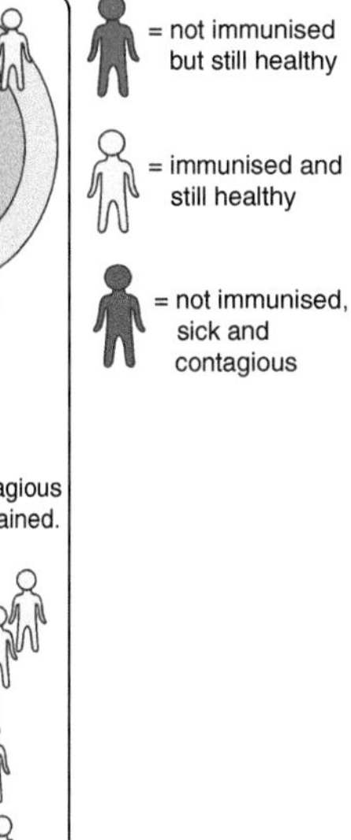

Figure 3.3.22 Mass vaccination programmes reduce the spread of disease

3 **Look at the graph in Figure 3.3.21.**

a **Why were antibodies produced more quickly after the second infection?**

b **Will the person be ill after the second infection?**

4 **How do mass vaccination programmes reduce the spread of disease?**

Vaccination and populations

Some viruses frequently mutate into new strains. This means that lymphocytes do not recognise the new strains. As a result, new vaccines are made regularly for some diseases, such as flu.

Some antibodies give lifelong protection, such as those that are active against measles. Sometimes antibodies fall below the critical level needed for immunity, as happens in the case of tetanus. Booster doses are needed to increase the antibodies again.

5 **Explain why new flu vaccines are made each year.**

6 **Many people now travel widely across the world. Discuss the global importance of vaccination.**

DID YOU KNOW?

One in three people who caught smallpox in the 1600s died from it. Smallpox is now almost eradicated globally.

DID YOU KNOW?

Flu vaccine is stored in eggs because it needs to be in a living organism to survive.

ADVICE

Remember: lymphocytes make *antibodies*. Pathogens have antigens.

3.3g Medicines

Learning objectives:

- explain the use of antibiotics and painkillers in the prevention and treatment of disease
- recall that the emergence of antibiotic-resistant bacteria is a serious threat
- explain that medicines are formulations of mixtures.

KEY WORDS

antibiotic
aspirin
drug
formulation
opiates
penicillin

When you have an infection or communicable disease, there are medicines that can help your body kill the pathogen or relieve symptoms to make you feel better.

Useful drugs

A **drug** is any chemical that alters how the body works. Medicines contain useful drugs. Many medicines do not affect the pathogen that makes you ill. They just relieve the symptoms caused by the infection, for example, painkillers and cough medicines. Other medicines work inside the body to kill bacterial pathogens: these are **antibiotics**, for example, **penicillin**.

Antibiotics work by interfering with the pathogen's metabolism; for example, with processes that make bacterial cell walls. Antibiotics do not affect human cells and they do not kill viral, protist or fungal pathogens. They only kill bacterial pathogens.

Most medicines are mixtures, made by combining ingredients in carefully measured quantities to create **formulations** with the required properties. At least one of the ingredients is the drug – the painkilling chemical or the antibiotic, for example – but other ingredients can be added to make it easier for a patient to take the drug, in solution or as a capsule or tablet.

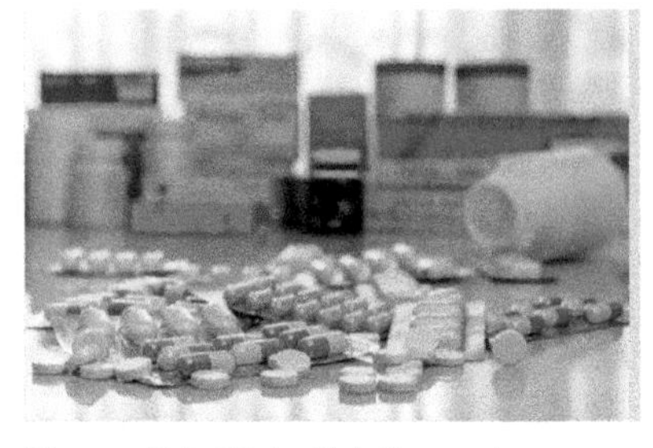

Figure 3.3.23 Antibiotics and painkillers are medicinal drugs

DID YOU KNOW?

Antiviral drugs specific to a particular virus are used to treat viral infections. While antibiotics can kill bacteria, antiviral drugs only slow down viral development. It is difficult to develop drugs that kill viruses without also damaging the body's tissues.

1. **What is a medicine?**
2. **How are painkillers and antibiotics similar and different?**

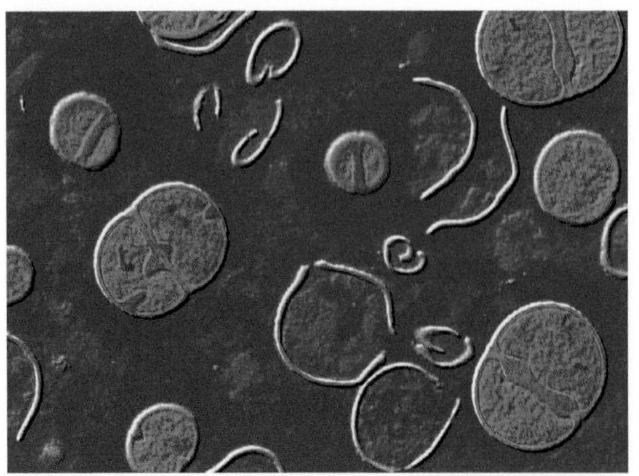

Figure 3.3.24 A false-colour electron micrograph showing whole bacteria and remnants of bacteria that have been affected by antibiotics

How do medicines work?

When you are injured, sensory nerve endings send pain messages to your brain. Painkillers stop these nerve impulses, so you feel little or no pain. Many painkillers are based on two natural drugs:

- **aspirin**, from willow bark
- **opiates**, from poppies.

There are many different antibiotics. Some antibiotics work against one type of bacterial infection. Others are used to treat many different bacterial infections.

A prescribed course of antibiotics has the correct amount to kill the bacterial pathogen completely. This is why you must always take the complete antibiotic course.

Antibiotics cannot be used to kill viral pathogens. This is because viruses live and reproduce inside cells. Antibiotics do not harm body cells and the viruses inside are protected.

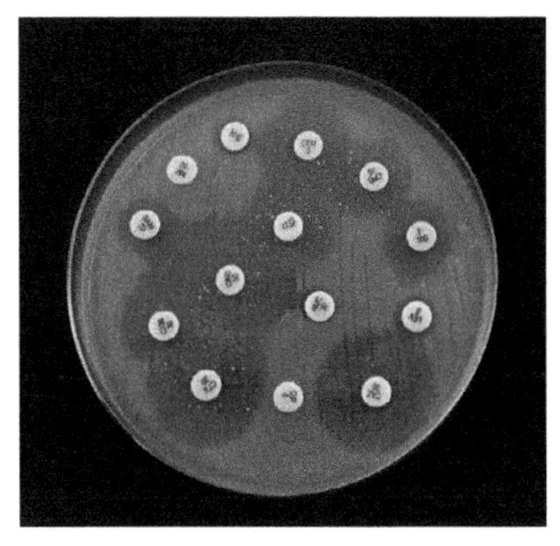

Figure 3.3.25 In an antibiotic test, white discs containing different antibiotics are placed on the surface of nutrient jelly with bacteria growing on it. Darker areas develop around the discs, showing where the antibiotics have stopped the bacterial growth

3. **Explain how painkillers work.**
4. **Compare the use of painkillers and antibiotics.**

MAKING LINKS

See topic 4.4d to find out more about antibiotic resistance in bacteria to find out how it provides evidence for evolution.

Antibiotic resistance

The use of antibiotics has greatly reduced deaths from infectious bacterial diseases. But misuse of antibiotics has led to the emergence of strains of bacteria that are antibiotic resistant – they cannot be killed by an antibiotic – which some believe is the worst public health problem facing us today.

Genes mutate in bacteria all the time, and by chance some mutations may make the bacteria resistant to an antibiotic. As they reproduce, the resistance genes spread through the bacterial population. Through the widespread use of antibiotics – in agriculture as well as for humans – populations of 'normal' bacteria are often reduced massively, leaving resistant strains to thrive.

When you are prescribed antibiotics, it is essential to complete the whole course so that *all* the infecting bacteria are killed. If you don't take all the antibiotics, any bacteria left will be more resistant than those killed first, and these resistant strains will then spread. It's also important to use the correct antibiotic for a specific bacterium, so that it is as effective as possible in killing all the bacteria.

DID YOU KNOW?

Alexander Fleming discovered penicillin, the first antibiotic, in 1928.

KEY INFORMATION

Remember: painkillers relieve infection symptoms. Antibiotics kill bacteria.

5. **Explain how widespread use of antibiotics is speeding up the evolution of antibiotic-resistant strains of bacteria.**
6. **Why is the development of bacteria with antibiotic resistance such a threat to public health?**

Qu: How are new drugs and medicines developed? ⟶

3.3h Testing new drugs

Learning objectives:

- describe how new medicines are discovered, developed and tested
- explain that research is published only after evaluation by peer review.

KEY WORDS

dose
efficacy
placebo

New and better drugs to treat diseases are developed and tested all the time, although the process takes many years. The starting point is often chemicals extracted from plants.

Making new drugs from old

Researchers sometimes use traditional medicines to start developing new drugs. Drugs are often based on extracts from plants and microorganisms, for example:

- the heart drug digitalis comes from foxgloves
- the painkiller aspirin was developed from a natural substance that comes from willow trees
- penicillin was discovered by Alexander Fleming in *Penicillium* mould.

Chemists working for pharmaceutical companies formulate most new drugs.

New drugs are tested and trialled before being prescribed to ensure they are:

- effective – able to prevent or cure a disease, or make you feel better; this is the drug's **efficacy**
- safe – not too toxic or with any undesirable side effects.

1. **Name two drugs and state the source of each.**
2. **Why do new medicinal drugs need to be tested and trialled?**

Figure 3.3.26 Research into new drugs may start with natural substances and traditional medicines

Developing new drugs

Stages in drug development are:

- preclinical testing in laboratories (using cells, tissues and live animals) to establish side effects and efficacy
- clinical trials, which use healthy volunteers and other patients. Low **doses** of the drug are given at the start. Then, if the drug is deemed safe, further trials are performed to find the optimum dose. Clinical trials are split into phases, as shown in Figure 3.3.27.

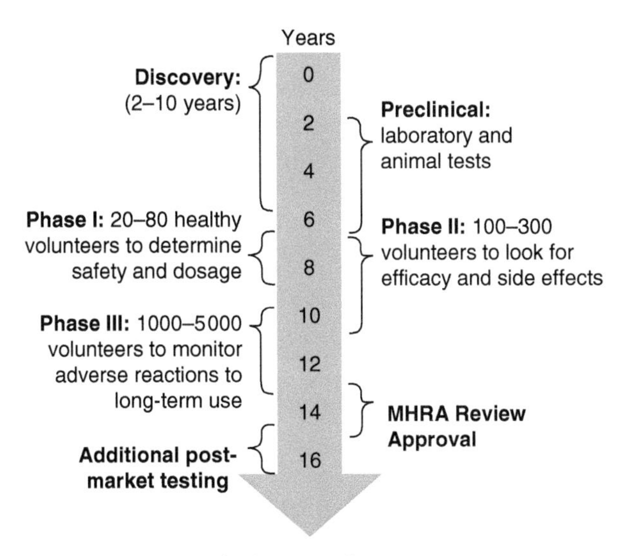

Figure 3.3.27 Developing new drugs

Trial results are peer reviewed by medical professionals in the MHRA (Medicines and Healthcare products Regulatory Agency) and only then are they published. Clinical trials involve some risk because unexpected side effects can occur.

KEY INFORMATION

The results of scientific testing and research are always examined by experts in the same field, before a paper is published describing the results. This is called peer review.

3. **Describe the stages in the development of a drug.**
4. **Why are potential drugs tested on human cells, animals and healthy human volunteers before being tested on patients?**

Double-blind trials

Double-blind trials are when patients are allocated randomly to groups, so that doctors and patients do not know, until the trial is complete, if they are taking:

- the new drug (this is the test group)
- a **placebo** (a treatment that does not contain the drug). This is the control group.

Clinical trials must not be influenced by the people involved in them, whether patients, doctors or employees of pharmaceutical companies. Sometimes, people feel better just because they think they will if they take a medicine. This is the 'placebo effect'.

5. **Why are double-blind trials conducted?**
6. **What is the placebo effect?**
7. **In a blind trial, patients don't know which drug they are receiving but the doctors do. Results from such trials are not always reliable. Suggest why this might be the case.**

DID YOU KNOW?

Before drug development was regulated by law, a drug called thalidomide was used to treat morning sickness in pregnant women. It caused severe limb deformities in foetuses.

REMEMBER!

A placebo is a treatment that does not contain the drug.

3.3i Genetic modification

Learning objectives:

- explain and evaluate some gene technologies used in medicine, taking into account benefits, risks, and the practical and ethical issues raised.

Genetic modification is a rapidly advancing area of research, bringing medical innovations that could help millions. But, as with all scientific advance, there are ethical issues to consider too.

Human insulin from bacteria

People with Type 1 diabetes need regular injections of the hormone insulin. From the 1920s, insulin was extracted from the pancreas tissue of pigs or cattle. But these types of insulin differ slightly from human insulin and caused some side effects.

Now it is possible to genetically modify the bacterium *Escherichia coli* so that it produces 'human' insulin identical to the protein produced by the human body.

Enzymes are used to remove the insulin gene from human DNA. The gene is transferred into the *E. coli* bacteria. The bacteria containing the gene are grown in fermenters. The protein is removed, processed and purified to make a human insulin product for people with Type 1 diabetes.

Figure 3.3.28 Human insulin production in India. This photograph shows the purification process

1. **How is a genetically modified *E. coli* bacterium different to a normal *E. coli* bacterium?**
2. **What are some advantages of producing insulin for humans from bacteria rather than from tissue from pigs or cattle?**

Proteins from sheep and goats

Sheep and goats have been genetically modified to produce chemicals in their milk that can be used to treat disease.

Tracy was the first transgenic farm mammal, created in 1990. She was a sheep genetically modified to produce a human protein, which it was hoped could treat cystic fibrosis. But clinical trials in 1998 showed that the protein caused breathing problems, so its development as a drug was not continued.

The first successful drug produced using genetically modified animals was approved in the USA in 2009. It comes from transgenic goats that produce a protein called human antithrombin in their milk. The drug is used to help people

DID YOU KNOW?

Transgenic organisms are also called genetically modified organisms (GMOs). Transgenic mammals are usually created by micro-injecting foreign DNA into the nucleus of a fertilised egg, which is then implanted into a surrogate mother.

with blood clotting problems during surgery or childbirth. One genetically modified goat can produce the same amount of antithrombin in a year as 90 000 human blood donations.

3 What are some pros and cons of using genetically modified goats to produce antithrombin rather than donated blood?

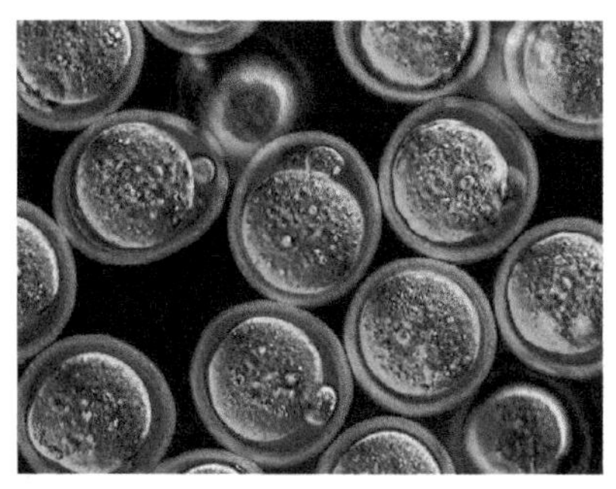

Figure 3.3.29 These are 'pronuclear' egg cells from a mouse, each measuring around 100 μm in diameter, and photographed using a polarising light microscope. They are fertilised but haven't yet formed zygotes. These cells will be injected with modified DNA, to create genetically modified organisms

Tissues for transplants

Research is also exploring the possibility of providing tissues and organs for transplants from animals like pigs that have been genetically modified so that the tissues are not rejected by the human immune system.

Using animal tissues for transplant is not a new idea. It could help save thousands of human lives each year, but rejection by the immune system and the possibility of infection by animal viruses have always been a big problem.

Now advances in gene-editing technologies mean that scientists can modify pig genes more quickly and accurately than in the past and reduce the risks of rejection or infection.

Figure 3.3.30 Biotechnology companies have ambitious plans to produce gene-edited pig lungs and other organs for human transplant on an industrial scale. Their huge pig farms could produce a thousand organs for transplant per year

4 Evaluate the practical risks and benefits of using animal tissues for human transplantation.

5 From an ethical standpoint, explain arguments for and against the large-scale production of animal organs for transplant.

3.3j Stem cells

Learning objectives:

- describe some uses of stem cells in medicine
- evaluate possible uses of stem cells in medicine taking into account benefits, risks and the ethical issues raised.

Stem cell research could provide almost limitless, patient-specific medical treatments that could transform health services. But the use of living stem cells raises moral and ethical concerns.

Stem cell research

Stem cells are unspecialised cells that can differentiate into many different cell types. Because of this potential, stem cells are useful in many areas of scientific and medical research.

Embryonic stem cells can develop into any of the many types of cells in the body. Adult stem cells can only give rise to certain types of cells. For example, blood stem cells in bone marrow can develop into any kind of red or white blood cell, but no other type.

Stem cells for research may be based on:

- stem cells from embryos that are a few days old
- adult stem cells from certain parts of the body, such as bone marrow
- fetal stem cells taken from blood in the umbilical cord.

Figure 3.3.31 A computer-generated illustration of embryonic stem cells (×1500)

1. **List the possible sources of stem cells for research.**
2. **What do you think is the function of adult stem cells in the body?**

Using stem cells in medicine

Patients with cancers such as leukaemia often receive high doses of chemotherapy. As well as destroying leukaemia cells, the treatment kills healthy stem cells in the bone marrow. These can be replaced in a stem cell transplant, which helps stimulate new bone marrow growth, supplies new blood cells and restores the immune system.

Most medical uses of stem cells are still experimental. Treatments based on stem cells, or specialised cells grown from them, are being investigated to treat conditions such as:

- spinal injuries leading to paralysis
- diseases in which certain tissues degenerate, like Alzheimer's disease, Parkinson's disease, multiple sclerosis, and Type 1 diabetes – using embryonic or fetal stem cells
- heart disease – using the patient's own stem cells from bone marrow.

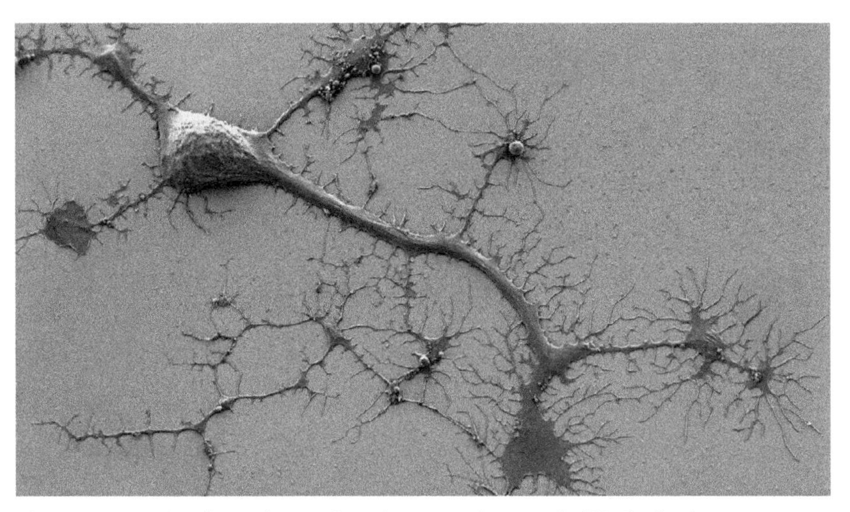

Figure 3.3.32 A coloured scanning electron micrograph (SEM) of a human neuron that has been developed in the laboratory from an embryonic stem cell (×12 000). Cells like this could be used to repair damaged tissue in diseases like Parkinson's and Type 1 diabetes

3. **Name two conditions that could be treated with stem cells in the future.**
4. **Why are stem cell transplants important for people who have had chemotherapy?**

Stem cell use is controversial

Research is needed to find out more about stem cells and their use in treatments. Scientists hope to be able to culture stem cells in limitless numbers.

However, using stem cells removed from a living human embryo is especially controversial. Until recently, the embryos used were those left over from *in-vitro* fertilisation treatment (IVF). These spare embryos would be destroyed if they had not been donated by the IVF couples for research.

British law now allows embryos to be created purely for scientific research. Some people object to this. Some argue that life begins at conception, so an embryo has rights. Who should decide when a human life ends?

There is also concern that scientists do not yet fully understand how stem cell differentiation is controlled. So there are fears that their ability to proliferate could lead to cancer when they are transplanted into a patient.

5. **Write down one ethical objection to stem cell research.**
6. **Evaluate the potential benefits and drawbacks of using stem cells in medicine.**

DID YOU KNOW?

Stem cell transplants are not new. Bone marrow transplants containing stem cells have been carried out since 1968. In a stem cell transplant, the patient receives the stem cells through a drip. They enter the bloodstream, and then travel to the bone marrow where they start to make new blood cells.

KEY INFORMATION

Stem cell use raises moral and ethical questions. A moral question looks at whether something is right or wrong. An ethical question discusses the reasons why something might be right or wrong.

3.3k Interactions between different types of disease

Learning objectives:

- describe the interactions between different types of disease.

We live in complex environments where many different factors can affect our physical and mental health at one time. These factors may interact, leading to new effects.

Interactions

Our bodies are constantly bombarded with a variety of pathogens, environmental substances and life experiences, so it's not surprising that their effects sometimes interact. For example:

- severe or prolonged physical ill health – for example, obesity or cancer – can lead to depression and other mental illnesses
- defects in the immune system mean that an individual is more likely to suffer from infectious diseases
- immune reactions initially caused by a pathogen can trigger allergies
- viruses living in cells can be the trigger for cancers.

1. **Suggest how having a serious illness might affect mental health.**
2. **Explain one other example of how different diseases interact.**

Immune system malfunctions

A person's immune system can be suppressed by disease treatments such as chemotherapy for cancer, or by drugs taken following transplants to reduce the chances of tissue rejection. Infections such as HIV, which causes AIDS, also interfere with the immune system. In later stages, people with AIDS have an increasing risk of opportunistic infections, because their bodies are not able to fight off invading pathogens. They may succumb to common infections like tuberculosis, and tumours that rarely affect people with properly working immune systems.

Over-reaction of the immune system can also result from interaction of disease-causing agents. Immune reactions initially caused by a pathogen can later cause allergies, such as skin rashes. Severe viral respiratory infections in early childhood can trigger asthma as children grow.

3. **Describe how interaction of the HIV virus and the tuberculosis bacterium may affect a person with AIDS.**

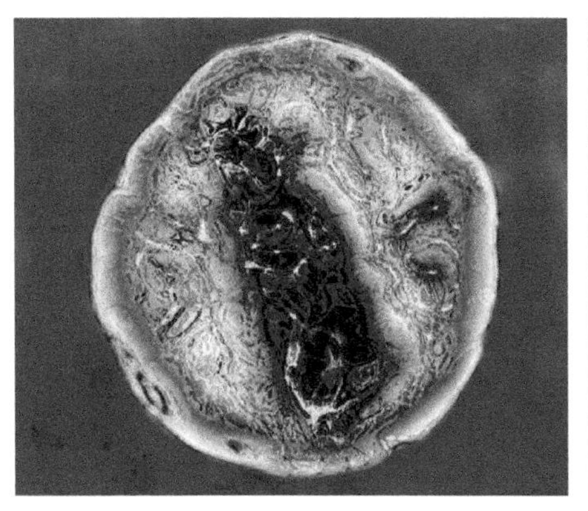

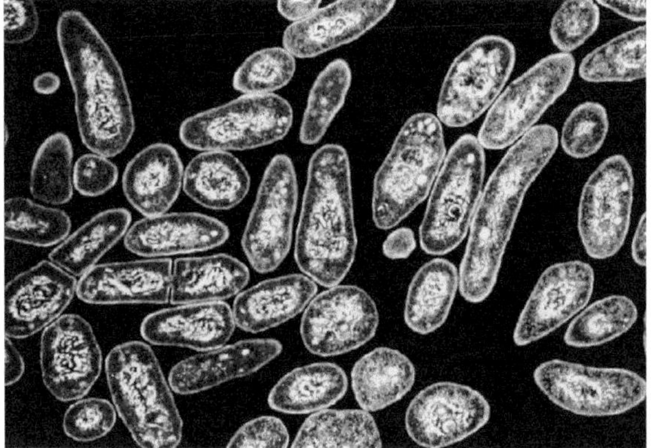

Figure 3.3.33 Infection with the human immunodeficiency virus, HIV (left); false-colour transmission electron micrograph, ×414 000, can make it more likely that a person will later contract infections like *Mycobacterium tuberculosis* (right); false-colour transmission electron micrograph, ×3675

Viruses and cancer

Human papilloma virus (HPV) is a virus that can trigger cancer in different body parts, in women and men. People with weak immune systems are less able to resist HPV, but not everyone who gets HPV develops cancer.

Worldwide, cervical cancer is the second most common cancer in women, and the vast majority of cases are attributable to HPV. It often takes many years for cervical cancer to develop after getting HPV and it does not usually show symptoms until it is quite advanced. It is hard to treat.

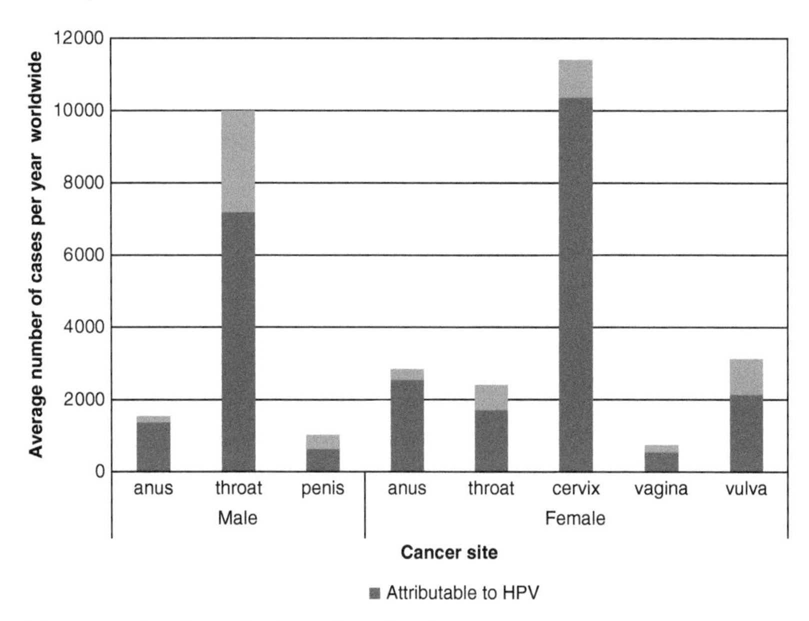

Figure 3.3.34 What do these data show?

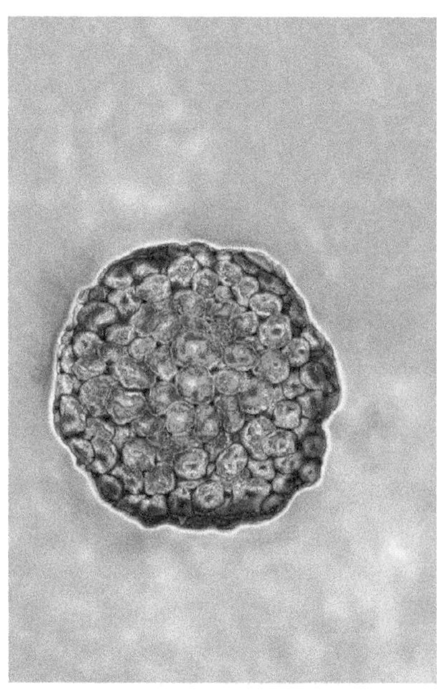

Figure 3.3.35 Infection with the human papilloma virus, HPV, can lead to cancer

4. **Describe what is shown by the data in the graph in Figure 3.3.34.**
5. **Cancer is a non-communicable disease, but it can be caused by an infection. Explain this statement.**

MATHS SKILLS

3.3l Sampling and scientific data

Learning objectives:

- understand why sampling is used in science
- be able to explain different sampling techniques.

KEY WORDS

bias
systematic

When undertaking clinical investigations, scientists are unable to test all people, so sampling is used.

Populations and samples

We cannot study every individual in a population, so we use a sample of the population.

Choosing the right size of sample is important. Carrying out studies costs money so scientists don't want to use sample groups that are unnecessarily large. However, the risk with small groups is that they aren't representative enough of the whole population. This is known as a 'sampling error' and it means that the results of the study don't indicate well what the population as a whole is like.

For example, for a new antibiotic to be effective it should cure 95–100% of the cases of an illness. Imagine that the population taking the antibiotic is represented by a container of coloured balls. Each ball is a patient. Red balls represent patients who are not cured by the antibiotic. The blue balls represent the patients who are cured.

A scientist takes a small sample, of 10 balls. There is 1 red ball and 9 blue balls:

$$\frac{\text{Patients not cured (red balls)}}{\text{All patients (red and blue balls)}} \times 100 = \text{\% chance of not being cured}$$

So

$$\frac{1}{10} \times 100 = 10\% \text{ chance of not being cured.}$$

This suggests that the antibiotic is only effective for 90% of patients treated.

However, in another sample, this time of 100 balls (patients), there are only two red balls:

$$\frac{2}{100} \times 100 = 2\% \text{ chance of not being cured.}$$

This larger sample suggests that the antibiotic is 98% effective.

When using a small sample size, a sampling error can easily be made, which makes the outcome unreliable. To avoid this

DID YOU KNOW?

A drug called thalidomide relieved morning sickness in pregnant women. Pregnant women were not in the sample used to test the safety and effectiveness of thalidomide, so scientists did not discover its effects on unborn babies, until too late. Many babies born to mothers who took thalidomide had limb abnormalities.

REMEMBER!

A small error in a small sample can mean a large error.

sampling error, large-scale clinical trials that test new drugs in humans will likely involve thousands of patients.

1 Why do scientists use samples for investigations?

2 Figure 3.3.36 represents more samples taken from the same population in the antibiotic investigation.
  a Calculate the percentage of people who were cured in each sample.
  b How does sample size affect the results of the test?

Sampling techniques

Research aims to find consistent patterns in repeat samples, which show relationships. Scientists select samples for clinical trials by:

- selecting randomly from the whole population
- grouping individuals by characteristics, then randomly selecting people from each group
- focusing on a particular subset of a population.

The sampling technique used can affect the outcomes.

Sample size needs to be considered as part of the research process. For a small population, 5–10% is a large enough sample size. For a larger population of size '*n*', sample size = $\sqrt{n}$. One way of choosing a good sample size is to find the square root of population number.

The symbol '$\sqrt{\ }$' means square root, so

$\sqrt{25}$ means the 'square root of 25'.

The square root is the inverse of the square of a number.

So, for a population of 900, a good sample would be $\sqrt{900} = 30$.

Looking at bias

Bias means that results don't represent the true situation. Bias results from systematic errors in procedure, which can give false outcomes. For example, if a sample is not chosen randomly, it might contain factors that are not representative of the whole population, which would bias the results.

Bias must be eliminated wherever possible. In clinical drug trials, a control group is used (people with similar diet, weight etc. to the test group), in double-blind tests (topic 3.3h). Neither the patients nor the doctors can tell which are the test or the placebo drugs, so they don't know which patients get the test drug. This eliminates the chance of patients' responses being biased by expectations.

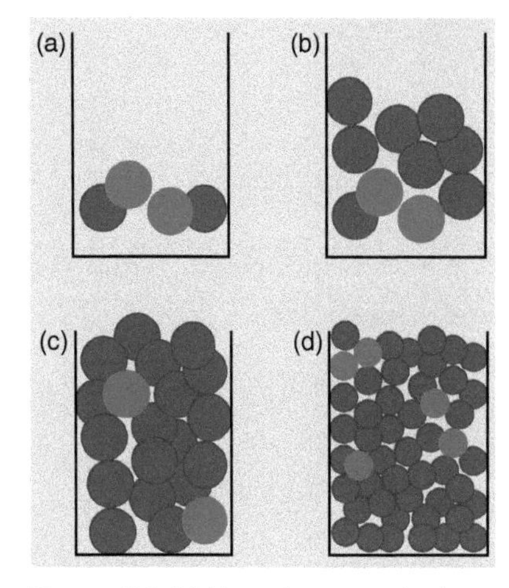

Figure 3.3.36 How does sample size affect results?

3 Why must care be taken when selecting the sampling method to be used?

4 What sample size would you suggest for trials involving these total populations?
  a 36 patients
  b 169 patients
  c 729 patients
  d 136 161 patients

5 Explain fully how bias in clinical trials can be minimised.

6 Scientists studied 31 000 heart disease patients. Their database of medical data was arranged according to the hospital where the patients were treated. The scientists chose the first 50 patients in the database as their sample.
  a Give two reasons why this sample might not be representative of the whole population.
  b Suggest how the sampling method could be improved.
  c Suggest a suitable sample size.

Check your progress

You should be able to:

describe the symptoms of some viral, bacterial, fungal and protist diseases.	→	describe the transmission and control of different diseases.	→	explain applications of science to prevent the spread of diseases.
describe how the body protects itself from pathogens.	→	explain the role of the immune system.	→	explain the specificity of antibodies.
recall why vaccinations are used.	→	explain how vaccinations trigger an immune response.	→	evaluate the global use of vaccinations.
describe the use of antibiotics and painkillers.	→	explain how antibiotics and painkillers treat disease.	→	explain the limitations of antibiotics. explain the impact of antibiotic-resistant bacteria.
recognise the need for potential new medicines to be tested and trialled extensively.	→	describe the process of development of potential new medicines, including preclinical and clinical testing.	→	explain how double-blind trials form a vital part of clinical testing.
recall that genetic modification of organisms is used to produce medically useful proteins.	→	describe some uses of gene technology in modern medicine.	→	evaluate gene technologies, in terms of benefits, risks, and ethical issues involved.
recall where stem cells are found.	→	understand the potential of stem cell therapies.	→	evaluate scientific and ethical issues involved with stem cell therapies.
recall that different diseases can interact in patients.	→	describe some ways in which diseases can interact, to produce new effects.	→	explain how diseases may interact.

Worked example

Viruses are pathogens that cause many diseases. Babies and children are vaccinated to protect them from many viruses.

1 a Why are antibiotics not effective against viral infections?

Antibiotics cannot kill viruses.

Correct. A fuller answer would explain that antibiotics kill or inhibit the growth of bacteria, but viruses survive and replicate very differently (inside host cells) and so are not affected.

b What is in a vaccine?

A weak form of the pathogen.

Correct, although 'inactive' or 'dead' would be better words than 'weak'.

c What does the vaccine do in the person's body?

It makes antibodies.

This is inaccurate. White blood cells (lymphocytes) make antibodies.

2 a Name one disease for which a stem cell transplant often forms part of the treatment.

Leukaemia.

Correct. High dose chemotherapy for other forms of cancer may also be followed up with stem cell transplants, to replace damaged bone marrow tissue.

b Outline one practical and one ethical concern about using embryonic stem cells in medical treatments.

We don't fully understand how stem cell growth is controlled, so some people worry that they might cause cancer when they are transplanted into a patient.

Embryonic stem cells come from living human embryos, which then die. Some people think it's wrong to end a human life like this, even if it helps save other people's lives.

These are good, full answers.

3 The graph shows the bacteria present in the body during an infection.

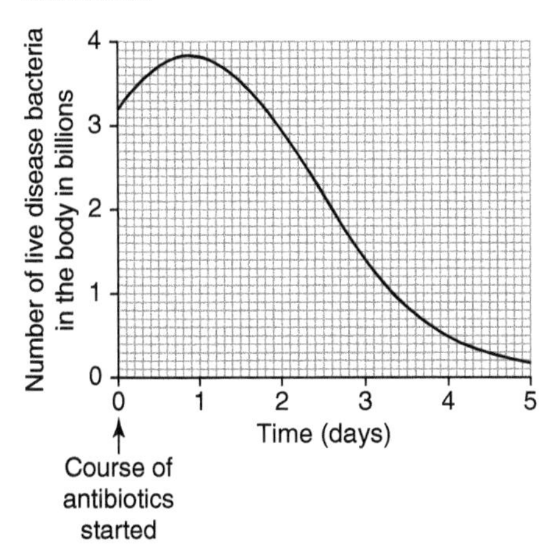

a Suggest why the numbers increase after antibiotics were taken.

The antibiotics didn't work to start with.

Incorrect. The antibiotic is working – the growth rate of the bacterial population starts to slow down almost right away. After a day the growth rate has fallen to zero, and then bacterial numbers start to decline.

b Explain why the full course of antibiotics needs to be taken.

Not taking the full course can lead to antibiotic-resistant bacteria.

Correct, but more detail could improve this answer. If you don't take all the antibiotics, any bacteria left will be more resistant than those killed first, and these resistant strains will then spread.

End of chapter questions

Getting started

1. What is a disease 'vector'? — 1 Mark
2. Describe how the stomach helps defend the body from pathogens. — 1 Mark
3. What type of pathogen causes athlete's foot? — 1 Mark
4. Name a disease that is caused by a protist pathogen. — 1 Mark
5. What type of blood cells destroy bacterial pathogens? — 1 Mark
6. A blood test reveals that a patient has a white cell count of 4 000 000 000. Write this number in standard form. — 1 Mark
7. a What is phagocytosis? — 1 Mark

   b Give an example of phagocytosis. — 1 Mark
8. Why must new drugs be tested before they are prescribed? — 2 Marks

Going further

9. What do vaccines contain? — 1 Mark
10. Give two examples of physical defences that help prevent pathogens entering your body. — 2 Marks
11. Why can a person with AIDS die from a simple infection? — 2 Marks
12. Give one ethical objection to genetic modification. — 1 Mark
13. How can the spread of measles be controlled? — 3 Marks
14. Outline how painkilling drugs work. — 1 Mark

More challenging

15. Why is an antibody specific? — 1 Mark
16. Explain the effects of antibiotics on bacteria and on viruses. — 2 Marks
17. Why is it difficult to develop drugs that kill viruses? — 2 Marks
18. For an antibiotic to be considered effective it should cure 95–100% of the cases of an illness. Out of a sample of 568 patients given a new antibiotic, 483 were cured.

    a Calculate the chance of not being cured. — 1 Mark

    b The patients sampled were all university students aged between 18 and 25. Suggest how this may have affected the results of the investigation. — 2 Marks

    c Explain what we mean by bias in a clinical trial, and suggest one way in which it can be avoided. — 2 Marks

Most demanding

19 **Some students see a newspaper article on a European stem cell clinic. The clinic uses stem cell therapy to treat diabetes and other conditions.**

The article contains some information on the clinic's treatment of diabetes. It includes data on the 55 patients the clinic has treated so far, which it claims is a success.

Type of diabetes	One month after treatment, number of patients who ...		
	showed an improvement	showed no change	became worse
Type 1	8	13	2
Type 2	20	9	3
Total	28	22	5

a **What is a stem cell?** 1 Mark

b **Calculate the overall percentage of patients who:**

(i) **showed an improvement** 1 Mark

(ii) **showed no change** 1 Mark

(iii) **became worse** 1 Mark

c **Does the data support the newspaper article's claims? Explain your answer.** 2 Marks

20 **MRSA is an antibiotic-resistant strain of bacteria. The graph shows the number of cases of patients with MRSA infections from 1993 to 2005 in the USA.**

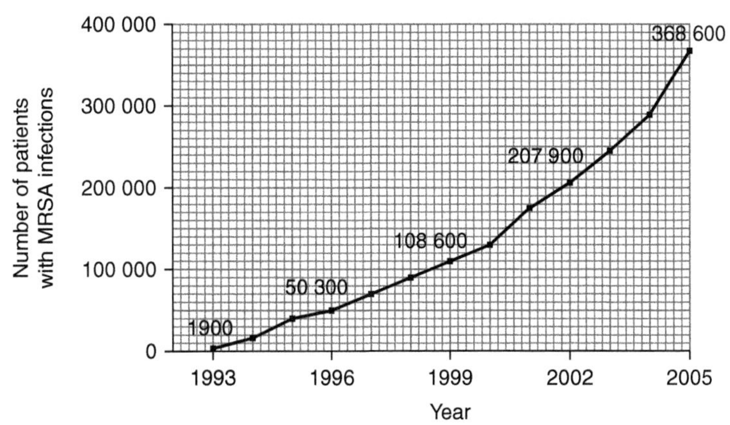

Explain the trend in the graph, even though the patients were treated with antibiotics. 4 Marks

Total: 40 Marks

How the ideas in this topic link together

Living and non-living systems change over time. They are affected by changes at a global level as well as changes at a molecular level in the cells of living organisms.

The study of how the Earth's atmosphere developed and its current situation depends on very complex – and sometimes incomplete – data, and so scientists set up and test models that are based on assumptions. Despite the uncertainty, almost all scientists agree that humans affect the environment around them and that we can evaluate our impact at a local and global level.

Our growing knowledge of genes, chromosomes and the inheritance of characteristics has far-reaching consequences. It has enabled us to manipulate the reproduction of other living organisms through selective breeding and genetic engineering. It also enabled us to understand how bacterial mutations can give rise to antibiotic resistance.

Humans affect the living and non-living systems around them. This can raise important ethical questions.

Working Scientifically Focus

- Interpret evidence and evaluate different theories
- Explain the importance of scientists publishing their findings and theories so that other scientists can critically evaluate them
- Describe, explain or evaluate ways in which human activities affect the environment

4 Explaining change

Contents

THE EARTH'S ATMOSPHERE

IDEAS YOU HAVE MET BEFORE:

PLANTS USE CARBON DIOXIDE IN PHOTOSYNTHESIS

- Plants photosynthesise using carbon dioxide and water.
- Plants make oxygen, which is used by animals to stay alive.
- Plants need sunlight for photosynthesis.

GLOBAL WARMING MELTS ICE CAPS

- There are different climates in different parts of the world.
- The Earth's surface is constantly changing.
- Animals have difficulty surviving because of habitat loss.

USING LESS FOSSIL FUEL CUTS DOWN POLLUTION

- Cars, ships and planes use fossil fuels to transport us.
- We use fossil fuels to heat buildings and drive machinery.
- Burning fossil fuels causes air pollution, which is bad for health.

THE WATER CYCLE

- Water evaporates from oceans and seas to form clouds.
- Clouds move the condensed water inland where it precipitates.
- Rain fills the rivers, lakes and reservoirs to provide fresh water.

4.1

IN THIS CHAPTER YOU WILL FIND OUT ABOUT:

WHAT WAS THE EARTH'S EARLY ATMOSPHERE?

- The early atmosphere arose from gases from volcanoes.
- Water vapour condensed to form the oceans.
- There was a high percentage of carbon dioxide and no oxygen.

WHY DID THE EARLY ATMOSPHERE CHANGE?

- Algae used carbon dioxide for photosynthesis.
- Photosynthesis produced oxygen so levels increased.
- Oceans dissolved carbon dioxide making them acidic.

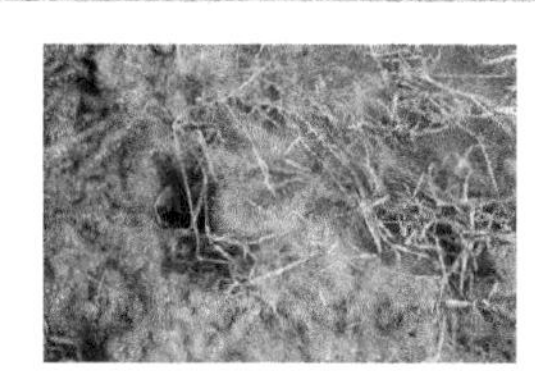

WHAT ARE THE CONSEQUENCES OF THE GREENHOUSE EFFECT?

- Greenhouse gases are essential for keeping temperatures stable.
- If we had no greenhouse effect the Earth would not sustain life.
- The balance of gases changes the effect and could be damaging.

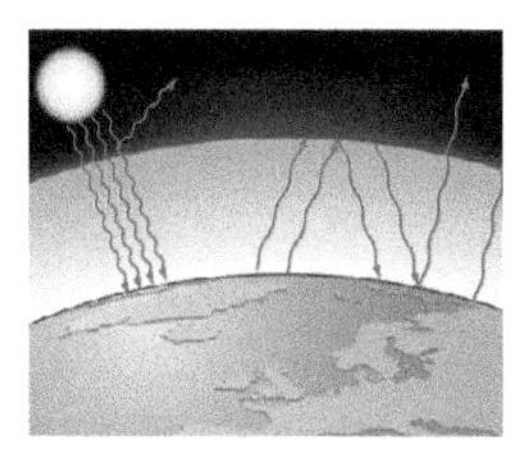

HOW CAN WE REDUCE THE EFFECT OF HUMAN ACTIVITY?

- We can reduce the use of fossil fuels and use renewable energy.
- We can use sunlight more directly for energy needs.
- We can use resources more efficiently and fairly.

4.1a The early atmosphere

Learning objectives:

- describe ideas about the Earth's early atmosphere
- interpret evidence about the Earth's early atmosphere
- evaluate different theories about the Earth's early atmosphere.

KEY WORDS

atmosphere
sediments
volcanic

The age of the Earth is currently estimated to be between 4.5 and 4.6 billion years. That is a little over four and a half thousand million years (often written as 4.5×10^9 years). During that time the atmosphere has changed from possibly hydrogen and helium at first, probably followed by gases from volcanic activity to finally an atmosphere rich in oxygen, which is what we have now.

REMEMBER!

You only have to remember this one theory, but there are more theories you can look up and evaluate against evidence.

The Earth's early atmosphere

We *definitely know* we now have an **atmosphere** rich in oxygen but we also *think* we know, from evidence, that oxygen was not in the Earth's early atmosphere. From the evidence we have, scientists have developed theories about the change in the atmosphere over the last 4.6 billion years.

One theory suggests that during the first billion years of the Earth's existence there was intense **volcanic** activity that released gases. This process is called degassing and the idea is based on the composition of gases vented out in present-day volcanic activity.

At the start of this period the Earth's atmosphere may have been like the atmospheres of Mars and Venus today, consisting of mainly carbon dioxide with little or no oxygen gas.

Volcanoes also produced nitrogen, which gradually built up in the atmosphere, and there may have been small proportions of methane and ammonia.

These gases formed the early atmosphere, along with water vapour, which condensed to form the oceans.

When the oceans formed, carbon dioxide dissolved in the water and carbonates were precipitated producing **sediments**, reducing the amount of carbon dioxide in the atmosphere.

Figure 4.1.1 Volcanic activity releases gases

Figure 4.1.2 The surface of Mars is being explored by robots but the atmosphere cannot support human life

1. **Explain the scientists' theory of how the first gases became part of the atmosphere for the first billion years.**
2. **Suggest why there was no oxygen in the early atmosphere.**
3. **There are several different theories about the Earth's early atmosphere. Suggest why.**

Evidence for the theories

The theories about what was in the Earth's early atmosphere and how the atmosphere was formed have changed and developed over time.

One set of evidence is gained by measuring carbon and boron isotope ratios in sediments under the sea.

Other models use the composition of gases given out by volcanoes today as evidence. An assumption is made that the same proportion of gases is given out today as was given out by volcanoes billions of years ago.

The question we need to ask is how well these assumptions fit with evidence gained through other means. We need to *evaluate evidence* to develop new models.

Figure 4.1.3 Water vapour was produced, which condensed to form the oceans

4. **Scientists use the composition of gases given off by volcanoes to develop models of the Earth's ancient atmosphere. Suggest one reason why.**

Evaluating the theories

The evidence for the early atmosphere is limited because of the timescale of 4.6 billion years, which means no direct measurements can be made. Models need to be developed using proxy evidence or by assuming that what happened in the past is still happening today and using that evidence from today (for example the composition of volcanic gases).

Proxy evidence is evidence from one source gathered from ancient times that can be used to make an assumption about another related ancient effect. An example would be counting the number of stomata on fossils of ancient leaves to make assumptions about the levels of carbon dioxide in the atmosphere.

Evidence from direct measurements would seem more valuable but there is not always agreement that the measurements were taken correctly. For example, the measurements taken from isotopes by one team are seen as more reproducible if their data compares well with the data from a different site by a different team. The more measurements taken by different teams that provide data that coincide, the more reproducible the data. This is why scientists publish data in scientific journals.

5. **Scientists have only indirect evidence for the atmosphere of 2 billion years ago. Suggest, with reasons, whether the current composition of volcanic gases gives better or worse evidence than the evidence from counting stomata.**
6. **Stomata are openings that allow gas exchange in plants. Predict a relationship between number of stomata and atmospheric carbon dioxide levels.**

DID YOU KNOW?

Some scientists are now rejecting the early models of the Earth's atmosphere and are studying zircon gemstones for new evidence.

4.1b Changes in the atmosphere

Learning objectives:

- identify the processes allowing oxygen levels to increase
- explain the role of algae in the composition of the atmosphere
- recall the equation for photosynthesis.

KEY WORDS

algae
evolved
photosynthesis

Algae first produced oxygen about 2.7 billion years ago and soon after this oxygen appeared in the atmosphere. Over the next billion years, plants evolved and the percentage of oxygen gradually increased to a level that enabled animals to evolve.

The early production of oxygen

We have already seen that the Earth is 4.5 billion years old. In the first billion years the Earth's atmosphere may have been like the atmospheres of Mars and Venus today, consisting of mainly carbon dioxide with little or no oxygen gas.

The first life forms appeared over 3.5 billion years ago, but these could live anaerobically, that is without oxygen. **Algae** first produced oxygen about 2.7 billion years ago.

Soon after the appearance of life forms that used **photosynthesis**, oxygen appeared in the atmosphere. Over the next billion years, plants **evolved** and the percentage of oxygen gradually increased to a level that enabled animals to evolve.

Figure 4.1.4 Photo of green alga. Algae started to produce oxygen about 2.7 billion years ago

1 When did algae first produce oxygen?

2 Explain the difference between living anaerobically and living aerobically.

Figure 4.1.5 These fossils are of early ferns that produced oxygen by photosynthesis. Ferns first arose sometime during the Devonian period, 410–365 million years ago

Algae and photosynthetic production of oxygen

Algae and plants produced the oxygen that is now in the atmosphere. This was produced by photosynthesis.

This process can be represented by the equation:

$$6CO_2 + 6H_2O \rightarrow C_6H_{12}O_6 + 6O_2$$

carbon dioxide + water → glucose + oxygen

You can see that the process uses up carbon dioxide. So, as more plants evolved the levels of carbon dioxide went down and the levels of oxygen went up. More about photosynthesis can be found in topic 2.2h.

KEY INFORMATION

You can show by experiment that aquatic plants produce oxygen.

Over billions of years, the percentage of oxygen in the atmosphere increased and the percentage of carbon dioxide decreased, until today's levels were reached.

The percentage of nitrogen also slowly increased, but because nitrogen is very unreactive, very little nitrogen was removed from the atmosphere.

3 Oxygen is a product of photosynthesis. What is the other product?

4 Suggest why the levels of oxygen increased.

5 Explain the role of algae in the increased levels of oxygen.

REMEMBER!

You do need to remember the equation for photosynthesis but not the role of nitrogen in the atmosphere.

Trapping carbon dioxide

Algae and plants decreased the percentage of carbon dioxide in the atmosphere using the process of photosynthesis.

They used carbon dioxide to make glucose, and they also released oxygen.

As the plants and algae grew and proliferated over time the amount of carbon dioxide decreased and the amount of oxygen increased.

As the plants died and decayed, they formed a thick layer of plant deposits, which was **compressed** first to peat and then eventually formed coal. See topic 4.2f for more information about peat.

Figure 4.1.6 Plant remains were compressed, forming coal, which can be seen as the black layers in this photo

Coal is formed from the thick plant deposits that were buried and compressed over millions of years. Coal is mostly carbon that became locked in the ground. This resulted in lower levels of carbon dioxide.

Another organism that used up carbon dioxide was plankton. The remains of plankton were deposited in muds on the sea floor and were covered over and compressed over millions of years. This produced crude oil and natural gas, which became trapped in the rocks.

Both these **fossil fuels** trapped carbon. Coal traps carbon as a **sedimentary** rock and crude oil/natural gas traps carbon between layers of other rock.

Figure 4.1.7 Peat, which has been harvested and dried (top), and bituminous coal (bottom)

6 Describe how coal was formed.

7 Describe two ways in which carbon dioxide levels were decreased over billions of years by the formation of fossil fuels.

4.1c The carbon cycle

Learning objectives:

- recall that plants take in carbon as carbon dioxide
- explain how carbon is recycled
- interpret a diagram of the carbon cycle.

KEY WORD

carbon sink

Materials are cycled through ecosystems continually. As one process removes them, another releases them into the ecosystem again. This helps to provide building blocks for future organisms.

Looking at carbon

The amount of carbon on Earth is fixed. Most carbon is found combined with other elements, for example in fossil fuels and carbonate rocks. Carbon is also found dissolved in water in rivers, lakes and oceans. A small amount is in the air in carbon dioxide.

Life depends on photosynthesis in producers, such as green plants, which use carbon dioxide from the air to photosynthesise. The carbon is used to make carbohydrates, proteins, fats and DNA that form new biomass, which is eaten by animals (consumers).

All organisms respire to release energy for cellular processes. They release waste carbon dioxide back into the environment, which is then used by plants.

The atmosphere contains about 0.04% carbon dioxide, which is enough for every plant to produce biomass (food), by photosynthesis. The process uses sunlight to create the energy store of food and oxygen that living things need for respiration.

When animals eat plants, they absorb carbon from them. Carbon passes along food chains, even when organisms die and decay. Energy is transferred along the food chain and to the environment.

Carbon dioxide is returned to the atmosphere by:

- plants, animals and decomposers respiring:
  glucose + oxygen → carbon dioxide + water
- burning (combustion) of fossil fuels and wood:
  fossil fuel/wood + oxygen → carbon dioxide + water

The continual cycling of carbon is shown in Figure 4.1.9.

Burning fossil fuels and wood releases more carbon into the atmosphere. Increased levels of carbon dioxide in the atmosphere cause global warming.

1. **Name two processes that release carbon dioxide.**
2. **Use Figure 4.1.9 to describe the carbon cycle.**

Figure 4.1.8 Burning fossil fuels releases carbon, as carbon dioxide, into the air

MAKING LINKS

See topic 4.2a for more information about food chains (feeding relationships between organisms).

REMEMBER!

The carbon cycle returns carbon from organisms to the air for plants to use in photosynthesis.

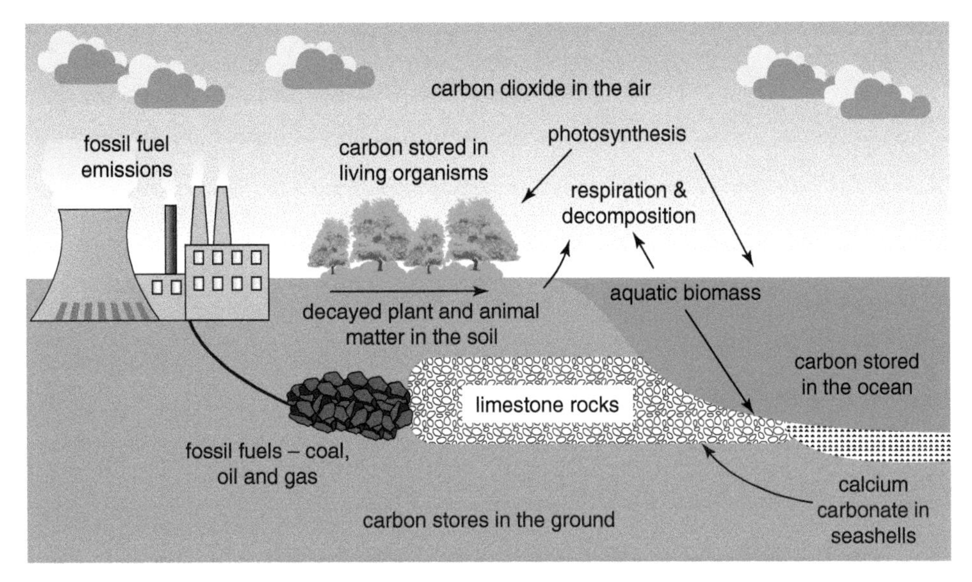

Figure 4.1.9 The carbon cycle showing the main stores of carbon and the flows of carbon between them

Carbon stores

Shells of marine organisms contain carbonates. Corals and microscopic algae cover themselves with calcium carbonate. Shells of dead organisms fall to the sea floor. Over millions of years they are compressed to form limestone. Limestone rocks are eroded by water in the soil that contains dissolved carbon dioxide, which comes from the respiration of soil organisms.

Carbon dioxide can be absorbed by oceans and held in a **carbon sink**. During volcanic eruptions and forest fires, massive amounts of carbon dioxide are released into the atmosphere.

3 Why is the carbon cycle important to all living organisms?

4 Explain the processes involved in the carbon cycle.

Explaining decay

Worms, woodlice and maggots break down waste and dead material into smaller pieces for decomposers to digest.

Decomposers are microorganisms (bacteria and fungi). They break down the smaller pieces of dead material. Decomposers release waste carbon dioxide, water, heat and nutrients (that plants use).

Decomposers decay waste in compost heaps and sewage works. Factors that speed up decay are:

- plenty of microbes
- warmth
- plenty of oxygen
- some moisture.

Life processes depend on enzymes. Factors that increase enzyme action speed up decay.

DID YOU KNOW?

Between 1000 and 100 000 million metric tonnes of carbon pass through the carbon cycle every year.

Figure 4.1.10 A woodlouse is pictured. Woodlice live on rotting wood and help to break it down for decomposers

DID YOU KNOW?

The Body Farm is a research site in the USA where scientists study human decay to support forensic analysis.

5 Explain the role of microorganisms in cycling materials.

6 Why does warmth speed up the process of decay?

KEY CONCEPT

4.1d The greenhouse effect

Learning objectives:

- describe the greenhouse gases
- explain the greenhouse effect
- explain these processes as interaction of short and long wavelength radiation with matter.

KEY WORDS

absorbed
greenhouse
radiation
wavelength

The average surface temperature of the Earth is 14°C. This is because we have a blanket of gases as an atmosphere that protects us by keeping the temperatures relatively stable. The greenhouse effect is a natural phenomenon that is beneficial for us.

Why are greenhouse gases important?

Water vapour, carbon dioxide and methane are the **greenhouse** gases in our atmosphere. These gases allow **radiation** from the Sun to pass through and warm the Earth. We all feel this on a bright sunny day.

What is less obvious, but very important, is that the warm Earth radiates energy back into the atmosphere and out into space. If this did not happen, the Earth would get warmer and warmer.

The greenhouse gases trap some of this radiation and help to keep us relatively warm.

So greenhouse gases in the atmosphere maintain the temperatures on Earth high enough to support life.

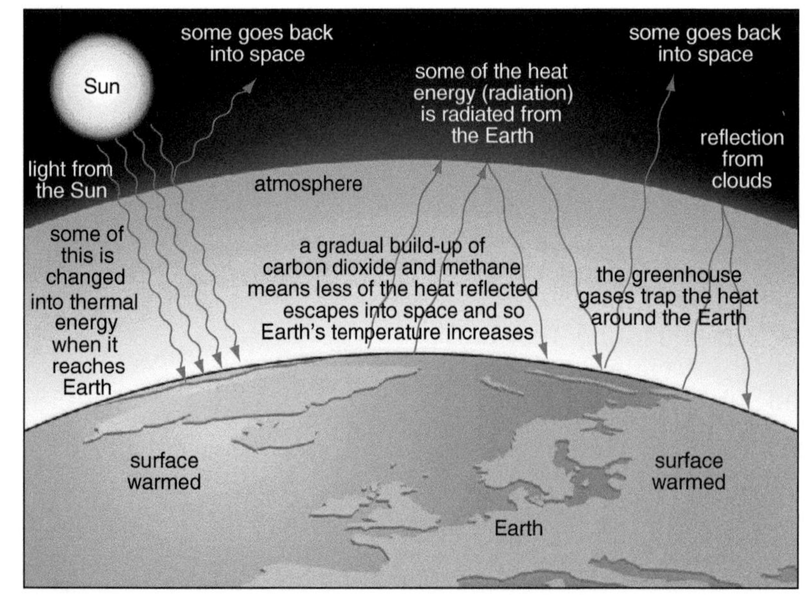

Figure 4.1.11 A diagram representing the greenhouse effect. Solar radiation reaches Earth. Energy radiated back from the Earth is trapped by the greenhouse gases

DID YOU KNOW?

The idea of a greenhouse effect was first put forward in 1824 by Joseph Fourier. It is estimated that the greenhouse effect makes the Earth 33°C warmer than it would be otherwise. What is the temperature where you are today? Now reduce that by 33°C ...

1. **Name two greenhouse gases.**
2. **Explain why greenhouse gases are important to life on Earth.**

How does the greenhouse effect work?

Radiation from the Sun is of short **wavelength**. It can pass through the atmosphere and not be **absorbed** by the atmospheric gases.

The gases allow *short wavelength* radiation to pass through the atmosphere to the Earth's surface. The gases then absorb the outgoing *long wavelength* radiation from the Earth, causing an increase in temperature. This temperature increase is beneficial to us as the temperature of the Earth is kept relatively stable.

Even though the temperature at the polar ice caps gets very low and the temperature at the equator can get very high, between these limits life can be sustained. Both our Moon and other planets experience temperatures way outside these limits. Life, as we know it, would be difficult to sustain elsewhere.

DID YOU KNOW?

The surface temperature on the Moon varies between –173°C and +127°C. The range on Earth is –89°C to +58°C.

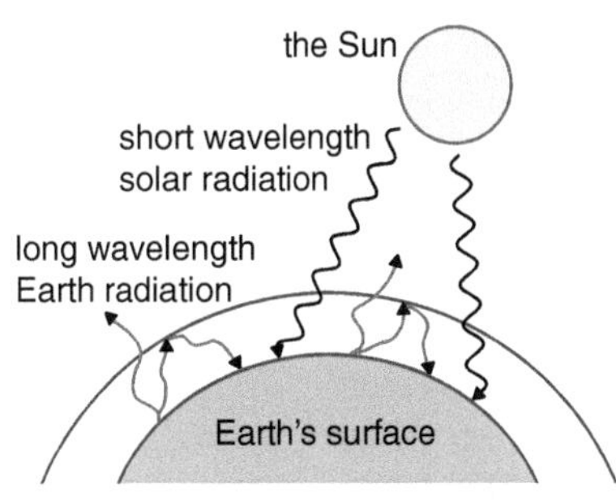

Figure 4.1.12 Short wavelength radiation from the Sun enters the atmosphere. Long wavelength radiation radiated back from the Earth is absorbed by the greenhouse gases

3 **Describe the wavelength of incoming solar radiation compared with outgoing radiation from the Earth. Use a diagram to help.**

4 **Venus has an atmosphere of carbon dioxide. Suggest whether the average temperature will be higher or lower than expected when considering its distance from the Sun. Explain your reasoning.**

What types of radiation are involved?

Radiation from the Sun is of short wavelength such as ultraviolet (UV), visible light and near infrared (near IR). It can pass through the atmosphere and not be absorbed by the gases. The radiation that comes back from the Earth is of a longer wavelength, which is far IR. This long wavelength IR radiation is absorbed by the molecules of water vapour, carbon dioxide and methane, which are the greenhouse gases.

The electromagnetic spectrum identifies radiation according to its wavelength and frequency.

This is an example of the spectrum to show the relative wavelengths of UV, visible light, near IR and far IR.

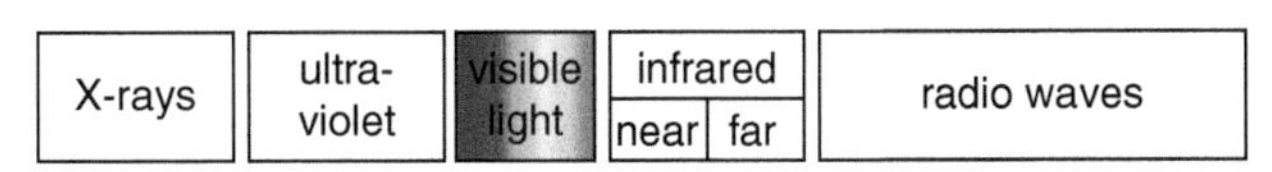

Figure 4.1.13 A representation of the relative wavelengths of UV, near IR and far IR. Wavelengths increase from left to right

MAKING LINKS

Look back at topic 1.4f to remind yourself about electromagnetic waves.

5 **Suggest what the Earth would be like if we did not have greenhouse gases.**

4.1e Human impacts on the climate

Learning objectives:

- describe human activities that increase the amounts of carbon dioxide and methane
- evaluate the quality of evidence about global climate change
- recognise the uncertainties in predictions about climate change.

KEY WORDS

average global temperature
climate
correlation
deforestation
global warming

There are many different causes of climate change, most of which are natural. Examples are changes in radiation from the Sun and volcanic eruption. There are certain human activities that are thought to be causes of recent climate change. If humans cause a change, it is called an anthropogenic change.

Human activities

We have seen that greenhouse gases are essential for maintaining stable temperatures on Earth so that life, as we know it, can be sustained. This is a natural phenomenon.

However, as far back as 1896 scientists were concerned that some human activities *increase* the amounts of greenhouse gases in the atmosphere and that we are altering the balance of the greenhouse effect. A Swedish chemist called Svante Arrhenius suggested that doubling the carbon dioxide level would cause the average temperature on Earth to increase by 5°C.

The original concern was over the burning of coal, producing carbon dioxide. As more fossil fuels are burned, more carbon dioxide is produced. You can see this in the graph.

The increase in the percentage of carbon dioxide in the atmosphere over the last 100 years is related to the increased use of fossil fuels – there is a **correlation**.

There are now concerns over many other activities that not only increase the amount of carbon dioxide but also increase the amount of methane.

The human activities that affect these greenhouse gases include more: combustion of fossil fuels and **deforestation** (carbon dioxide); and animal farming, through the digestion and decomposition of waste, and decomposition of rubbish in landfill sites (methane).

1. **Describe two reasons why carbon dioxide levels have increased in the last 200 years.**
2. **Explain why deforestation may lead to increased levels of carbon dioxide.**

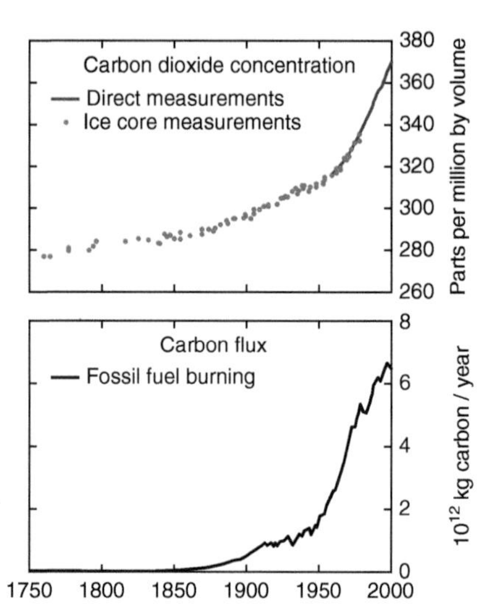

Figure 4.1.14 Graph showing that fossil fuel use has risen and carbon dioxide levels have risen at the same time – there is a correlation

Figure 4.1.15 Two examples of human activities that result in increased levels of carbon dioxide or methane in the atmosphere

The scale of global climate change

4.1e

The average global temperature of the Earth and its atmosphere is increasing. This is **global warming**. The scientific consensus is that the rise in greenhouse gas concentrations has caused the rise in temperature.

DID YOU KNOW?

The natural causes of increasing average global temperature are called 'forcings'.

(a)

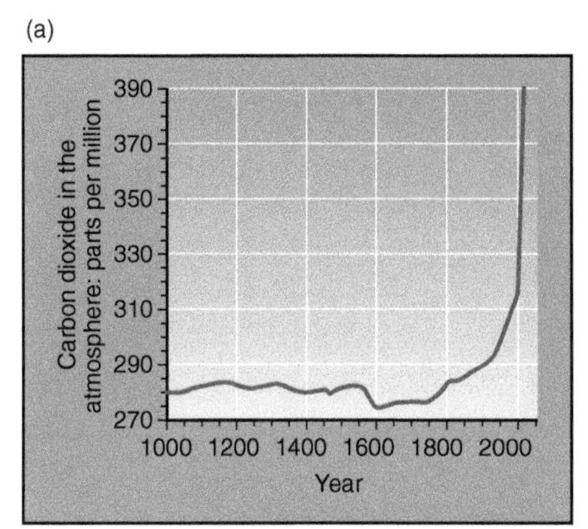

(b)

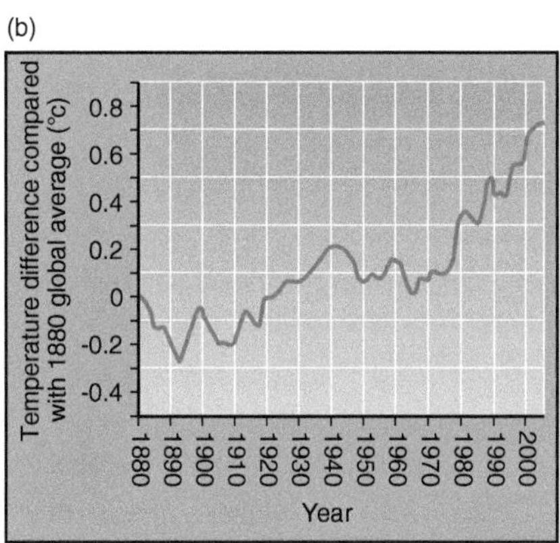

Figure 4.1.16 (a) Carbon dioxide in the atmosphere 1000–2000; (b) temperature difference from 1880 global average temperature, 1880–2000. What do these graphs tell you?

Climate describes the long-term patterns of weather in different parts of the world. Climate change is shown by changes to the patterns in air temperature, rainfall, sunshine and wind speed.

The major cause of global climate change is an increase in the **average global temperature**.

Human activities producing greenhouse gases add to the natural causes of the rise in average temperature.

These activities are burning fossil fuels, deforestation, increased agriculture and rubbish decomposition.

REMEMBER!

You do not need to know about the models of climate change or the work of the IPCC but you do need to know how to evaluate the quality of evidence.

3 **Describe which human activities may add to global warming.**

4 **Explain two things that might happen if global warming continues.**

DID YOU KNOW?

Petrol, used in car engines, comes from a fossil fuel. It burns in oxygen to make carbon dioxide. Increased carbon dioxide levels have been linked to climate change.

Modelling climate change

It is very difficult to model such complex systems as global climate change. This is why the Intergovernmental Panel on Climate Change (IPCC) was set up through the United Nations in 1988. This panel draws on evidence from hundreds of scientists who contribute to papers and review each other's evidence. They try to model what might happen and give probabilities to future events. For example, they *predict* that oceans will warm and that it is *very likely* that Arctic ice cover will continue to decrease.

Discussion of climate change presented in the media is often simpified, and speculation and opinions may be presented based on only parts of the evidence and part of the complex interpretations. At any stage of any explanation there may also be a biased viewpoint.

5 **There has been an increase in methane from agriculture over the last 50 years. Explain how you would search for evidence of its effect on the temperature of the Earth.**

6 **Explain why a simple model of predicted climate change seen on a 2-minute TV news item may not give quality information.**

4.1f Effects of climate change

Learning objectives:

- describe four potential effects of global climate change
- discuss the scale and risk of global climate change
- discuss the environmental implications of climate change.

KEY WORDS

distribution
erosion
food producing capacity

The increasing levels of carbon dioxide from the use of fossil fuels and other human activities is a concern, particularly as the global population has risen so much over the last few decades. This means more human activities, which will produce higher levels of greenhouse gases. This will have a number of consequences that will affect all species living on Earth.

Figure 4.1.17 The potential effects of global climate change by warming include:

- sea level rise, which may cause flooding and increased coastal **erosion**
- more frequent and severe storms
- changes in the amount, timing and distribution of rainfall

The risks of global climate change

Scientists are concerned that if the average global temperature rises by 2°C there will be irreversible changes taking place, caused by a change in climate. Climate is not a short-term weather pattern change that follows cycles, but a long-term change in the whole weather system. A large number of scientists contribute evidence to the Intergovernmental Panel on Climate Change (IPCC) so that the risks of climate change can be assessed. No government can do this on their own, because it is a global issue.

The scientists use a number of indicators to demonstrate how the climate is changing. One such sensitive indicator is the retreat of glaciers.

Glaciers are mountain reserves of fresh water that are frozen. The mass of water now held in glaciers continues to decline.

1. **Explain the difference between a weather pattern and a climate change.**
2. **Suggest why it is important that glaciers remain high in mass.**

Figure 4.1.18 Glaciers are sources of frozen fresh water that are diminishing. The sign here shows where the glacier was in 1908

Environmental concerns of climate change

There will be big changes in the environment while global warming continues. The impact will affect different regions in different ways. In some regions the impacts are already changing lives, for example for the people of the Arctic region where ice sheets are retreating.

The main effects are predicted to be:

- temperature stress for humans and wildlife (some areas will become too hot for people to make a living)
- water stress for humans and wildlife (fresh water supplies will reduce in some regions)

- changes in the **food producing capacity** of some regions (the production of wheat and maize has already been affected)
- changes to the **distribution** of wildlife species (migration patterns are changing)

Scientists are working on complex models to try to determine what may happen, as there are many interrelated factors to take into account.

3 Explain how each of the potential social effects of climate change listed in this section are related to the physical changes that are taking place.

REMEMBER!

You are not expected to remember all the details involved in the complex issue of global climate change but you should aim to be able to discuss the possible causes and implications.

Case studies

The white lemuroid possum is the first mammal in Australia to have become almost extinct due to global warming. There are just four known adults left. The white possum's habitat spans cooler areas of high-altitude rainforest. Possums are vulnerable to increases in environmental temperature because they cannot maintain their body temperature.

Little terns are vulnerable to high tides and storms. These are happening more often because of global warming. Little terns migrate to the UK each spring and make their colonies just above the high tide line. Their nests are vulnerable to flooding by stormy seas.

Coastal mangrove forests grow in equatorial regions. Increasing numbers of storms and typhoons are undermining the fine sediment that the mangroves grow in. Seedlings cannot root and essential nutrients for the mangrove ecosystems are washed away.

Increasing sea temperatures are causing bleaching of coral reefs. The algae living in corals cannot survive and the food source for the corals is affected. Around 20% of coral reefs have been destroyed in just 50 years.

REMEMBER!

Learn three ways that global warming affects biodiversity.

DID YOU KNOW?

If sea levels rose by 1 m, half of the world's important coastal wetlands would be threatened.

Figure 4.1.19 The coral in the centre of the image has been bleached, but the surrounding corals have not, or have recovered their colour

4 Explain how global warming affects biodiversity.

5 Evaluate the impact of climate change on ecosystems.

Figure 4.1.20 How does global warming affect little terns?

4.1g Mitigating the effects of climate change

Learning objectives:

- explain that carbon footprints can be reduced by reducing emissions of carbon dioxide and methane
- describe how emissions of carbon dioxide can be reduced
- describe how emissions of methane can be reduced.

KEY WORDS

alternative energy
carbon capture
carbon off-setting
carbon neutrality

There are two responses to increasing average global temperature – reduce it or adapt to it. Adapting to it means coping with the social changes and ways of living. Reducing it means reducing greenhouse gases. The reduction of carbon dioxide and methane emissions is called a mitigation action so that we do not have to adapt.

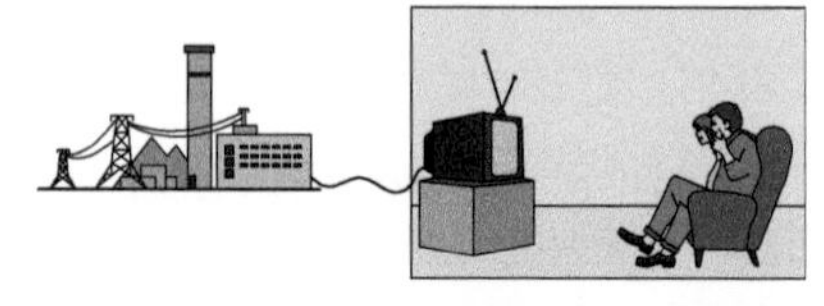

Figure 4.1.21 We do not burn fuel when watching TV – or do we?

What is a carbon footprint?

A carbon footprint is the total amount of carbon dioxide and other greenhouse gases emitted over the full life cycle of a product, service or event.

We can often see what we are doing when burning a fuel that gives out Carbon dioxide. Examples are when we ride in a car or bus. At other times we may not notice. For instance, when we turn on the light or electric shower we do not always realise that the electricity we are using was probably generated by burning fossil fuels. We also do not take into account how much energy was used in making the car or the shower in the first place, and so how much carbon fuel was used to make it.

There is now an Intergovernmental Agreement, called the Kyoto Protocol, that, in essence, states that we should all reduce our carbon footprint.

Figure 4.1.22 Alternative energy sources: using solar energy

1. **Explain how playing a computer game has a carbon footprint.**
2. **Explain why governments agree that carbon footprints need reducing.**

Figure 4.1.23 Alternative energy sources: using nuclear energy

Reducing our personal carbon footprint

Actions to reduce our personal carbon footprint include:

- increased use of **alternative energy** supplies
- energy efficiency in our homes
- energy efficiency by driving cars that use less fuel (i.e. have higher mpg figures).

3. **Explain how using solar energy reduces carbon dioxide emissions.**
4. **Explain whether or not using solar energy panels reduces the carbon footprint.**

Figure 4.1.24 Alternative energy sources: using wind energy

5 **Look at Figure 4.1.25. Explain how each of the methods labelled help to reduce the amount of energy we use in our homes.**

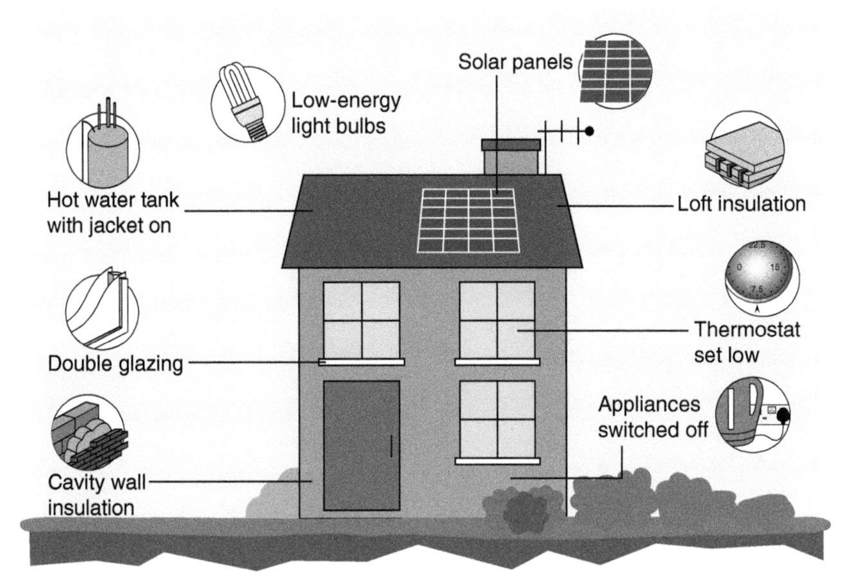

Figure 4.1.25 A diagram to show how households can use less energy by installing more insulation or use energy more efficiently

Governments reducing the carbon footprint

It is important that we reduce our own personal carbon footprint but there are some measures that governments or large companies or organisations will need to implement as we cannot do this on our own. These include:

- **Carbon capture** and storage – taking the emissions from a power station and depositing the carbon dioxide into an underground geological formation so that it will not enter the atmosphere. This is a new idea and so will need investment.
- Recycling carbon – many local governments have implemented waste recycling and composting schemes to help reduce greenhouse gas emissions from landfills and incinerators.
- **Carbon off-setting**, including increasing the carbon sink through tree planting and reforestation. Carbon off-setting can take place on a small local scale or a large scale such as the European Union Emissions Trading System. This trading market grew out of the agreements reached through the Kyoto Protocol.
- **Carbon neutrality** – zero net release – is the aim of carbon off-setting. People and organisations aim to take out of the atmosphere as much carbon dioxide as they put into it, for example, by planting trees that use the equivalent carbon dioxide quantity as their emission from their electricity use. The idea is to produce a zero carbon footprint.

6 **Explain two different ways by which forestation can help reduce the carbon footprint.**

DID YOU KNOW?

This idea could have been tried by carbon capture and storing the carbon dioxide in the ocean, but this would have made the acidification of the oceans worse.

REMEMBER!

You do not need to know all the details of this topic but it is important to know where to find current information and be able to discuss it.

4.1h Air pollution

Learning objectives:

- describe how carbon monoxide, soot, sulfur dioxide and oxides of nitrogen are produced by burning fuels
- predict the products of combustion of a fuel knowing the composition of the fuel
- predict the products of combustion of a fuel knowing the conditions in which it is used.

KEY WORDS

hydrocarbons
particulates

Burning a fuel produces energy, which is what we want. However, burning a fuel can also produce unwanted products. If a fuel contains sulfur, then acid rain can result. If a fuel is burned in a shortage of oxygen then soot (carbon particles) may be produced, which is dirty and bad for human health. Burning fuels is a major source of pollution in the atmosphere.

Common atmospheric pollutants

The combustion of fuels is a major source of atmospheric pollutants.

Most fuels, including coal, contain carbon and other elements such as hydrogen and may also contain some sulfur. The sulfur combines with oxygen to make sulfur dioxide.

When a fuel is burned the gases given off will include carbon dioxide and water vapour. However, they may also include unwanted gases such as carbon monoxide, sulfur dioxide and oxides of nitrogen.

Burning fuels may also release solid particles and unburned **hydrocarbons**. These form **particulates** in the atmosphere.

1 Write down three unwanted pollutants that are released when a fuel is burned.

DID YOU KNOW?

Carbon dioxide is not included in this group as, although it is a greenhouse gas and levels need to be reduced, it is not a harmful gas.

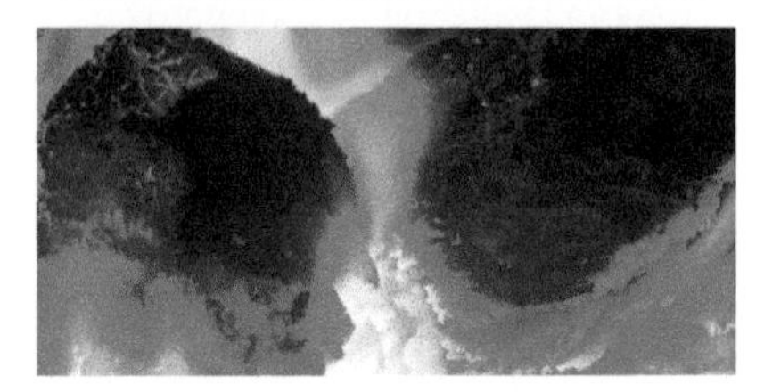

Figure 4.1.26 An image of coal burning. There are regulations in many urban areas on the types of fuels you can burn so that the air in cities is kept clean

Problems caused by pollutant gases

Carbon monoxide is formed by the incomplete combustion of hydrocarbon fuels when there is not enough air. Carbon monoxide is a toxic gas that combines very strongly with haemoglobin in the blood. At low doses it puts a strain on the heart by reducing the capacity of the blood to carry oxygen. At high doses it makes people lose consciousness or even die.

Sulfur dioxide is produced by burning fuels that contain some sulfur. These include coal in power stations and some diesel fuel burned in ships and heavy vehicles. Sulfur dioxide turns to sulfuric acid in moist air. Along with oxides of nitrogen, sulfuric acid dissolved in rain water is called acid rain, and it damages plants and buildings. It also harms living organisms in ponds, rivers and lakes.

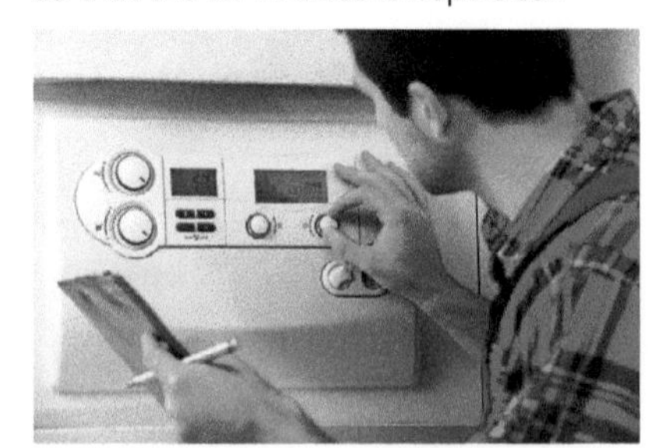

Figure 4.1.27 The control panel of a boiler. Regularly checking boiler emissions for carbon monoxide is important

Figure 4.1.28 Cars need to have an annual check for the levels of NO_x gases they emit

Oxides of nitrogen are produced by the reaction of nitrogen and oxygen from the air at the high temperatures involved when fuels are burned. Sulfur dioxide and oxides of nitrogen cause or worsen respiratory diseases such as emphysema and bronchitis.

2 **Explain the term 'incomplete combustion'.**

3 **Why is carbon monoxide such a dangerous pollutant?**

4 **How is acid rain formed?**

Particulates

Solid particles and unburned hydrocarbons form particulate matter (PM) in the atmosphere.

Particulates cause health problems for humans because of damage to the lungs. The particles have a range of diameters and some are more penetrating in the lungs than others. Particles with diameters of less than 10 micrometers (PM_{10}) can enter the airways to the lungs. With diameters of less than 2.5 micrometers ($PM_{2.5}$) they can penetrate the alveoli. Some have different shapes and so they are just the right shape to penetrate and remain in lung tissue. Some are small enough to get into the bloodstream. Particulates are deemed to be the most lethal pollution as they have the ability to go deep into lungs and enter the bloodstream. They can lodge in the body causing heart attacks and DNA mutations. There is no safe level of PM.

Figure 4.1.30 Dust and other particulates from natural causes or burning fossil fuels can cause damage to the lungs

5 **Particulates can cause health problems for humans.**

a **Give two different ways that particulates can cause damage in the human body.**

b **Governments have put measures in place to reduce particulate emissions. Explain the impact this may have on the climate.**

DID YOU KNOW?

Nitrogen dioxide is NO_2, but there are other oxides of nitrogen in exhaust gases, so they are grouped together as NO_x. The common name is then NO_x gases.

Figure 4.1.29 The damaging effects of acid rain can be seen on this statue

KEY INFORMATION

Remember rain is already slightly acidic because it dissolves carbon dioxide from the atmosphere. Sulfur dioxide and oxides of nitrogen make the rain *more* acidic to levels where damage is done.

4.1i The water cycle

Learning objectives:

- explain the stages in water cycles
- explain the importance of the water cycle to living organisms.

KEY WORDS

aquifer
precipitation
run-off

The water cycle provides fresh water for plants and animals on land. It describes how water moves on, above or just below the ground. Water molecules move between rivers, oceans and the atmosphere by precipitation, evaporation, transpiration and condensation.

Water cycles

Earth is known as the 'blue planet' because 71% of its surface is covered with water. Water also exists below the land surface and as water vapour in the air. Of this abundance of water, only a small percentage (about 0.3%) is usable by humans. The other 99.7% is in the oceans, soils and ice caps and floating in the atmosphere.

There is no new water on the planet; it is ancient water that is continually recycled.

The water cycle is a natural cycle in which water from the oceans evaporates, condenses to form clouds, moves over high ground and **precipitates** (rains, snows, sleets or hails) water back to the ground to rivers, lakes and aquifers. **Aquifers** are bodies of permeable rock that can contain or transmit groundwater. In England and Wales there are 11 main aquifers.

For thousands of years we have lived with this natural cycle: drinking water from wells, rivers and lakes and excreting waste into the ground, where the solids are broken down by bacteria. The water finds its way back to groundwater or rivers and evaporates to form clouds.

One problem associated with the water cycle is water contamination. Chemicals that go into the water often are very difficult, if not impossible, to remove. One potential source of contamination of water is **run-off**: the overland flow of water. Although precipitation naturally causes run-off, stripping vegetation from land can add to the severity of it. The soil from such areas, as well as any pesticides or fertilisers that are present, are washed into the streams, oceans and lakes.

Figure 4.1.31 The Perito Moreno Glacier, Patagonia, Argentina

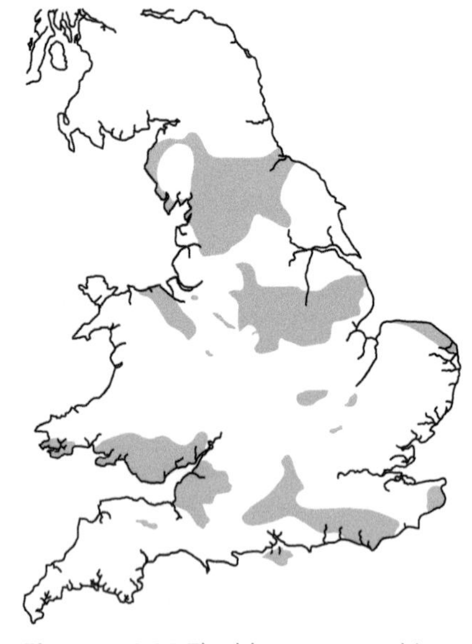

Figure 4.1.32 The blue areas on this map of England and Wales show the extent of Carboniferous Limestone aquifer – one of the main aquifers in England and Wales

1. **What are aquifers?**
2. **Explain why only 0.3% of total water is usable by humans.**

Explaining the water cycle

Water is continuously recycled by the following processes.

- Precipitation: water droplets in clouds get bigger and heavier and they fall as rain, snow or sleet.
- Evaporation: water evaporates as it is heated by the Sun's energy. Water vapour is carried upwards in convection currents.
- Transpiration: water vapour is released into the air through stomata in leaves.
- Condensation: water vapour rises, cools and condenses back into water droplets, which form clouds.

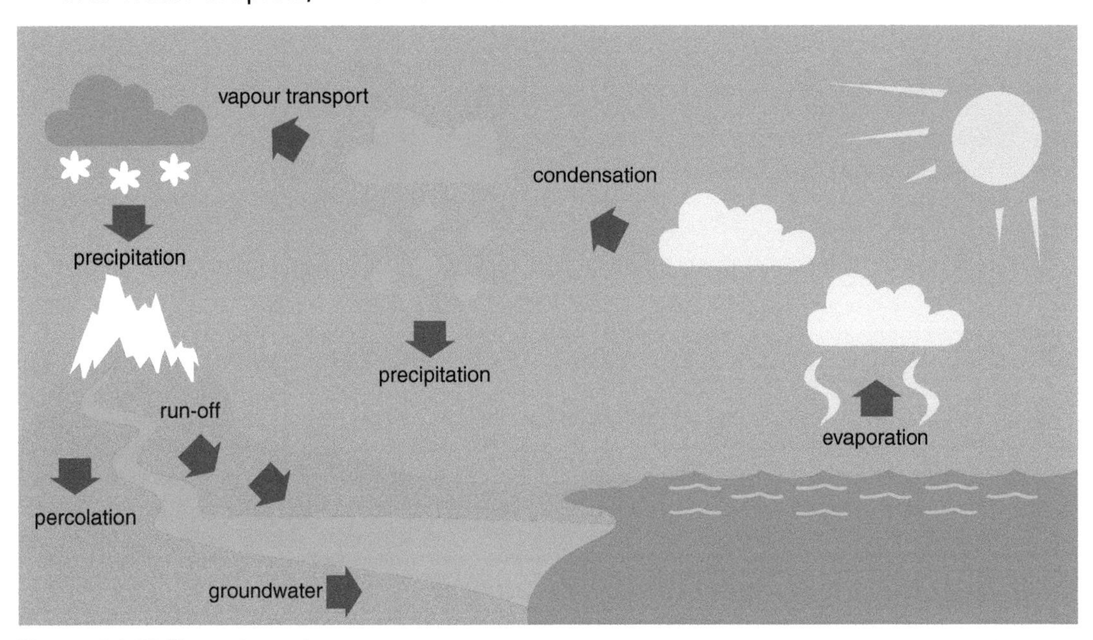

Figure 4.1.33 The water cycle

The water cycle is important because it circulates water that:

- maintains habitats
- maintains internal fluids and transport systems
- is needed for chemical reactions
- is a reactant in photosynthesis.

3 Explain the stages in the water cycle.

4 In which stage of the water cycle can water be a:

a gas

b solid?

DID YOU KNOW?

Clean water saves more lives than medicines do. That is why, after earlier disasters and issues in developing countries, relief organisations concentrate on providing clean water supplies.

REQUIRED PRACTICAL

4.1j Analysis and purification of water samples from different sources, including pH, dissolved solids and distillation

Learning objectives:

- describe how safety is managed, apparatus is used and accurate measurements are made
- recognise when sampling techniques need to be used and made representative
- evaluate methods and suggest possible improvements and further investigations.

KEY WORDS

boiling point
distillation
purity

These pages are designed to help you think about aspects of the investigation rather than to guide you through it step by step.

Fresh drinking water is essential for everyone. Water needs to be made safe, which means testing and treatment. River water can be filtered and bacteria removed or seawater can be distilled.

Analysing water samples

Various skills are needed to analyse water and carry out a **distillation**, including applying sampling techniques, manual dexterity and the safe use of heating devices.

Planning your sampling

One way of selecting samples is by section sampling. In this method, if 30 samples of river water are to be tested, they must be taken equally in each section over the whole stretch of river.

Imagine a new town is being planned that needs potable water. A river runs nearby. Your job is to test the water to find the best place to site a water treatment plant. The pH needs to be 6.5–9.5. The river is 1 km long. It can be divided into three sections:

A. Open countryside over chalk (0.5 km)
B. Past a large dairy farm (0.2 km)
C. Past an old lead mine (0.3 km)

Think about these questions:

1. **Thirty samples of river water are tested altogether. Calculate the number of samples needed for each selection.**
2. **Explain how you will test the pH of the samples.**

SAFETY

For this practical, a risk assessment is required, and eye protection must be worn.

DID YOU KNOW?

There are two steps to getting the most accurate results from a chemical analysis: i) collecting samples and ii) analysing samples. Errors need to be reduced in both steps.

MAKING LINKS

When answering question 1, it might help you to refer back to topic 3.3k, about scientific sampling.

The effects of changing pH are covered in more detail in topics 7.3f and 7.4l.

3. **Other than 'What is the pH?', what other questions will you ask about the sites that the samples are taken from?**

Some countries do not have access to fresh water, only seawater. This has dissolved salts in it that need to be removed. This is done by distillation. Figure 4.1.34 shows the apparatus for the standard technique.

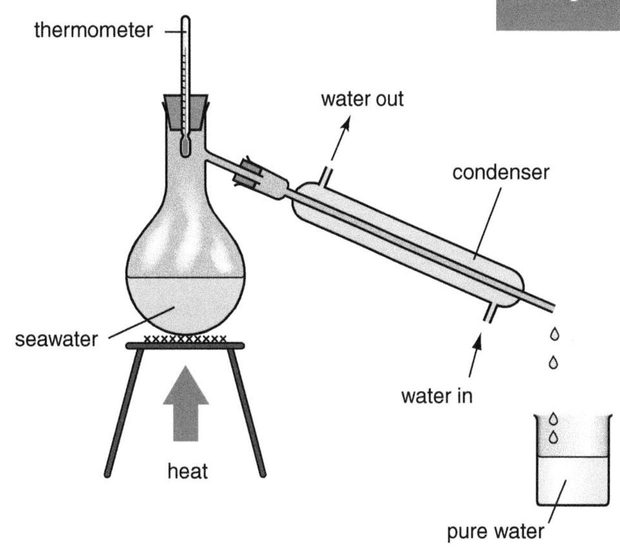

Figure 4.1.34 Distillation to remove salt from seawater

Think about these questions:

4. **Explain why the water in the flask is heated.**
5. **Why is a condenser attached to the arm of the heating flask?**
6. **Describe how distillation works.**
7. **Explain the purpose of the thermometer.**
8. **Why does the salt not evaporate?**
9. **Explain why the cold water flows from the *bottom* of the condenser out into the sink from the *top* of the condenser.**

Analysing the results

When three samples are gathered, they are tested for **purity** by finding the **boiling point**. Jo and Akira tested their three samples:

	1st sample	2nd sample	3rd sample
Boiling point before distillation in °C	102.1	102.0	101.9
Boiling point after distillation in °C	101.4	101.6	100.8

10. **Explain their results.**

The way the equipment was used, and how techniques can be improved, should be considered.

11. **Suggest ways to improve the sampling of river water above. For example, after answering question 3, would all the sites need to be included in the sampling?**
12. **Suggest ways to improve the measurement of pH or the distillation technique shown.**
13. **Suggest what other investigations could be done to analyse drinking water samples.**
14. **Taking averages is common in science.**
    - **a** **Calculate an average of the three sample boiling points both before and after the distillation.**
    - **b** **Explain whether separate or average boiling points are more useful.**
    - **c** **The pH for 50 samples of river water at different locations was measured. Suggest whether an average or separate values would be more useful.**

REMEMBER!

The water in the cooler jacket of the condenser does not mix with the water going through the centre of the condenser.

KEY INFORMATION

A simplified equation for section sampling could be: Sample number for each section = total sample number × (size of section ÷ total size of all sections)

Therefore, the sample size for section A is estimated to be:
30 samples × (0.5 km ÷ 1 km) = 15 samples

4.1k Sources of potable water

Learning objectives:

- distinguish between potable water and pure water
- describe the differences in treatment of ground water and salty water
- give reasons for the steps used to produce potable water.

KEY WORDS

desalination
potable
reverse osmosis
sedimentation

Safe drinking water is essential for life. Where does it come from and how is it made safe? For humans, the appropriate quality drinking water should have sufficiently low levels of dissolved salts and microbes. How do you get safe water if it does not rain?

Producing potable water

Water that is safe to drink is called **potable** water. In a chemical sense potable water is not pure water because it contains dissolved substances, which are needed by the body.

There are several methods used to produce potable water. The method chosen depends on available supplies of water and local conditions.

Traditionally, settlements grew up on river banks or close to lakes because there was an abundant source of water. Other settlements relied on water from groundwater that collected in aquifers. They extracted it from wells.

In the UK, as the population grew in cities, reservoirs were built to provide constant fresh drinking water for everyone.

The water is provided by rain. This water, with low levels of dissolved substances (fresh water) collects in the ground and in lakes and rivers and is then treated to make it potable.

There are three main stages in water treatment:

- **sedimentation** of particles so that solids clump together and drop to the bottom
- filtration of very fine particles using sand
- sterilising to kill microbes, which could otherwise cause diseases such as typhoid or cholera. Sterilising agents that are added to the water include chlorine, ozone or ultraviolet light.

1. **Explain the stages involved in rain becoming drinking water.**
2. **Suggest why the sterilisation stage of water purification is carried out last.**
3. **Some data for a drinking water sample are shown below. Explain whether the water is potable.**

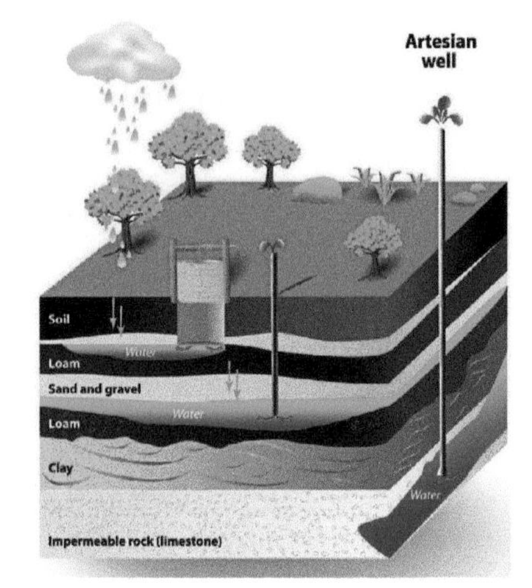

Figure 4.1.35 Groundwater is held in aquifers and is reached from the surface through artesian wells

Figure 4.1.36 Huge reserves of water are held to provide drinking water

Figure 4.1.37 A sedimentation tank at a water-treatment works

	Bacteria per 100 ml	Lead / µg per ml	Nitrate / mg per l
Sample of drinking water	2	2	15
Maximum allowed	0	10	50

Potable water from seawater

Seawater has so many substances dissolved in it that it is undrinkable. However, if supplies of fresh water are limited, **desalination** of salty water or seawater is necessary to remove the dissolved substances.

Desalination can be done by distillation or by processes that use membranes such as **reverse osmosis**. These processes require large amounts of energy, so are very expensive. They are only used when there is not enough fresh water.

Many countries with little fresh water rely on desalination processes using distillation or reverse osmosis. Spain is the largest operator of desalination plants in Europe.

4. **Suggest why Spain operates more desalination plants than the UK.**
5. **Explain which stage in distillation makes the process so costly.**

The urban water cycle

Urban lifestyles and industrial processes produce large amounts of waste water that require treatment before being released into the environment. This can be from domestic washing machines, dishwashers and showers. It can be from industrial processes such as cleaning cycles and solvent usage in manufacturing processes and from cooling systems.

The industrial waste water may contain pollutants such as organic matter and harmful chemicals that require extra treatment before it can go back into the fresh water cycle.

Agriculture is by far the biggest user of water. Agricultural waste water can cause problems to ecosystems by containing too many nutrients.

This is why sewage and agricultural waste water require treatment to ensure the removal of organic matter and harmful microbes.

Sewage treatment includes:

- screening and grit removal – at this initial stage, most of the inorganic material is removed from the sewage
- **sedimentation** to produce sewage sludge and effluent
- anaerobic digestion of sewage sludge – anaerobic bacteria are added to break down or 'digest' the organic matter in the sludge to water and minerals. This process releases methane, carbon dioxide and hydrogen gases.
- **aerobic** biological treatment of effluent – oxygen is pumped through the sewage treatment works to allow aerobic bacteria to break down any remaining organic matter in the effluent.

Figure 4.1.38 Aerial view of a sewage-treatment works, showing a number of tanks where sedimentation and treatments with bacteria take place

MAKING LINKS

Refer back to topics 2.1a and 2.1b for a reminder about aerobic and anaerobic respiration.

6. **Suggest how sedimentation works and predict one limitation.**
7. **Explain why urban lifestyles need organised waste water management.**

MATHS SKILLS

4.1l Use ratios, fractions and percentages

Learning objectives:

- extract information from charts, graphs and tables
- use orders of magnitude to evaluate significance of data.

KEY WORDS

approximately
fractions
orders of magnitude
percentages
segment

To support life as we know it there must be a proportion of oxygen in the atmosphere. The percentage of oxygen in the air was not always 20% and there was no oxygen until plants began to photosynthesise. How does a fifth of our atmosphere contain oxygen when animals are constantly using it? Why did the percentage composition of the atmosphere change?

The composition of air

Air is a mixture of gases. It contains nitrogen and oxygen and there are other gases present such as carbon dioxide, water vapour and noble gases like argon in small amounts.

If the other gases are taken out of the diagram it will show that the air is **approximately** $\frac{4}{5}$ nitrogen and $\frac{1}{5}$ oxygen. Approximately means 'about'. If the circle is divided into 5 equal parts, 4 parts represent nitrogen and 1 part represents oxygen. The **fractions** are $\frac{4}{5}$ and $\frac{1}{5}$, *respectively.*

If the circle is now divided into 10 equal parts (so double the number of parts) then 8 represent nitrogen and 2 represent oxygen. The fractions are $\frac{8}{10}$ and $\frac{2}{10}$, respectively.

Now imagine that each of the ten **segments** were divided into 10. There would be 100 segments in the circle. 80 of these would represent nitrogen and 20 segments would represent oxygen. The fractions are $\frac{80}{100}$ and $\frac{20}{100}$, respectively.

Fractions representing $\frac{1}{100}$ are **percentages.** So $\frac{80}{100}$ is 80% and $\frac{20}{100}$ is 20%.

Now as fractions, percentages and a ratio the summary is:

$\frac{4}{5}$	$\frac{8}{10}$	$\frac{80}{100}$	80%	Ratio 4:1
$\frac{1}{5}$	$\frac{2}{10}$	$\frac{20}{100}$	20%	

1 **The atmosphere of Mars is approximately $\frac{19}{20}$ carbon dioxide, $\frac{1}{20}$ other gases. What is: a the percentage of gases; b the ratio of gases?**

2 **If an ancient atmosphere had been 3:1 methane to ammonia, what would be the percentage of each gas?**

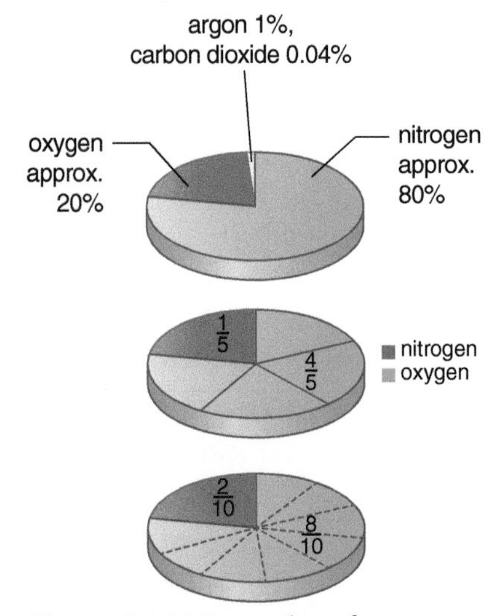

Figure 4.1.39 Proportion of gases

REMEMBER!

If you are drawing a pie chart, you draw the segments in order from largest to smallest clockwise.

Extracting and interpreting data from graphs

You can extract the average temperature for 1950 from Figure 4.1.40, but if you are asked to suggest why temperatures are rising, you will have to use your knowledge and understanding of the reasons for global warming.

Some graphs have two vertical axes. In Figure 4.1.41, the red line denotes total crop yield while the grey line denotes world population. You need to make sure you look at the correct axis when you are extracting data.

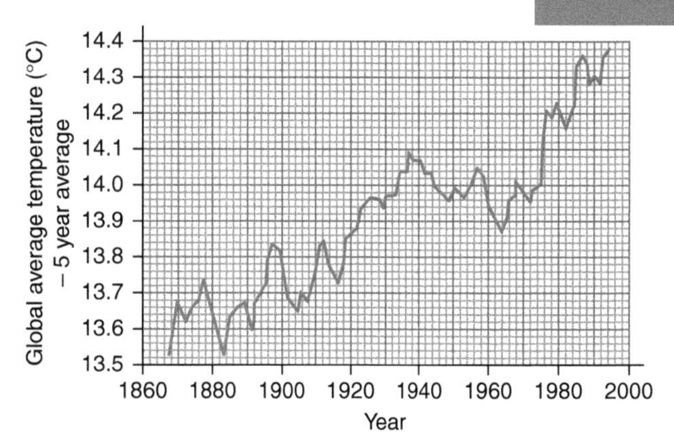

Figure 4.1.40 Changes in average global temperatures

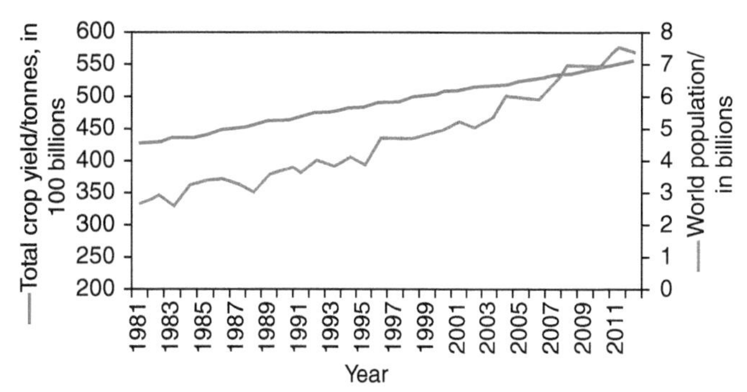

Figure 4.1.41 Graph showing world population and total crop yield

3 **What is the highest temperature shown in Figure 4.1.41? How much higher is it than the temperature in 1880?**

4 **Discuss whether or not the data in Figure 4.1.42 show that there has been a growing threat of malnutrition since 2009.**

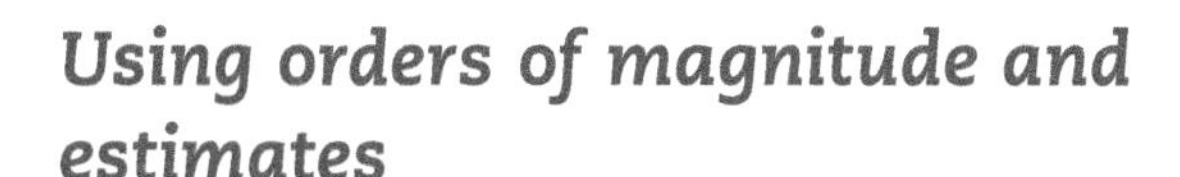

Using orders of magnitude and estimates

Orders of magnitude are often used to make very approximate comparisons between things that are very different sizes. The difference is expressed in factors of ten.

	Typical diameter
Grain of sand	1 mm
Grain of pollen	100 µm
Particle of soot from a diesel car	100 nm

Diameter of a grain of sand = 1 mm = 10^{-3} m

Diameter of a particle of soot = 100 nm = 10^{-9} m

So the difference in their order of magnitude is 6 (the difference between –3 and –9).

Scientists and mathematicians sometimes have to estimate values that they don't know exactly. Using orders of magnitude to estimate is really useful.

You can obtain surprisingly accurate answers from simple calculations if you make correct estimates of the order of magnitude for each quantity.

5 **What is the difference in the order of magnitude between a grain of pollen and a particle of soot?**

6 Using the correct order of magnitude, state how long it would **take you to walk from London to Glasgow.**

REMEMBER!

1 millimetre (mm) = 10^{-3} m
1 micrometre (µm) = 10^{-6} m
1 nanometre (nm) = 10^{-9} m

Check your progress

You should be able to:

describe the ideas about the Earth's early atmosphere.	→	interpret evidence about the Earth's early atmosphere.	→	evaluate different theories about the Earth's early atmosphere.
identify the processes allowing oxygen levels to increase.	→	explain the role of algae in the composition of the Earth's early atmosphere.	→	describe how photosynthesis contributed to the Earth's early atmosphere.
describe the main changes in the atmosphere over time.	→	describe some of the likely causes of these changes.	→	explain how levels of carbon dioxide are measured.
name the greenhouse gases.	→	explain the greenhouse effect.	→	explain these processes as interaction of radiation with matter.
evaluate the quality of evidence about global climate change.	→	describe uncertainties in the evidence base.	→	recognise the importance of a worldwide effort to bring about change.
discuss the scale of global climate change.	→	discuss the risk of global climate change.	→	discuss the environmental implications of global climate change.
identify the major source of air pollutants.	→	describe the health problems related to some products of combustion.	→	explain how pollutants harm health and the environment.
explain the importance of the water cycle to living organisms.	→	describe the process of distillation.	→	describe principal methods for increasing the availability of potable water.

Worked example

The air contains a mixture of gases. One gas occupies 80% of the total.

1 Identify the gas.

a oxygen b carbon dioxide

c nitrogen (circled) d helium

The answer nitrogen is correct.

2 Describe which gases formed the early atmosphere and where they came from.

They were carbon dioxide and nitrogen and they came from volcanoes.

This answer is correct. The process is called degassing.

3 Describe how in the early atmosphere:
a levels of carbon dioxide were reduced
b levels of oxygen were increased

a Plants used up the carbon dioxide.

b Plants gave out oxygen.

The answers given for a) and b) are not incorrect but they lack detail. Fuller answers would include: a) Plants used carbon dioxide for photosynthesis. Carbon dioxide was also used to make animal skeletons and shells, which formed sedimentary rocks.
b) Through photosynthesis plants produced oxygen. Over time levels increased.

4 Describe how the levels of oxygen and carbon dioxide are maintained in the atmosphere today.

Plants use carbon dioxide during photosynthesis and give out oxygen. Animals use oxygen for breathing in and breathe out carbon dioxide. It's a cycle.

The answer is correct for maintaining levels of oxygen. However, air is breathed in and out. Oxygen is used for respiration in cells, not for breathing.

5 Write the symbol equation for photosynthesis.

$CO_2 + H_2O \rightarrow C_6H_{12}O_6 + O_2$

The student has recalled the reactants and products but has not understood how to balance the equation, which should be $6CO_2 + 6H_2O \rightarrow C_6H_{12}O_6 + 6O_2$

6 Describe the steps needed to make river water into potable water.

The water needs to be left in a holding tank for solids to sink as sediment. The water is filtered through sand to take out any small pieces of waste and fine particles. The microbes are killed using chlorine.

The answer is full and correct. The last stage is called sterilisation.

End of chapter questions

Getting started

1. **Identify the gas that causes acid rain.**

   A sulfur dioxide **B** hydrogen
   C carbon monoxide **D** argon — 1 Mark

2. **Which of these gases is a greenhouse gas?**

   A hydrogen **B** oxygen **C** argon **D** methane — 1 Mark

3. **Suggest two factors that caused carbon dioxide levels to decrease in the Earth's early atmosphere.** — 2 Marks

4. **A species of bird has started to migrate to a different island further north than usual. What could this be due to?**

   A increased oxygen levels **B** global warming
   C acid rain **D** ozone depletion — 1 Mark

5. **Which stage is used to destroy microbes when treating water?**

   A filtration **B** sedimentation **C** chlorination **D** distillation — 1 Mark

6. **Peter is building his own new house. Suggest two things he could include in the building to reduce his carbon footprint.** — 2 Marks

7. **Match the readings to the change.**

Readings	Change
ppm CO_2 27 31 35 37	increase in global warming
ppm NO_x 3.6 4.4 5.2 5.6	increase in greenhouse gas
°C 12 12.5 12.9 13.1	increase in acid rain

2 Marks

Going further

8. **Describe how potable water is made from seawater.** — 2 Marks

9. **Identify the atmospheric pollutants given off from the incomplete combustion of fuels.** — 1 Mark

10. **Compare the wavelength of radiation entering the Earth's atmosphere to that being given out at the Earth's surface.** — 1 Mark

11. **Describe how respiration increases the level of carbon dioxide.** — 2 Marks

12. **Explain the role of anaerobic bacteria in the treatment of sewage.** — 2 Marks

13. **These average winter temperatures at a glacier have changed each year. Explain what may happen in 10 years' time.**

    Data table for Years 1–7: — 2 Marks

−2.0°C	−1.8°C	−2.5°C	−1.5°C	−1.5°C	−1.1°C	−0.8°C

More challenging

14 Describe the difference in wavelength between radiation reaching the Earth from the Sun and radiation reflected back from the Earth's surface. How does this affect how much radiation is absorbed by greenhouse gases in the atmosphere?

2 Marks

15 **The diagram shows the carbon cycle.**

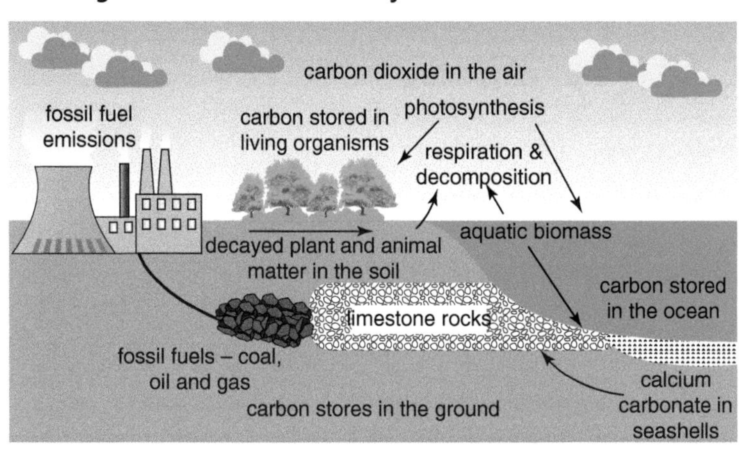

Name three different stores of carbon shown on the diagram. Use the diagram to explain, as fully as you can, how carbon is cycled.

6 Marks

16 **The amount in millions of tonnes of NO_2 and SO_2 produced in the UK is shown in the table from 1970 to 2010, every 5 years.**

Year	1970	1975	1980	1985	1990	1995	2000	2005	2010
NO_2	2.67	2.57	2.67	2.60	2.88	2.31	1.79	1.58	1.12
SO_2	6.32	5.17	4.76	3.68	3.68	2.36	1.22	0.71	0.43

Describe and explain the trend for NO_2 and SO_2 produced.

2 Marks

Most demanding

17 **Explain what is meant by the term 'carbon off-setting'.**

2 Marks

18 **Life on Earth depends on water, on land and in the seas. Explain, as fully as possible, the importance of water in supporting life on Earth.**

4 Marks

19 **Earth's early atmosphere was probably similar to Venus's atmosphere now.**

	Earth's atmosphere / %	Venus's atmosphere / %
Nitrogen	78.1	3.5
Oxygen	21.0	trace
Carbon dioxide	0.039	96.5

a **Explain why planets such as Venus are used as a model for the early atmosphere on Earth.**

b **Work out how many times more carbon dioxide there is on Venus compared to Earth.**

c **Explain the difference in oxygen concentration.**

4 Marks

Total: 40 Marks

ECOSYSTEMS AND BIODIVERSITY

IDEAS YOU HAVE MET BEFORE:

ORGANISMS IN AN ECOSYSTEM DEPEND ON EACH OTHER FOR SURVIVAL

- Food chains show how energy is transferred between living organisms.
- Food webs are a series of interconnected food chains.
- A change in the population of one organism affects other organisms in the food chain.
- Insects pollinate plants to provide food and shelter for other species.
- Some organisms survive by living in or on another species.

ORGANISMS AFFECT AND ARE AFFECTED BY THEIR ENVIRONMENT

- Organisms can affect populations of other plants and animals in their habitat.
- If one organism is removed from a food web, the whole food web can be affected.
- Changes in the environment can affect populations.

HUMAN ACTIVITIES IMPACT ON THE ECOSYSTEM

- Increased carbon dioxide emissions have caused global temperatures to increase.
- Humans produce toxic substances that can accumulate in food chains.
- Humans frequently destroy ecosystems to use the land for other purposes.

4.2

IN THIS CHAPTER YOU WILL FIND OUT ABOUT:

WHAT FACTORS AFFECT THE SIZE AND DISTRIBUTION OF POPULATIONS?

- What the different levels of organisation are in an ecosystem.
- What the difference is between producers and consumers.
- How large and small predator populations will affect a prey population.
- Why predator–prey cycles are out of phase with each other.
- How distribution and numbers of species are sampled.

HOW DO PLANTS AND ANIMALS IN A COMMUNITY INTERACT?

- How a change in an abiotic (non-living) factor affects a community.
- How changes in biotic (living) factors affect a community.
- Why animals compete for resources in a habitat.
- Why interdependence is important for communities.
- Why certain environmental changes affect populations more than others.

WHAT ARE THE POSITIVE AND NEGATIVE IMPACTS OF HUMAN ACTIVITIES ON BIODIVERSITY?

- What biodiversity is and why it is important to maintain biodiversity.
- How human population growth has impacted on the use of land and how this affects biodiversity.
- How waste from human activities has affected the environment and biodiversity.
- Which human activities are promoting biodiversity.

4.2a Habitats and communities

Learning objectives:

- describe different levels of organisation in an ecosystem
- describe the differences between producers and consumers
- describe predator and prey cycles.

KEY WORDS

community	population
consumer	predators
ecosystem	prey
habitat	producer

Ecology looks at how organisms survive, how they relate to other organisms and their physical environment, and the features that make them successful in their habitat.

Ecosystems

Living organisms are affected by their environment. For example, if plants cannot absorb enough water for their needs, they wilt and may die. But plants affect their environment, too. Plant roots hold soil particles together and stop the wind from blowing the soil away.

An **ecosystem** is made up of all the living organisms in a particular environment together with the non-living components such as soil, air and water.

Ecosystems can be:

- natural, e.g. oceans, lakes, puddles and the rainforest
- artificial, e.g. fish farms and planted forests.

Artificial ecosystems have low biodiversity: the variety and number of plants is limited, so only a small number of animal species can survive. An example of an artificial ecosystem is a garden pond.

A **population** is the total number of one species in an ecosystem. A **community** is all the plants and animals living in an ecosystem. The place where a living organism lives in the ecosystem is its **habitat**.

Figure 4.2.1 Fish swimming on a coral reef. An example of a natural ecosystem

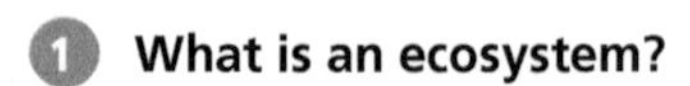

1. **What is an ecosystem?**

2. **Describe the different populations in the marine ecosystem in Figure 4.2.1.**

REMEMBER!

A community is all the different plants and animals in an ecosystem, and a population is the total number of one species.

Feeding relationships

Feeding relationships within a community can be represented by food chains. All food chains begin with a **producer** – an organism that synthesises molecules. It is usually a green plant, which absorbs light to make glucose molecules. Organisms that eat other organisms are **consumers**.

Primary consumers eat producers and are eaten by secondary consumers. Some animals, such as humans, are primary and secondary consumers because they eat both plants and animals.

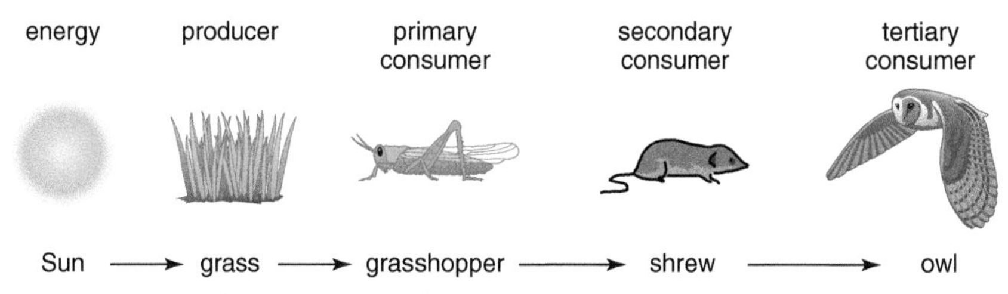

Figure 4.2.2 Food chains describe feeding relationships

3 **Explain why humans are primary and secondary consumers.**

Explaining population cycles

Consumers that hunt and eat other animals are **predators**. The animals that they eat are **prey**. In a community, the numbers of predators and prey rise and fall in cycles.

In predator–prey relationships the size of each population is dependent on the other. The Canadian lynx eats mainly snowshoe hares. Data from a study between 1845 and 1937 were used to produce the graph showing their relationship (Figure 4.2.3). The snowshoe hare population rises and falls in a 10-year cycle. Because the lynx population is dependent on the snowshoe hare, it also rises and falls but lags behind that of the hare population cycle by about 2 years.

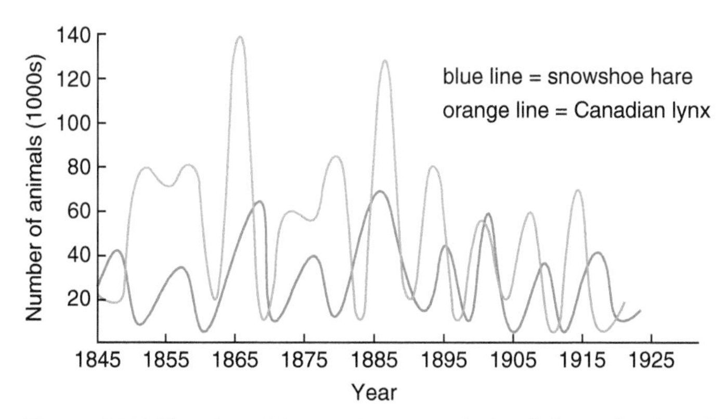

Figure 4.2.3 The size of the predator population follows the size of the prey population

4 **Describe what happens to the predator population when:**

a prey numbers are high

b prey numbers are low.

5 **Explain how large and small predator populations will each affect the number of prey.**

6 **Explain why the predator cycles and the prey cycles are out of phase with each other.**

4.2b Interdependence and competition

Learning objectives:

- describe how competition impacts on populations
- explain why animals in the same habitat are in competition
- describe how one population affects another in a community
- explain the importance of interdependence.

KEY WORD

competition
interdependence

If plants and animals are to survive, they need certain resources from their habitat. However, this may put them in competition with other species as well as their own. This will affect the size and distribution of populations.

Why do organisms compete?

Organisms only survive if they have sufficient resources for their needs. They compete for the available resources in their habitat.

Plants in a community may compete for:

- light
- space
- water
- mineral ions.

Animals often compete for:

- food and water
- mates
- territory.

Figure 4.2.4 Photos of three different organisms in their natural habitats. Elephant seals on the seashore; giraffes, gazelles and wildebeest at a waterhole; and a cactus growing in the desert

Animals and plants that get more of the resources are more successful than those that get less. Successful organisms are more likely to survive and reproduce so the size of their population is more likely to increase. For example, dolphins feed on several foods and are more likely to survive than marine animals called dugongs, which feed only on seagrass. Animals will travel to where food is available.

1. **How does competition affect the distribution and number of organisms?**
2. **Describe how organisms are competing in Figure 4.2.4.**

Competition

Competition is the conflict between organisms for a limited essential resource. When species compete, if they are not perfectly matched one will eventually become more successful than the other. A less successful species may:

- try to compete, unsuccessfully, and become extinct
- stay in its habitat but adopt new survival strategies
- move to another area looking for resources.

Humans are very successful organisms. We compete with animals and plants all over the world.

3 **What happens when two organisms compete?**

4 **Explain why organisms in the same habitat are in competition.**

Figure 4.2.5 A photo of sweetcorn growing in a field. How do farmers compete with other organisms on their farmland?

Interdependence

Within a community each species depends on other species for their needs, such as food, shelter, pollination and seed dispersal. This is called **interdependence**. If one species is removed it affects the whole community. For example, bees depend on flowers for nectar while flowers depend on bees for pollination. If the bee population dies out, there is less chance of flowers being pollinated.

Food webs can be used to understand the interdependence of species within an ecosystem, in terms of food sources. Arrows are used to show feeding relationships. For example, in Figure 4.2.6, the two arrows from 'vole' show that voles are a food source for both kestrels and stoats.

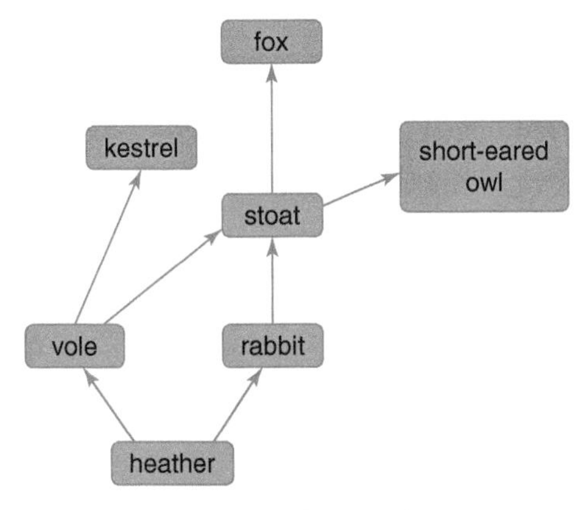

Figure 4.2.6 A simple food web

5 **What is interdependence?**

6 **Look at the food web in Figure 4.2.6. Describe the changes that would occur if the population of rabbits were to die out**

4.2c Factors that affect communities

Learning objectives:

- identify factors that affect ecosystems
- explain how biotic and abiotic factors affect communities
- describe the effect of interacting factors on species distribution.

KEY WORDS

abiotic factor
biotic factor
distribution
stable community

Living organisms are affected by many different factors as they grow and try to survive.

What are abiotic factors?

Many factors affect where an organism lives. **Abiotic factors** are physical (that is, non-living) conditions that affect the **distribution** of an organism. These factors include:

- temperature
- light intensity
- oxygen levels for animals that live in water
- carbon dioxide levels for plants
- moisture levels
- soil pH and mineral content for plants
- wind intensity and direction.

Biotic factors are caused by living organisms affecting other populations in their ecosystem. Biotic factors include:

- food availability
- new pathogens
- new predators
- competition between species.

Abiotic and biotic factors change over time. For example temperature varies daily, monthly, seasonally and over many years.

Figure 4.2.7 Grassland growing alongside a woodland area

Figure 4.2.8 Tropical fish swimming in the ocean

Figure 4.2.9 A panda bear eating bamboo shoots

1. **Choose one abiotic and one biotic factor. Explain how a lack of each factor would affect a plant and an animal.**
2. a **What abiotic factors affect the organisms shown in Figures 4.2.7 (the grass), 4.2.8 (tropical fish) and 4.2.9 (panda bear)?**

   b **Apart from blocking light, how else might trees affect the grass that grows beneath them?**

Looking at changes

The numbers and types of organisms can gradually change across an ecosystem. These changes are easy to see on the seashore, where there are distinct zones of organisms due to changing tides (see Figure 4.2.10).

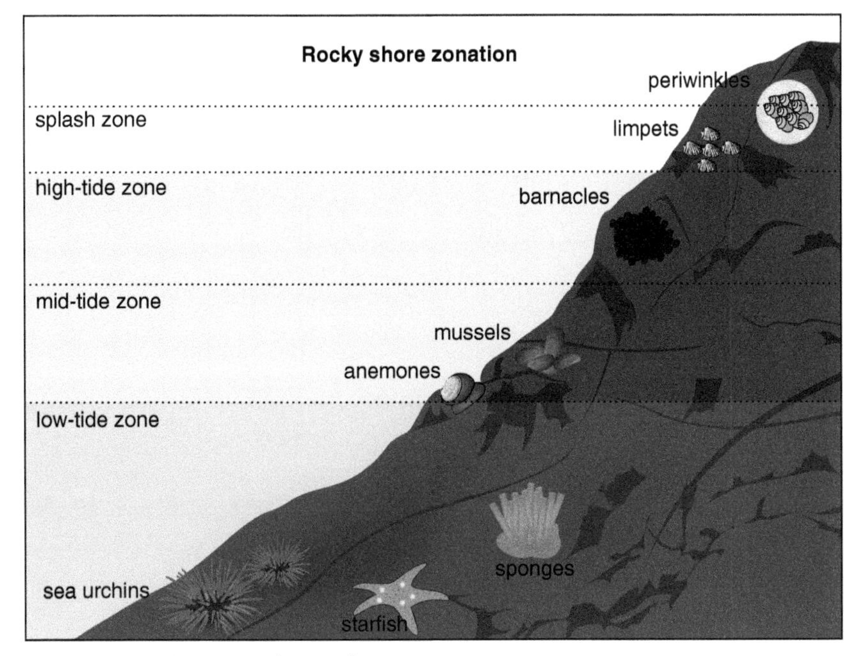

Figure 4.2.10 Zones on the seashore

3 **Describe the abiotic factors at each tidal zone.**

4 **Use your answer to question 3 to suggest why limpets are found higher up the shore than anemones.**

DID YOU KNOW?

Although plants are rare in the desert, their distribution changes after it rains. When water is available, desert plants flower and produce seeds quickly. This increases their numbers rapidly.

REMEMBER!

Abiotic factors are non-living. Biotic factors are living organisms. Changes in these factors cause changes in the distribution of organisms.

Interacting factors and species distribution

Factors that affect the distribution of organisms do not work alone. Two or more factors can interact to form very different environments within a habitat.

Animals such as rabbits and sheep graze on plants. The amount of grazing affects the numbers of plant species found.

Little grazing allows a few plants to out-compete others. As grazing increases more plant species grow because dominant plants are controlled by the animals, allowing weaker species to grow. Only specially adapted plants can resist the effect of intensive grazing and survive.

A **stable community** is where the biotic and abiotic factors are in balance so that population sizes remain fairly constant.

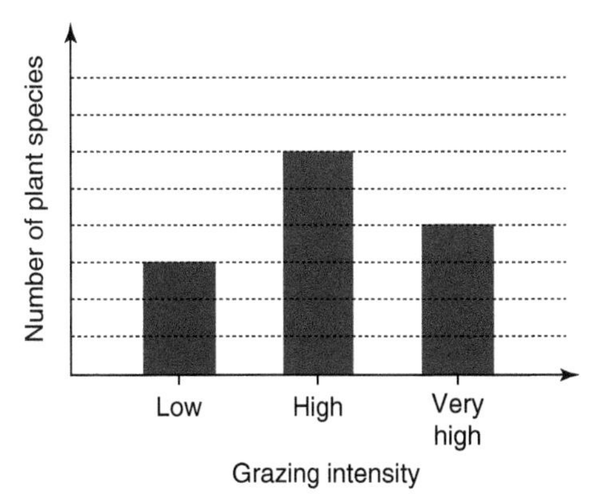

Figure 4.2.11 How does grazing affect plants?

5 **What is a stable community?**

6 **The water content of soil is calculated as a percentage by comparing the original mass of a soil sample after it has been dried.**

In a soil sample taken from a pond, the mass of soil before drying was 43.20 g. After drying, it was 21.89 g. What is the percentage water content of the soil?

DID YOU KNOW?

Tropical rainforests and ancient oak woodlands are stable communities.

REQUIRED PRACTICAL

4.2d Investigating the population size of a common species in a habitat

Learning objectives:

- describe abundance and distribution
- plan experiments to test a hypothesis
- explain the apparatus and techniques used to sample a population
- explain how a representative sample is taken
- explain how to use quadrats and transects.

KEY WORDS

abundance
distribution
hypothesis
quadrat
transect

It is impossible to count every plant or animal in a habitat, but numbers can be estimated by taking samples of the organisms from the habitat. The larger the sample, the more accurate your estimate of the population size is likely to be. This allows population sizes to be compared between different areas.

Plants can be sampled more easily than animals because they cannot move around.

These pages are designed to help you think about aspects of the investigation rather than to guide you through it step by step.

Developing a hypothesis

It is summertime. Asha and Jalan are going to investigate the population of daisies in trampled and un-trampled areas of the school field. The centre of the school field is used to play cricket and tennis and to train for athletic events. The students run, jump and lie on the grass. Not many students go on the parts of the field away from this area.

Figure 4.2.12 Leaf, root and flower of a daisy

Asha and Jalan are going to investigate the numbers or the **abundance** as well as the spread or **distribution** of the daisies. They need to form a **hypothesis** before they do the investigation. They have made some observations on the field and now know that:

- trampling on soil compacts it
- trampling on plants can crush delicate leaves and flowers
- daisy leaves are not very delicate
- daisy plants have very long fibrous roots and thick leaf cuticles.

1 **How do you think compaction may affect**

a the soil?

b the daisies growing in the soil?

2 **If meristems, leaves and flowers can be crushed by trampling how might this affect the daisies?**

KEY INFORMATION

Sampling many small sections of an area gives a representative sample of the whole area.

Planning an investigation

The teacher has given Asha and Jalan a **quadrat**. The quadrat is a square grid that is placed on the ground and the species within it are identified and recorded. They can use a tape measure to lay a straight line across the field to count and record daisy plants. This is called a **transect**. They know that it is important to take random samples, make repeat observations and count whole daisy plants.

SAFETY

A risk assessment will be required.
The area of land to be used should be checked for hazards such as broken glass and toxic plants (e.g. ragwort).

3. **Suggest a hypothesis for the fieldwork that Asha and Jalan are going to do.**
4. **Suggest how Asha and Jalan can use this equipment to collect representative data to test the hypothesis from Question 3.**
5. **Look at Figure 4.2.12. Why is it important to count whole daisy plants and not just daisy flowers? How many daisies are in the quadrat?**
6. **How can Asha and Jalan calculate the mean number of daisy plants in each area?**
7. **Another student has asked Asha and Jalan for a copy of their results. Suggest why they asked for it.**

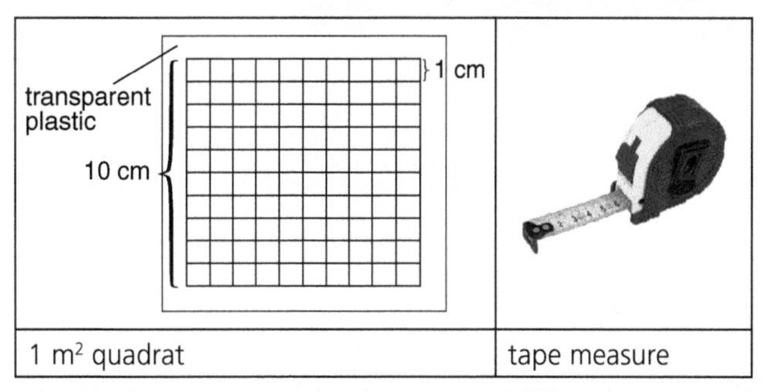

Figure 4.2.13 Quadrat and tape

Developing explanations

Asha and Jalan did their investigation. They placed their tape across the school field from the trampled area across to the uncompacted area. Here are their data.

Quadrat position on tape (m)	No. of daisies per 1 m^2 quadrat	
	Un-trampled area	Trampled area
0	3	6
3	4	6
8	3	7
11	2	6
18	2	4
22	4	7
30	3	6

Figure 4.2.14 Quadrat placed over daisies

Scientists always use evidence to explain the data they collect. The evidence can be from knowledge or observations. They also relate their data to their original hypothesis.

REMEMBER!

To get an accurate representation of distribution of a species, it is important that samples are taken at random. This is so that your results are not biased.

8. **Suggest what Asha and Jalan can conclude from their data.**
9. **Explain the data using the information that Asha and Jalan had before they started the investigation.**

4.2e Biodiversity

Learning objectives:

- recall that biodiversity is the range of different plants and organisms living in an ecosystem
- describe the benefits of maintaining biodiversity
- explain some ways of maintaining local and global biodiversity.

KEY WORDS

biodiversity
self-supporting

Every species depends on others for food, shelter, pollination, and so on. Because of interdependence, if one species is removed the whole community is affected.

Biodiversity

In every ecosystem there are different living organisms. **Biodiversity** is the range of different plant and animal species living in an ecosystem. Some natural ecosystems, such as the rainforest or shallow tropical coral reefs, have high biodiversity. They provide a wide variety of food throughout the year and shelter, above and below the ground or water. Other natural ecosystems, such as the Arctic tundra or deep-sea thermal vents, have low diversity, with species adapted to survive those extreme environments.

Artificial ecosystems have low biodiversity: the variety and number of plants there is limited, so only a small number of animal species can survive. An example of an artificial ecosystem is a garden pond.

Figure 4.2.15 A waterfall feeding into a river in the Andes mountain range in South America

Figure 4.2.16 Bluebells growing in an area of woodland in the UK

Figure 4.2.17 Water lilies and various grasses growing in a pond in the UK

1. **What is high biodiversity?**
2. **Look at the photos in Figures 4.2.15 to 4.2.17. How are these ecosystems similar and different?**

Benefits of local and global biodiversity

Biodiversity is greater in ecosystems that provide a bigger range of different habitats, which are home to larger populations of a variety of organisms. Small populations are in greater danger of dying out if an ecosystem is disrupted in some way.

High biodiversity is important in an ecosystem because:

- it allows a wide variation of food sources, reducing the dependence of a species on a particular food source
- it provides us with food, medicines, the atmosphere and water.

All ecosystems are **self-supporting**: all the requirements for living organisms to grow and survive are present. They do need an external energy source, which is usually the Sun.

DID YOU KNOW?

Siberian tigers are endangered because of deforestation, hunting and poaching.

All animals depend on plants for oxygen and food. Plants depend on animals for carbon dioxide, pollination and seed dispersal. This interdependence is of mutual benefit to all organisms in an ecosystem.

3 Explain why high biodiversity is important.

4 Explain what is meant by ecosystems being self-supporting.

Challenges to local and global biodiversity

The future of the human species on Earth relies on us maintaining a good level of biodiversity, yet we are threatening it by our actions.

Huge areas of tropical forest are being destroyed. This deforestation is happening to:

- provide land for cattle and rice fields
- grow crops, such as oil palm and sugar cane to make biofuels.

Forests are often destroyed by burning. Mass destruction of trees has:

- increased the release of carbon dioxide into the atmosphere due to burning
- reduced the rate that carbon dioxide is removed from the atmosphere by photosynthesis.

The average global temperature of the Earth and its atmosphere is increasing due to global warming caused by increasing atmospheric levels of carbon dioxide.

Scientists think that global warming is changing the climate. The average world temperature rise is small (about 0.8°C since 1880), but some species, such as coral reefs, are sensitive to this. This is because warmer than normal sea temperatures cause the corals to eject the algae that live in their tissues and enable them to survive. The corals go completely white (called coral bleaching) and die, sometimes causing the reef itself to crumble and other reef creatures to lose their valuable habitat.

Figure 4.2.18 A coral reef, showing a bleached coral in the foreground surrounded by corals that still contain their algae

DID YOU KNOW?

About 13 million hectares of forest have been cleared or lost through natural disasters. By 2030, there may only be 10% of our forests left.

5 How do human actions affect global warming?

6 Explain how global warming might affect biodiversity.

MAKING LINKS

Look back to topic 4.1g to remind yourself about the impacts of a rise in global temperatures.

4.2f Negative human impacts on ecosystems

Learning objectives:

- describe how humans interact negatively with the ecosystem
- explain how this impacts on biodiversity.

Human life is dependent on a rich biodiversity for survival. Our actions are decreasing biodiversity, and are doing so at an alarming rate.

Figure 4.2.19 Photo of the sea shaping the rocks on the coastline

Figure 4.2.20 Snowstorm on a mountain

Figure 4.2.21 Land being used for farming

Environmental changes

Environmental changes affect the distribution and behaviour of organisms. Changes may be short-lived or long-lasting. These changes may cause migration, enable the survival of some well-adapted organisms or lead to the death of organisms.

Environmental changes include temperature changes, and changes in the availability of water and atmospheric gases. Environmental changes can be caused by natural phenomena such as seasonal changes or volcanic activity. Humans also cause environmental change.

Examples of environmental change caused by humans include:

- deforestation to provide land for farming and housing
- destruction of peat bogs and other areas of peat
- polluting streams, rivers and lakes with chemicals.

1. **Look at the photos in Figures 4.2.19 to 4.2.21. In each one, identify the environmental change shown and suggest what caused it.**
2. **Give an example of:**

   a a long-lived environmental change

   b a short-lived environmental change.

Effects on land and water

The world's population has increased rapidly, from 1 billion (1000 million) in 1880 to about 7 billion in 2012. People use increasing amounts of the Earth's resources, resulting in a decrease in the land available for other organisms.

As our population increases, biodiversity decreases:

- More land is needed for homes, shops, factories and roads. Building sites destroy habitats. Roads divide habitats, making it harder for organisms to find food and mates.
- New quarries are mined to provide stone, slate and metal

REMEMBER!

Humans reduce land availability for other organisms by building, farming, quarrying and dumping waste.

ores for building materials. Habitats are destroyed.
- More farmland is needed and fertilisers are used. Many farms grow one crop over huge areas. This affects food availability for insect pollinators. There are fewer available nesting sites for birds.
- More waste is sent to landfill, and more sewage and industrial waste are produced. This can pollute the land.

Figure 4.2.22 Intensive farming using fertilisers increases crop yields

3 **Suggest what will happen if human populations keep rising.**

4 **Describe the impact of:**

a land-use change

b fertiliser run-off.

5 **Look at Figure 4.2.23. Why are there no fish in the water?**

If water contains a lot of fertilisers caused by run-off from farmers' fields or sewage, the nitrates and phosphates in the water increase and then algal growth increases. Algae cover the water surface and prevent light from reaching water plants.

The plants and algae die. Bacteria respire as they break down dead plants and use up oxygen in the water. The other living organisms in the water die.

Toxic chemicals from household and industrial waste taken to landfill sites can spread into soil and enter waterways. Pesticides and herbicides are also washed into waterways. Toxins build up in food chains, kill organisms and affect feeding relationships.

Figure 4.2.23 Photo of a stagnant waterway at the mouth of a tunnel, situated near agricultural land. The water is covered in algae

Peat bogs

Decomposers cannot completely break down plant material in acidic conditions with little oxygen, so peat forms. Peat is an important carbon store. Peat is used as a fuel and as compost by gardeners. Compost improves soil quality to increase food production.

Peat bogs form over thousands of years in marshy areas. Peat is being destroyed faster than it is being made. The loss of peat bogs reduces the variety of different plants, animals and microorganisms that live there.

6 **What is the impact on the environment of using peat as a fuel?**

7 **How might using peat-free compost reduce this impact on the environment?**

Figure 4.2.24 Peat cut to be used for fuel

4.2g Positive human impacts on ecosystems

Learning objectives:

- describe positive human interactions on biodiversity
- describe some conservation measures
- describe the impact of breeding programmes
- explain how habitats are regenerated.

KEY WORDS

conservation
field margins
monoculture
regeneration
sustainable

Increasing awareness of the problems caused by urbanisation and global warming have led to interventions that reduce the negative effect we have on ecosystems.

Protecting ecosystems

Programmes have been developed to reduce the negative effects on ecosystems and biodiversity that are caused by our actions. These measures include the following.

- Introducing breeding programmes for endangered species.
- The protection and **regeneration** (restoring) of rare habitats.
- Introducing wider **field margins** to provide habitat for many wild species. Field margins are grass strips between hedges, and the crop edge in a field with **monocultures** – one type of crop grown year after year.

Figure 4.2.25 Hedgerows are being replanted to encourage more wildlife to use the habitat between fields

- Reducing deforestation and carbon dioxide emissions. **Sustainable** strategies include replanting trees.
- Recycling resources instead of dumping waste in landfill. Many materials are recycled in the UK.
- Cloning plant species. Cloning can be done quickly and economically using stem cells found in the meristems. This protects plant species from extinction.

Figure 4.2.26 Why must we recycle waste?

1. **How are ecosystems and biodiversity protected?**
2. **How do hedgerows increase biodiversity?**

Conservation

Conservation programmes are introduced:

- because we have a moral responsibility to protect endangered species
- so more plant species may be identified for medicines
- to minimise damage to food chains and webs
- to protect future food supplies.

Figure 4.2.27 Arabian oryx populations are growing

Captive breeding programmes are planned to ensure genetic diversity is maintained, such as for the Arabian oryx and for giant pandas. Successful programmes allow species to be reintroduced to the wild. The wild Arabian oryx became extinct in the wild in the 1970s, but a successful breeding programme means the animal has now been reclassified as 'vulnerable'.

Many endangered species, for example rhinos and tigers, are hunted and poached, despite legal protection. Seed banks store seeds carefully to protect plant species for the future.

REMEMBER!

Learn the different programmes for protecting ecosystems and biodiversity.

3. **Why are conservation programmes introduced?**
4. **How do breeding programmes help endangered species?**

Protecting rare habitats

There are many difficulties and issues associated with organising conservation programmes. Some of these include:

- ensuring long-term funding
- having qualified scientists who understand the issues
- animals and plants do not recognise boundaries
- many organisations and governments may be involved, working locally, nationally and internationally
- lack of 'policing' of protected areas.

Figure 4.2.28 Many species depend on sustainable mangrove forests

Mangrove forests are rich ecosystems that prevent coastal erosion and reduce carbon emissions. They are declining rapidly due to land development, and their use as a fuel and building material. In Abu Dhabi, however, mangroves are increasing due to massive planting programmes over the last two decades. The local environment agency works with land developers and the public to maintain healthy, litter-free, sustainable forests. The forests protect seagrass beds, which are the sole food of the endangered manatee.

5. **What difficulties are involved in managing conservation programmes?**
6. **Explain how mangrove forests are protected and regenerated.**
7. **Explain why it is important to protect biodiversity.**

Check your progress

You should be able to:

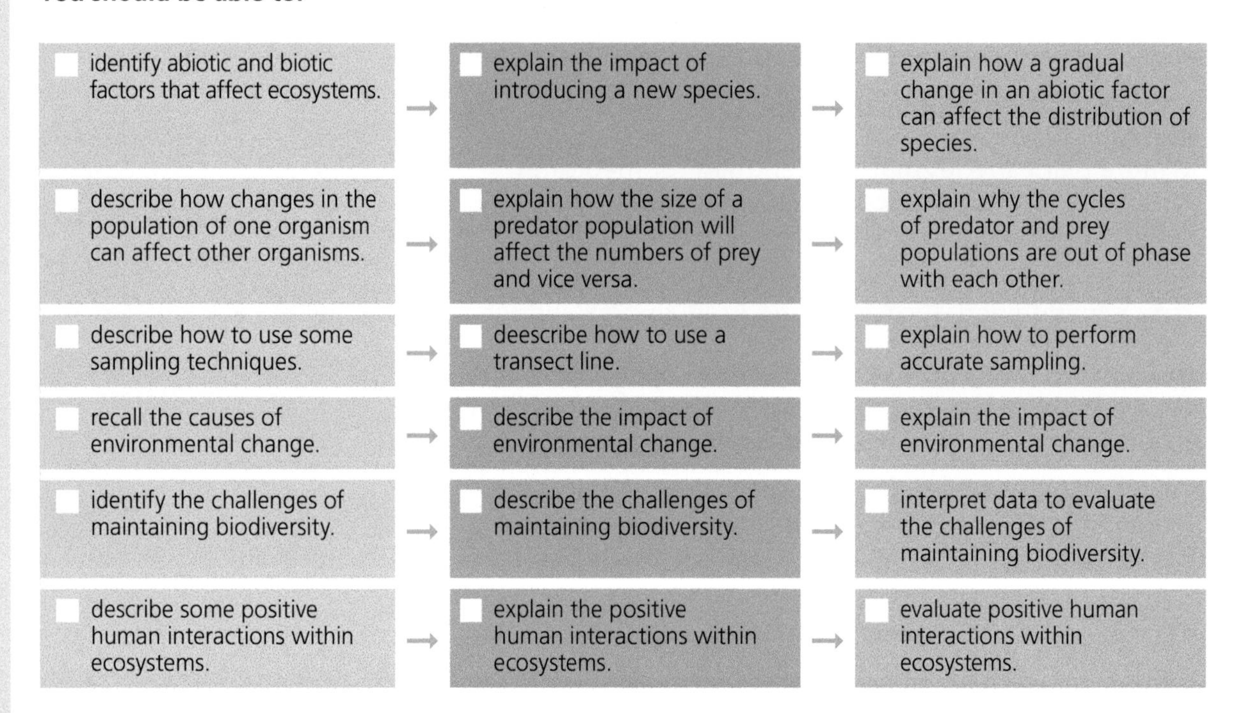

identify abiotic and biotic factors that affect ecosystems. →	explain the impact of introducing a new species. →	explain how a gradual change in an abiotic factor can affect the distribution of species.
describe how changes in the population of one organism can affect other organisms. →	explain how the size of a predator population will affect the numbers of prey and vice versa. →	explain why the cycles of predator and prey populations are out of phase with each other.
describe how to use some sampling techniques. →	deescribe how to use a transect line. →	explain how to perform accurate sampling.
recall the causes of environmental change. →	describe the impact of environmental change. →	explain the impact of environmental change.
identify the challenges of maintaining biodiversity. →	describe the challenges of maintaining biodiversity. →	interpret data to evaluate the challenges of maintaining biodiversity.
describe some positive human interactions within ecosystems. →	explain the positive human interactions within ecosystems. →	evaluate positive human interactions within ecosystems.

Worked example

Some students investigated the average number of plant species growing in a habitat. They investigated different sized areas of the habitat. Look at the graph.

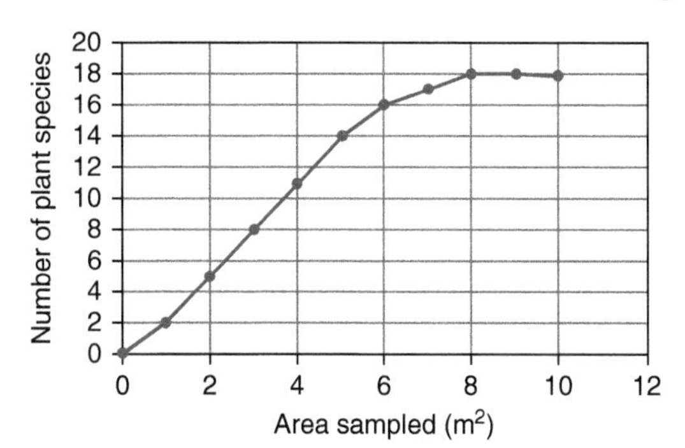

1. **Describe what the graph shows.**

The number of different plants found increases from 0 to 18.

Maximum number of plants found has been correctly identified.

Fuller responses would include details such as: as area sampled increases the number of species found increases; the ideal area to sample is $8\,m^2$.

2. **Amy and Laura decide to investigate how the number and type of plant species growing in a salt marsh change with distance from the sea.**

What equipment should they use to do this?

A quadrat and a tape measure.

Correct answer.

3. **Describe how Amy and Laura should use this apparatus.**

They make a straight line and place the quadrat at random along it. Then they count the different species and their numbers.

Good answer. The student has remembered the importance of taking random samples.

4. **Why will the results of Amy's and Laura's investigation be an estimate of the population sizes?**

They have only sampled the habitat.

Correct answer identified. A more detailed response would include that it is impossible to count every plant.

5. **Amy and Laura repeat the sampling process in the salt marsh twice. Explain why.**

To improve the accuracy of their results.

This is a correct response.

Overall, there is some good understanding of the topic shown but the answers need to be more detailed.

End of chapter questions

Getting started

1. What is the difference between a population and a community? 1 Mark
2. What are quadrats used for? 1 Mark
3. What do food chains always start with? 1 Mark
4. Why can badgers be primary and secondary consumers? 1 Mark
5. Name two abiotic factors. 2 Marks
6. Explain why bluebells grow before the trees in a wood are in leaf. 1 Mark
7. What is interdependence? 1 Mark
8. Name two human activities that are destroying habitats. 2 Marks

Going further

9. Name one way by which the destruction of peat bogs could be reduced. 1 Mark
10. What is a monoculture crop? 1 Mark
11. The mass of pond soil is 58 g before drying and 24 g after drying. What is the percentage water content of the soil? 2 Marks
12. Jessica is carrying out an investigation into how many dandelion plants there are in her school playing field. The field is 20 metres by 30 metres. She selects nine spots at random and samples each spot with a 1 m^2 quadrat. Jessica's results are shown in the table.

Sample number	Number of dandelion plants
1	5
2	4
3	2
4	4
5	3
6	1
7	3
8	2
9	4

Jessica calculates the mean number of dandelions to be 3.1.

a Calculate the median and mode values for the number of dandelions across the areas sampled. 2 Marks

b Use the data to estimate the number of daisy plants in the playing field. 1 Mark

c Why is it important to place quadrats in a random way rather than choose an area? 1 Mark

13 The graph shows human population growth and the number of extinctions over time.

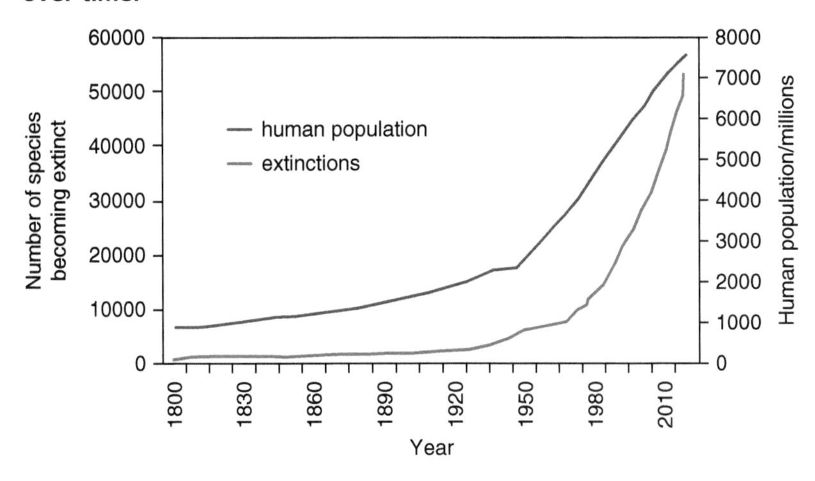

Describe what the graph shows.

2 Marks

More challenging

14 The diagram shows a simple food web. Explain the effects if the population of foxes were to increase.

2 Marks

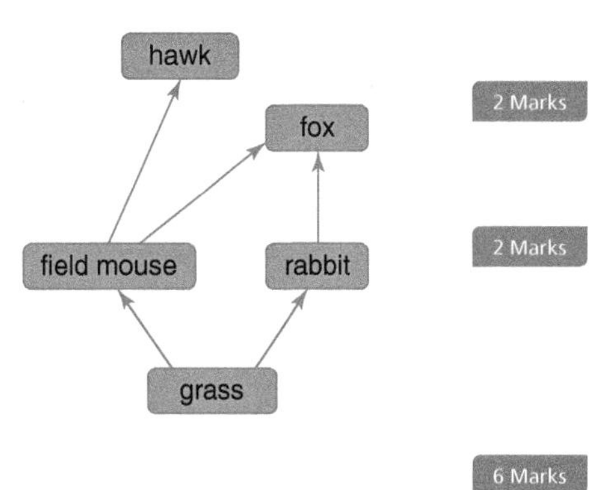

15 Explain how destruction of rainforests without replanting has helped increase the concentration of carbon dioxide in the atmosphere.

2 Marks

16 Alistair notices that the pond in his garden has turned green and all the fish have died. He thinks it might be because he used too much inorganic fertiliser in his garden. Explain how using inforganic fertiliser on his garden may have caused the fish in his pond to die.

6 Marks

Most demanding

17 Describe how recycling our waste rather than dumping it in landfill has a positive impact on biodiversity.

2 Marks

18 Ken has lots of moss growing in his lawn. He wants to estimate what percentage of his lawn is moss. Describe how Ken could do this, making sure the data he collects is random.

4 Marks

19 Sewage is accidently allowed to leak into a river.

Based on the data shown in the graph, predict the changes to the environment that might be observed as a result of the leak of sewage.

How would you investigate whether or not your predictions are correct?

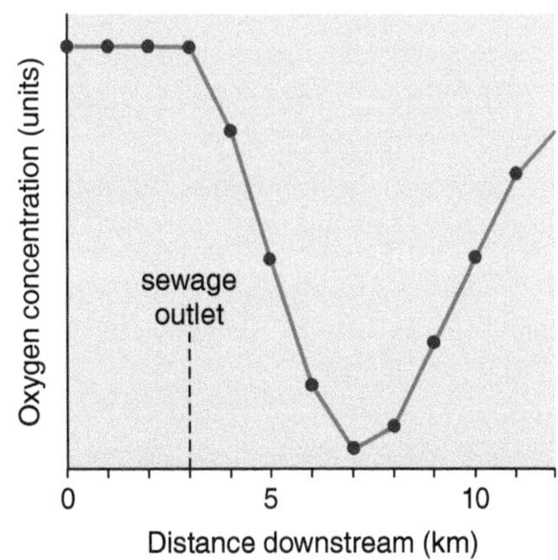

4 Marks

Total: 40 Marks

INHERITANCE

IDEAS YOU HAVE MET BEFORE:

OUR CELLS CONTAIN GENETIC INFORMATION.

- Genetic information is contained within genes on our chromosomes.
- Our chromosomes and genes are found in the nucleus of our cells.
- A change to our DNA, called a mutation, can lead to cancer.

IN MULTICELLULAR ORGANISMS, CELLS HAVE TO DIVIDE.

- Cells divide as we're growing, and to replace cells that are injured, worn out or have died.
- This type of cell division is called mitosis.
- When a cell divides by mitosis, two daughter cells are produced, each with an identical number of chromosomes and identical DNA.

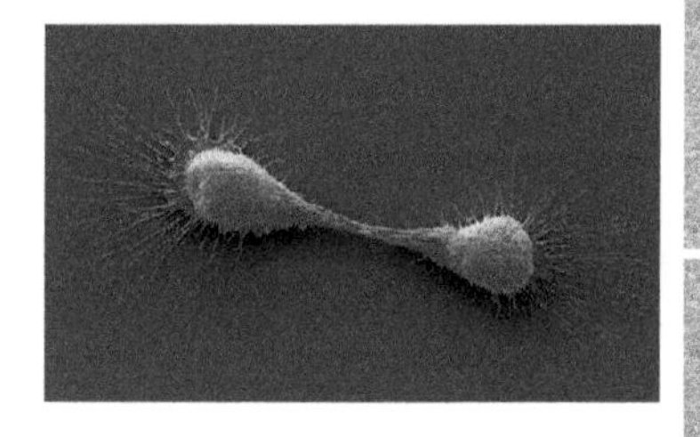

OUR CHARACTERISTICS ARE INHERITED AND PASSED ON.

- Passing on genetic information is called heredity.
- Sexual reproduction in humans leads to similarities and variation between individuals.
- Cells in reproductive organs divide by meiosis to produce egg and sperm cells (gametes).

4.3

IN THIS CHAPTER YOU WILL FIND OUT ABOUT:

OUR UNDERSTANDING OF DNA AND THE WAY GENES WORK

- DNA is a polymer made of two strands that twist around each other in a double helix.
- A gene is a short section of DNA that codes for the production of a particular protein.

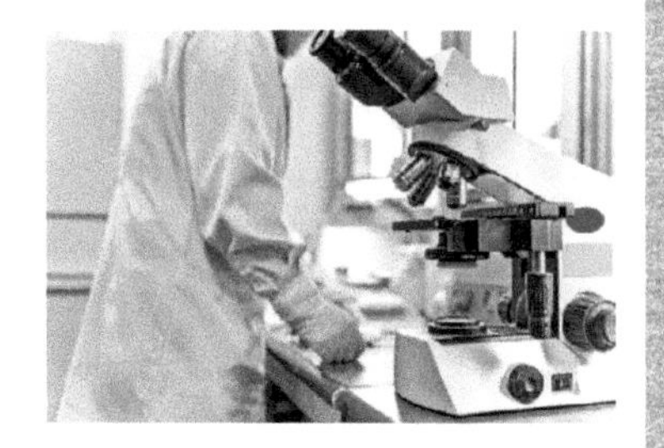

PRODUCTION OF SEX CELLS FOR REPRODUCTION

- During sexual reproduction, a cell divides by meiosis to produce four gametes, each with half the number of chromosomes.
- Our sex is determined by the 23rd pair of chromosomes.
- The way chromosomes are inherited means that the number of boys who are born is roughly the same as the number of girls.

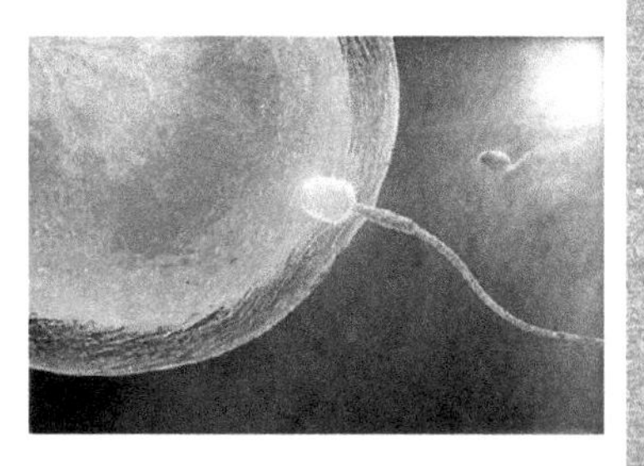

CHARACTERISTICS ARE INHERITED FROM ONE GENERATION TO THE NEXT

- Genetics allows us to understand the inheritance of certain characteristics in humans and in many other organisms.
- We can use genetic terms and predict the outcome of crosses.

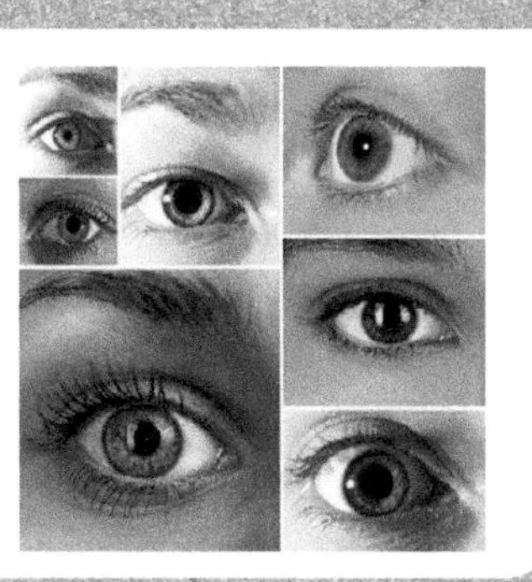

4.3a Chromosomes and genes

Learning objectives:

- describe DNA, chromosomes and genes
- describe the structure of DNA
- explain what the genome of an organism is.

KEY WORDS

chromosome
gene
genome
nucleotide

The genetic information of all organisms is contained in the nucleus as long, thin strands of DNA, deoxyribonucleic acid.

The genome of an organism

The **genome** of an organism is the entire genetic material of that organism.

The genetic information is carried by a chemical called deoxyribonucleic acid (DNA). DNA is made of very large molecules in long strands, twisted to form a double helix. The DNA is found within structures called **chromosomes**. The chromosomes are long, thin strands of DNA found in the nucleus.

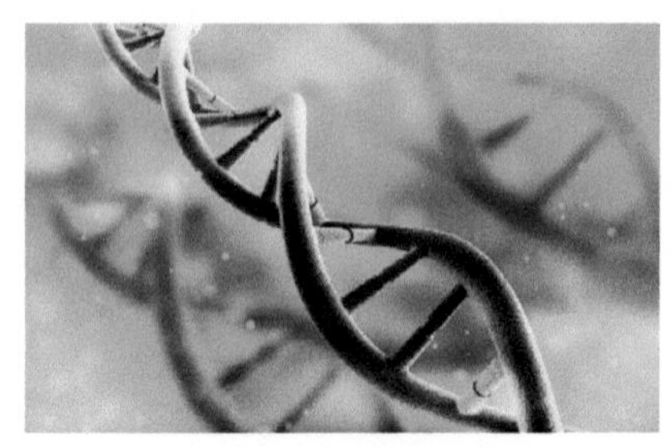

Figure 4.3.1 A schematic representation of DNA to illustrate the double helix shape. The molecule is like a ladder that has been gently twisted

1 What is the name of the chemical that makes up our chromosomes?

2 What is the shape of this molecule?

3 What is a genome?

MAKING LINKS

You first came across chromosomes when learning about mitosis and the cell cycle in topic 1.3j.

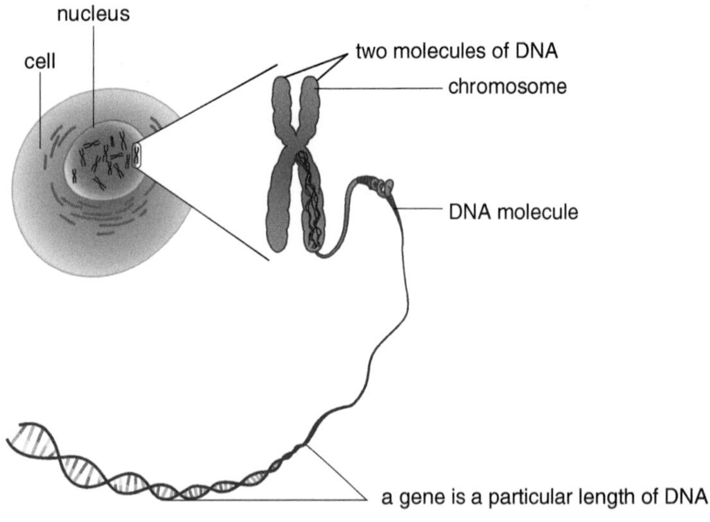

Figure 4.3.2 Diagram to explain the relationship between DNA and chromosomes in the nucleus of a cell. Each half of a double chromosome is a molecule of DNA

Genes and chromosomes

A **gene** is a short section of DNA on a chromosome. Each gene contains the code for a particular combination of amino acids to make a specific protein. These proteins determine all our inherited characteristics, such as blood group, hair colour and eye colour, including the minutest detail of our cells.

Our genes are distributed across our chromosomes. The chromosomes in a chromosome pair have the same types of genes, in the same order, along their length.

Human cells have 46 chromosomes: that's 23 pairs. This number varies from one organism to another. It's the chromosomes contained within a cell and their genetic make-up that determine the type of organism.

Figure 4.3.3 The female jumper ant has just one pair of chromosomes; the male just one chromosome. One species of fern has over a thousand chromosomes

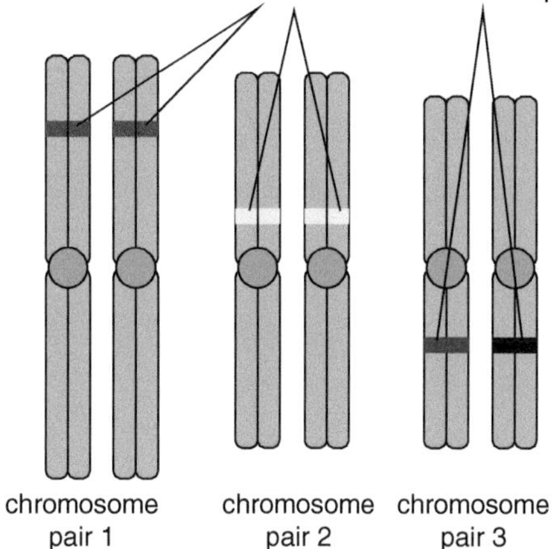

Figure 4.3.4 The chromosomes of an organism contain the genes. One gene pair per chromosome is shown here. In reality, there are thousands on each chromosome

Every person has two copies of each gene – one inherited from each parent.

COMMON MISCONCEPTIONS

Don't think that the amount of DNA or number of genes in a cell is linked to the complexity of an organism.

4. **Where are genes located on a chromosome pair?**
5. **How many copies of a gene does an organism have?**
6. **What is the difference between a gene, DNA and a chromosome?**

The structure of the gene

The DNA molecule is a long chain of nucleotides. A **nucleotide** is a 5-carbon sugar molecule joined to a phosphate and an organic base. There are four organic bases: adenine (A), thymine (T), cytosine (C) and guanine (G). The nucleotides are joined by their phosphate groups to form a long chain, often thousands of nucleotides long.

The sequence of bases down the length of the DNA molecule forms a code, which instructs the cell to make specific proteins. It is the sequence of bases in the DNA molecule that determines which proteins are made.

7. **How many organic bases are there?**
8. **How does the cell know which proteins to make?**

4.3b Sex determination in humans

Learning objectives:

- explain how meiosis halves the number of chromosomes for gamete production
- explain how fertilisation restores the chromosome number
- describe how the sex chromosomes determine the sex of the offspring.

KEY WORDS

gamete
sex determination
X-chromosome
Y-chromosome

Most organisms reproduce sexually. Two parents are involved. The parents produce sex cells or gametes by meiosis. Once the gametes have fused, sex is determined by the sex chromosomes.

Sexual reproduction

Sexual reproduction provides advantages over asexual reproduction:

- Genetic material comes from both parents, producing variation. Offspring are different from each other, and their parents.
- If the environment changes, because of their genetic differences, some offspring are more likely to survive than others.

1 Give two advantages of sexual reproduction.

2 How are sex cells produced?

MAKING LINKS

Refer back to topic 1.3k, where you studied meiosis.

Genes and chromosomes in sexual reproduction

Before sexual reproduction can occur, special cells are produced by the male and female sex organs. These cells have half the number of chromosomes and are called **gametes**. Gametes are produced in a form of cell division called meiosis.

	Female gametes	Male gametes
animals	egg cells	sperm
plants	egg cells	pollen grains

In sexual reproduction, the male and female gametes (sperm and egg cells in animals) join or fuse together. There is mixing of genetic information, which leads to variety in the offspring.

3 Describe what happens to the chromosomes when cells divide to produce gametes.

What determines our sex?

Egg and sperm cells (gametes) are produced during meiosis (reduction division). Each gamete contains one of the two sex chromosomes – from pair 23. These carry the genes for **sex determination**. Female humans have two X-shaped chromosomes (XX). Males have an X- and a Y-shaped chromosome (XY).

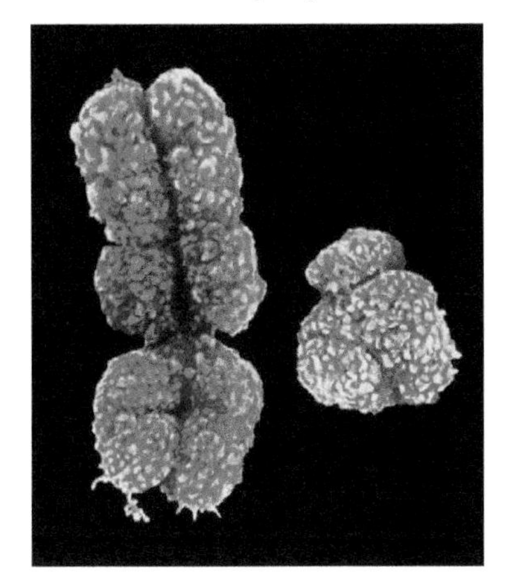

Figure 4.3.5 A coloured scanning electron micrograph (SEM) of an X-chromosome (on the left) and a Y-chromosome (on the right) at a magnification of x10 200

Look at Figure 4.3.6. It is a Punnett square, which shows how sex is determined in humans. All the eggs produced by the mother contain **X-chromosomes**. These are shown at the top of the diagram in pale pink. Gametes that come from the father are shown in pale blue on the left of the diagram. Half of the father's sperm contain the X-chromosome; half the **Y-chromosome**. Therefore, a child will always inherit an X-chromosome from their mother but they can receive either an X- or a Y-chromosome from their father. The four different possible combinations of X- and Y-chromosomes are shown in the dark purple squares. In theory, 50% of live-born children are female and 50% are male.

		Mother (XX) gametes	
		X	X
Father (XY) gametes	X	XX female	XX female
	Y	XY male	XY male

Figure 4.3.6 A Punnett square showing how sex is determined in humans

4 **In the UK, the proportion of live births is in the proportion of 1.05 males:1.00 females. Suggest why it isn't exactly 1:1.**

4.3c Single gene inheritance

Learning objectives:

- explain single gene inheritance
- predict the results of single gene crosses
- explain the difference between dominant and recessive characteristics
- explain homozygous and heterozygous characteristics.

KEY WORDS

allele, carrier, dominant, expression, genotype, heterozygous, homozygous, phenotype, recessive

Chromosomes are made of sections of DNA called genes. These genes code for proteins, which determine our characteristics.

An organism's genetics

Many characteristics are controlled by a single gene. Examples are red–green colour blindness in humans and fur colour in mice. Each gene has different versions called **alleles**. Every person has two alleles – one inherited from each parent.

The translation of the genetic code is called **expression** and the genes that are expressed make up the characteristics of the organism. Not all genes are expressed. A **dominant** allele is always expressed, even if only one copy is present. A **recessive** allele is only expressed if two copies are present (which means that no dominant allele is present).

To describe the genetics of an organism, we give the alleles letters. A dominant allele is given a capital letter; a recessive allele is written in lower case. The red–green colour blindness gene is a recessive gene and is represented by 'b'; the dominant form is non-colour blindness and is represented by 'B'. A person who is not colour blind will be either Bb or BB, but someone who is colour blind can only be bb.

When an organism has two alleles of the same type – either dominant or recessive – they are said to be **homozygous** for that characteristic. If the alleles are different, they are described as **heterozygous**.

REMEMBER!

It is important that you know, understand and can use the genetics terms on these pages.

DID YOU KNOW?

1 in 12 males in the UK are colour blind, compared to just 1 in 200 females.

1. **What is meant by a dominant allele?**
2. **How many copies of a recessive allele must be present for it be expressed?**

Cystic fibrosis

Cystic fibrosis is an example of an inherited disorder caused by recessive alleles. Sufferers produce mucus that is thicker and stickier than normal, making it difficult to breathe. In the 1980s, the gene for cystic fibrosis was mapped to chromosome 7. We can now diagnose cystic fibrosis quickly and predict its inheritance.

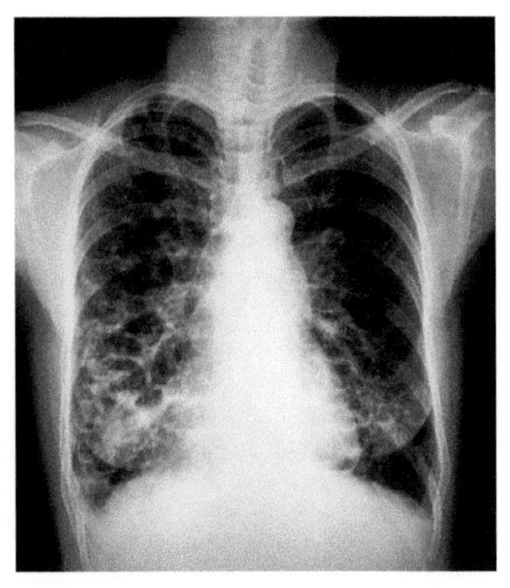

Figure 4.3.7 This X-ray shows the lungs of someone with cystic fibrosis. The airways are wider than they would normally be, and the light coloured patches near the bottom of the dark coloured areas represent the build-up of mucus in the lungs

If a mother with cystic fibrosis has children with a normal father we can draw genetic crosses to predict whether the children will have cystic fibrosis. The mother's genes must be cc, as she has the condition. The father could be CC or Cc.

Let's assume he's CC, so all his sperm will be C. The eggs produced by the mother will be c. So at fertilisation, all the children will be Cc. They will all be normal.

3 **What is the genotype of a person who is homozygous dominant for a gene, B?**

4 **What types of gametes will the person produce?**

Genetic diagrams

If the father is Cc – he's normal but a **carrier**. This time, the cross is more complicated. It's easier to show the possible outomes in a diagram called a Punnett square (see Figure 4.3.8).

Figure 4.3.9 shows the possibilities when a couple are both carriers of the cystic fibrosis allele. As carriers, they both have Cc genes. They will produce both C- and c-carrying gametes.

The probable ratio of normal children to children with cystic fibrosis in this family is 3:1. The ratio gives an overall probability, but of course which allele is in the sperm and which allele is in the egg is completely random.

		Mother (cc) gametes	
		c	c
Father (Cc) gametes	C	Cc normal (but a carrier)	Cc normal (but a carrier)
	c	cc cystic fibrosis	cc cystic fibrosis

Figure 4.3.8 There is a 50%, or 1 in 2, chance that a child will be born with cystic fibrosis when one parent has cystic fibrosis and the other is a carrier

		Mother (cc) gametes	
		C	c
Father (Cc) gametes	C	CC normal	Cc normal (but a carrier)
	c	Cc normal (but a carrier)	cc cystic fibrosis

Figure 4.3.9 The probability of the couple having a child with cystic fibrosis is 1 in 4 when both parents are carriers

5 **The ability to taste a chemical called PTC is controlled by a single gene that codes for a taste receptor on the tongue. A man who is heterozygous for gene T has children with a woman who is homozygous recessive for the gene. Draw a Punnett square showing the possible genotypes of the children.**

6 **For an eye colour gene in parrots, the brown allele is dominant to red. Draw a Punnett square showing the genotypes and phenotypes of a mating between two heterozygous parrots.**

REMEMBER!

You should able to interpret Punnett square diagrams, and to construct them if working at a higher level. For most of the questions you'll encounter, you'll find a 1:1 ratio, a 3:1 ratio, or – if crossing a homozygous dominant individual with a homozygous recessive – the offspring will all have the dominant form.

4.3d Genotype and phenotype

Learning objectives:

- compare the terms genotype and phenotype
- explain how the genome interacts with the environment to influence the phenotype
- describe how most phenotypic features are the result of multiple gene inheritance.

KEY WORDS

genotype
phenotype

The variation in the characteristics of individuals of the same kind may be due to differences in the genes they have inherited, the conditions in which they have developed or a combination of genes and the environment.

Genotypes and phenotypes

The alleles present for a particular gene make up the organism's **genotype**. In the example of red–green colour blindness, the genotype of a person with the condition is bb. All the genes present in an individual organism develop its observable appearance and character. These characteristics are its **phenotype**. Thus, the phenotype for someone with a bb genotype would be red–green colour blindness.

Sometimes the genotype interacts with the environment in which the organism grows, and this influences the phenotype.

1. **What is the difference between a genotype and a phenotype?**
2. **Write down the phenotype of someone who has the genotype Bb for red–green colour blindness.**

KEY INFORMATION

Remember that most of our characteristics are controlled by multiple genes. Those controlled by a single gene are rare.

Multiple gene inheritance

So far, we have only discussed characteristics determined by a single gene. However, some characteristics are determined by the combined effect of more than one pair of genes. These are referred to as **polygenic** characteristics.

Human height is an example of a characteristic determined by many genes, each with different alleles. The combined size of all the body parts from head to foot determines the height of the individual. The sizes of all of these body parts are determined by numerous genes to give an additive effect. The set of alleles that determine the height of a person contain all the genes involved and therefore, involve a more complex genotype for that characteristic. Human skin, hair and eye colour are also polygenic characteristics because they are influenced by more than one allele. The phenotype of these characteristics result in continuous variation with an unlimited number of possible values.

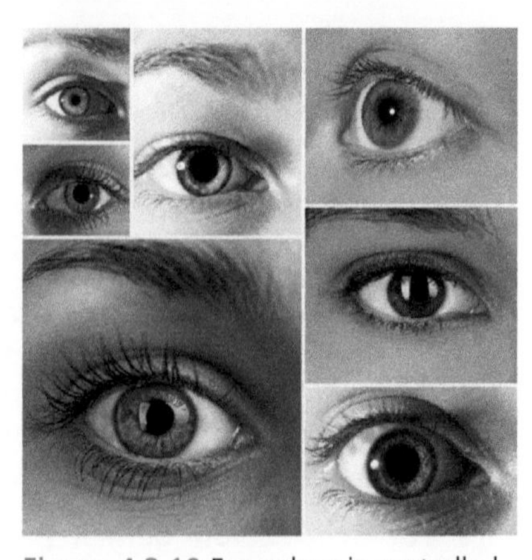

Figure 4.3.10 Eye colour is controlled by many genes. While we can write down the phenotype, it's not possible to write a genotype

3. **What effect does environment have on genotype in the case of height?**
4. **Which other traits are determined by multiple genes?**

Twin studies

There are two types of twins – 'identical' and 'non-identical'. Identical twins are created from one fertilised egg, which splits into two embryos, meaning that all the genes in each embryo are identical. Non-identical twins are formed when two separate eggs are fertilised at the same time. Identical twins therefore share all of their genes, whereas non-identical twins only share about 50% of them, the same as non-twin siblings.

If identical twins are compared for a particular trait, the effect of environment can be determined. Identical twins have exactly the same genes; as they grow, they can develop subtle differences such as a difference in height or weight. As they share the same genome, scientists can assume the differences are due to environment.

In other studies, identical and non-identical twins are compared to determine how strongly the environment influences a trait. If both identical twins show more similarity on a given trait compared to non-identical twins, this provides evidence that genotype significantly influences that trait. If both sets of twins share a trait to an equal extent, it is likely that the environment influences the trait more than genotype.

Figure 4.3.11 The two girls are identical twins; the two boys are non-identical twins

5. **How can identical twins help identify the contribution of genotype to a particular trait?**
6. **Interpret the graph in Figure 4.3.12 to explain the influence of genes on these traits.**

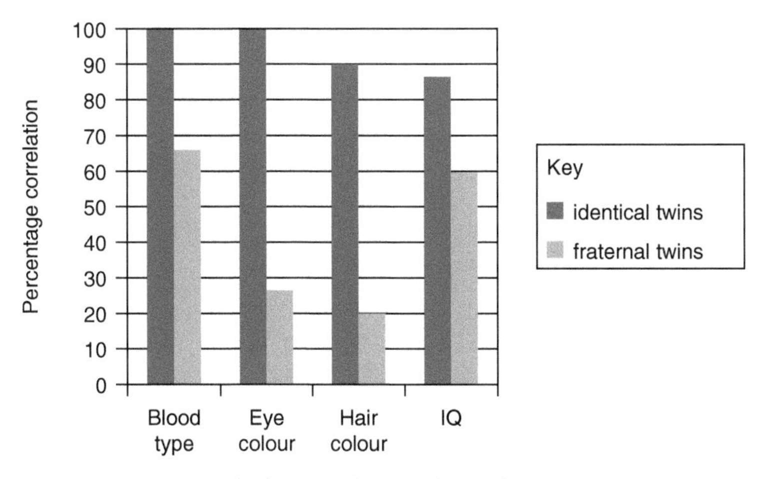

Figure 4.3.12 A graph showing the correlation for various traits

Check your progress

You should be able to:

recall the definition of the gene.	→	describe a gene as a section of DNA that controls a particular characteristic.	→	understand that genes work by coding for the production of a particular protein.
describe how some characteristics are controlled by a single gene.	→	explain how a gene may have different forms called alleles.	→	describe the expression of a gene in terms of alleles.
recall that females have two X-chromosomes and males have an X- and a Y- chromosome.	→	describe the process of sex determination in humans, using genetic diagrams.	→	explain why approximately 50% of live-born children are female and 50% are male.
recall that genes exist in different forms called alleles, and know that they exist as dominant and recessive alleles.	→	complete Punnett squares to show the inheritance of characteristics controlled by single genes.	→	construct Punnett squares to predict the outcome of genetic crosses.
explain the terms genotype and phenotype.	→	describe how genes and their interaction with the environment influence the phenotype.	→	recall that most phenotypic features are the result of multiple genes.

Worked example

A couple's second child was born with a condition called phenylketonuria (PKU). An example of single gene inheritance, the condition is caused by a recessive allele. The letter P is used to represent the dominant non-PKU allele, while 'p' is used to represent the recessive PKU-causing allele.

The family tree shows that there is a history of PKU in the mother's (8) family. The father (7) was unaware of the condition in his family.

The couple would like another child. A genetic counsellor draws a family tree.

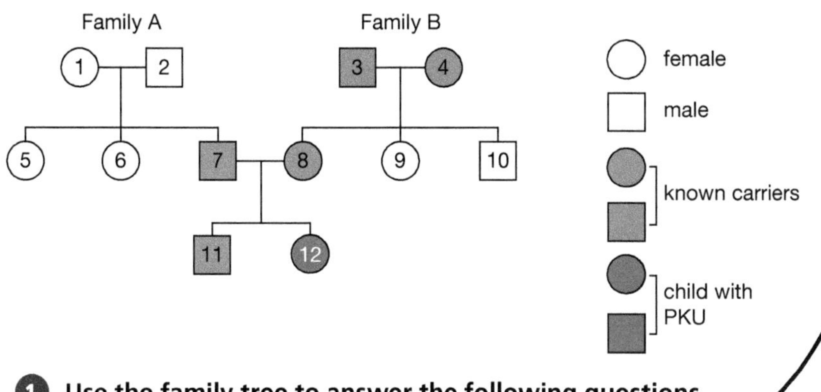

1 **Use the family tree to answer the following questions.**

What is the genotype of the father (7)?

Pp

The student has identified the genotype correctly.

2 **What can be deduced about the genotype of the father's parents, 1 and 2? Explain your answer.**

One parent (or both) must be carriers. For 7 to be a carrier, he must have inherited one recessive allele. So, one parent must have passed on a recessive allele. The other parent would have passed on a dominant allele (otherwise 7 would have PKU).

A good explanation. The student could have added that, their other allele *could* have been the recessive allele, which was not passed down to any of the three children.

3 **Draw a genetic diagram to show how 12 came to have PKU.**

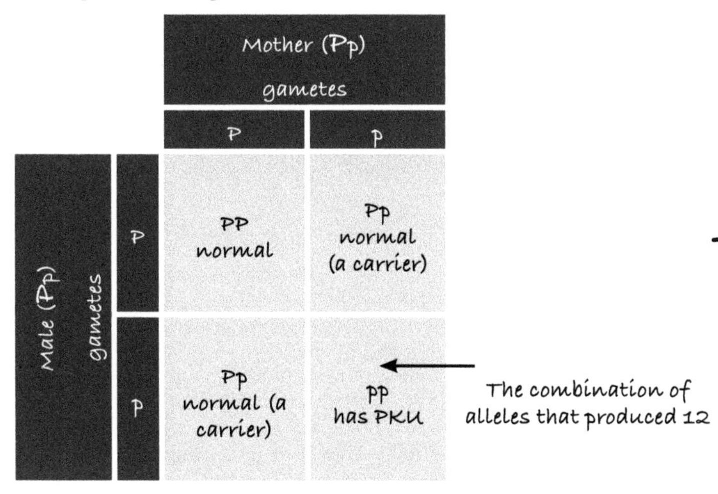

The student has included the parental genotypes, the gametes produced, the possible genotypes and phenotypes produced, and has highlighted the combination of alleles that must have produced 12.

End of chapter questions

Getting started

1. The genetic term used to describe the entire DNA of an organism is: 1 Mark

   A phenotype B genome C double helix D allele

2. The process of cell devision that results in the formation of gametes is: 1 Mark

   A sex determination B mitosis C meiosis D cloning

3. Use the words provided to complete the sentences.

alleles	double helix	chromosomes	different positions
DNA	each chromosome	identical positions	one chromosome

Alleles for a particular characteristic are located at_____ on_____ of the chromosome pair. 1 Mark

The structure of _____ is like a twisted ladder. The shape of this molecule is described as a . 1 Mark

4. How do genes control inherited characteristics? 2 Marks

5. People can either roll their tongue into a U-shape, or are unable to roll their tongue.

   a The diagram shows a pair of chromosomes.

   Is the allele for tongue-rolling dominant or recessive? Explain your answer. 2 Marks

   b Give the other possible genotypes related to tongue rolling. 1 Mark

   c Give the phenotypes for these genotypes. 1 Mark

genotype TT
can roll tongue

Going further

6. One of the advantages of sexual reproduction is: 1 Mark

   A less energy required

   B produces large number of offspring

   C variation

   D no need for a mate

7. **a** Describe the process of sex determination in human offspring. 2 Marks

   You can use a diagram to help with your description.

8. On the diagram you drew for question 7, circle the parts that represent an egg cell, and a sperm cell. 2 Marks

9. In a species of mouse, black coat colour is dominant to white.

   Two black mice mate.

   a Complete the Punnett square below to show the genotypes and phenotypes of the offspring. 2 Marks

b Calculate the probability of producing a white mouse. Express your answer as a fraction.

1 Mark

10 Some characteristics are controlled by single genes, while most are the result of multiple genes interacting. Give an example of each type of inheritance.

2 Marks

More challenging

11 The diagram represents meiosis in an animal cell.

Copy and complete the diagram by drawing in the remaining cells, to show the combination of chromosomes which would produce two genetically different cells.

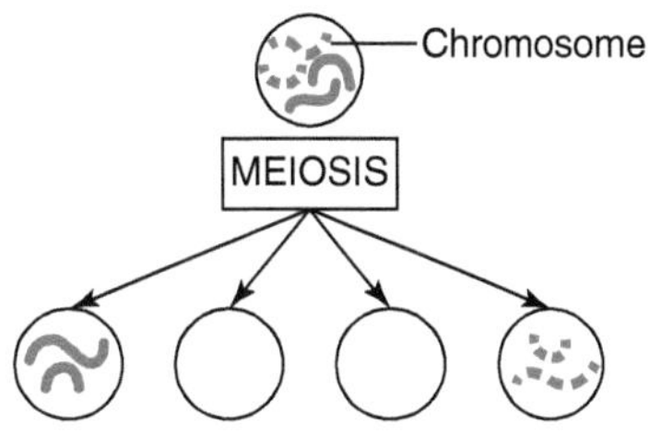

2 Marks

12 A couple have three children. Their genotypes are XX, XY and XX. What are their sexes?

1 Mark

13 A man and a woman have two sons. The woman is pregnant with a third child. What is the chance that this child will also be a boy?

1 Mark

A 0% B 25% C 50% D 100%

14 Copy and complete the genetic diagram below and identify any children with cystic fibrosis (CF).

6 Marks

Give the probability of any children having CF.

The following symbols have been used:

C = dominant allele for **not** having CF

c = recessive allele for having CF

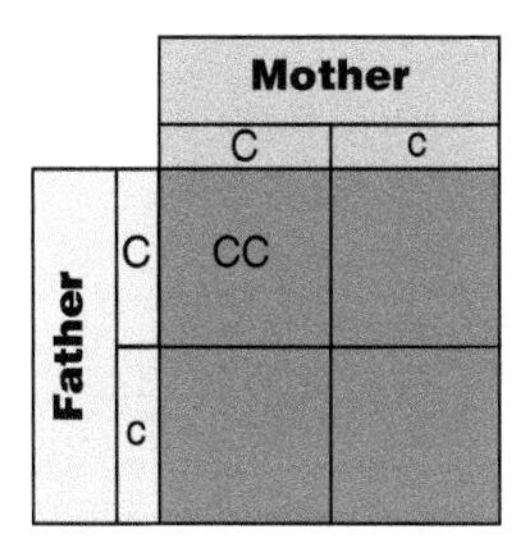

Most demanding

15 Using an appropriate example, explain the difference between the terms heterozygous and homozygous; phenotype and genotype; dominant and recessive.

3 Marks

16 In Estonia, a large percentage of the population has blue eyes. In an Estonian town with a population of 14 000, a survey reveals that 12 600 have blue eyes. Assuming direct proportionality, how many people would you expect to have blue eyes in another Estonian town with a population of 20 000?

2 Marks

17 Red–green colour blindness is controlled by a single gene. The recessive allele, which causes colour blindness, is represented by 'b'; the dominant allele (non-colour blindness) is represented by 'B'. Using these symbols, draw two Punnett squares to show:

a a cross between a mother who has a heterozygous genotype and a father who has colour blindness;

2 Marks

b a cross between a mother and father who are both heterozygous for the condition.

2 Marks

c For each of the Punnett squares you've drawn in parts **a** and **b**, write down the probability of any children having red-green colour blindness.

1 Mark

Total: 40 Marks

VARIATION AND EVOLUTION

IDEAS YOU HAVE MET BEFORE:

THERE ARE DIFFERENCES BETWEEN AND WITHIN SPECIES.

- Variation between individuals is continuous or discontinuous.
- Variation can be measured and the data represented graphically.
- Hereditary material is passed from one generation to the next through the genetic material, DNA.
- Changes in our DNA might cause gene mutations that trigger uncontrollable cell division, leading to cancer.

VARIATION BETWEEN SPECIES CAN LEAD TO NATURAL SELECTION.

- Variation between individuals of the same species means that some organisms compete more successfully than others.
- When an animal or plant is more successful it passes more of its genes to the next generation. Less successful organisms will not pass on as many genes. This is natural selection.
- Genetics allows us to understand the inheritance of certain characteristics in humans and many other organisms.

CHANGES IN THE ENVIRONMENT CAN LEAD TO EXTINCTIONS.

- Some individuals within a species, and some species, are less well adapted than others.
- Changes in the environment can lead to the extinction of those individuals that are less well adapted.
- Gene banks can be used to preserve hereditary material.

4.4

IN THIS CHAPTER YOU WILL FIND OUT ABOUT:

WHAT CAUSES VARIATION AND WHAT ARE THE EFFECTS ON THE INDIVIDUAL?

- Variation has genetic and environmental causes.
- Mutation, sexual reproduction and meiosis are processes that lead to variation.
- Variation results in differences in phenotypes and genotypes in a population.

HOW DO VARIATION AND NATURAL SELECTION LEAD TO THE EVOLUTION OF NEW SPECIES?

- In any population, there is a struggle for existence.
- Because of variation, some individuals are better suited to the environment, so they reproduce and pass on their genes to the next generation, a process called natural selection.
- Natural selection acts on populations, and if the environment changes a new species may result, by the processes of evolution and speciation.
- The evidence for natural selection and evolution comes from observations of current organisms, the fossil record and biochemistry, including DNA.
- Organisms are named by the binomial system of genus and species.
- Modern classifications systems are based on theories about evolution developed from analysis of differences in DNA molecules.
- Humans can also change the genetic make-up of organisms by selective breeding and genetic engineering.

WHAT IS THE EVIDENCE FOR EVOLUTION?

- Evidence for evolution comes from the study of fossils, which show how much or how little organisms have changed as life developed on Earth.
- Evolution of bacteria can be observed in a relatively short time because they reproduce so fast.
- Bacteria that cause disease evolve by natural selection when exposed to antibiotics.

4.4a Mutations

KEY WORDS

mutation

Learning objectives

- describe how genetic variation arises in a population
- explain how variants arise from mutations
- describe the negative and sometimes positive effects of mutations.

Sexual reproduction usually generates a large amount of variation within a species. This is essential to survival. Another way that variation arises is through gene mutation. Organisms that differ through genetic mutations are called variants.

Mutations are changes to our DNA

Mutations, changes to our DNA, occur continuously in our bodies. These changes can be the result of chemicals or radiation, but they also just happen. They are errors that are made when a cell divides, or when the instructions to produce a protein are being copied.

Sometimes, mutations can affect the way an animal or plant functions, leading to the death of the organism. If a mutation occurs in a body cell, cancer may result, for example. If a mutation occurs in cells that produce eggs and sperm, it may lead to a genetic disorder in the offspring.

Mutations can also be beneficial, making individuals more likely to survive and reproduce. A good example of this is the peppered moth (Figure 4.4.1).

Figure 4.4.1 The light form of the peppered moth (on the left) is camouflaged on lichen-covered tree bark. In areas with heavy industrial pollution the genetically mutated darker moth (on the right) is better camouflaged on sooty trees, as can be seen in this photograph

Sometimes, mutations lead to extra or reduced numbers of whole chromosomes. More commonly, a change in the sequence of chemicals – or bases – in a gene will occur.

1. **What is a mutation?**
2. **When can mutations be beneficial?**

Not all mutations have serious effects

Because we have two copies of each gene, it's likely that, if one is faulty, the other will be normal and so the protein can be produced as usual. There are also instances where a change in a gene may not lead to a change in the function of a protein.

MAKING LINKS

You first came across mutations when learning about ionising radiation. Look back to topic 3.2h and to Figure 3.2.20, to remind yourself.

Most mutations have no effect on the phenotype of an organism. This can happen in situations where the mutation occurs in a stretch of DNA with no function. This ends up not affecting the function of the protein made by the gene. Some mutations influence phenotype, such as a mutation in the ear shape of cats.

Figure 4.4.2 The American Curl is a breed of cat characterised by its unusual ear shape

Very few mutations can determine phenotype. For example DDT resistance in insects is caused by a single mutation.

3. **Why would a mutation result in a possible change of phenotype?**
4. **Explain why most mutations are not harmful.**

Mutations that affect the phenotype

A few gene or chromosome mutations can change the phenotype. For example, particular changes in DNA may cause a different protein to be produced. A chromosome mutation may result in damage or loss of a part of a chromosome or even the gain of an extra chromosome. A change in a gene or chromosome may result in production of proteins that disrupt the complex reactions in the cells, or even prevent the protein being produced in cells.

Most of these types of mutation produce harmful effects to the organism, often leading to early death. In some rare cases, a gene or chromosome mutation produces a beneficial effect in the organism and leads to development of a phenotype that is successfully passed on to some of its offspring.

5. **How common are mutations that affect phenotype?**
6. **How do mutations lead to a change in the phenotype?**

RESEARCH

Look up the disorder phenylketonuria, or PKU. What are the consequences of mutation in the alleles of the parents?

4.4b Evolution through natural selection

Learning objectives:

- explain the theory of evolution by natural selection
- describe the process of natural selection
- understand that when natural selection operates differently on populations, a new species is produced.

KEY WORDS

hybrid
interbreeding
natural selection
speciation

Mutation leads to variation in the population of a species. In some cases, it may confer an advantage to the individual, helping it to survive in that environment and produce offspring with the same variation.

Evidence for Darwin's theory of evolution

From 1831 to 1836, Charles Darwin travelled around the world on board the HMS *Beagle*.

A short period of the trip, in 1835, was spent on the Galápagos Islands. Those 19 days were to shape the ideas for Darwin's theory. The Galápagos Islands are 600 miles off the coast of South America. They arose from the sea by volcanic action.

Darwin realised that the animals and plants that lived there must have originated from mainland South America.

Some species on the islands were similar to those on the mainland, but *not* the same. Others were very different. The giant tortoises were all different from island to island. Darwin also noted three species of mockingbird, each living on a different island.

He realised that species were not fixed: they can change. Because of variation in a species, some individuals survived better than others. Darwin called this **natural selection**.

Darwin recorded his observations and collected specimens of the animals and plants from the islands.

1. **Why did Darwin think that the animals and plants found on the Galápagos Islands originated from South America?**
2. **What evidence did Darwin use to suggest the theory of natural selection?**

DID YOU KNOW?

During the 18th century, an English economist called Thomas Malthus wrote about how the human population was growing so fast that the supplies of food would not be able to keep up and many people would starve, reducing the size of the population.

Darwin applied Thomas Malthus' work on human populations to other living organisms. Darwin suggested that within a species, more offspring are born than are able to survive to adulthood and reproduce. The individuals that are better suited (adapted) to their environment are the ones most likely to survive and produce offspring with the same beneficial adaptations. This is the basis of the theory of evolution by natural selection.

giant Pinta tortoise

marine iguana

flightless cormorant

Galápagos penguin

Figure 4.4.3 Animals of the Galápagos are very unusual

Isolation and natural selection lead to a new species

Imagine two populations of a species that have been separated for a long time. Environmental conditions will be different in the two locations.

Naturally occurring genetic mutations will lead to differences emerging in the characteristics – the phenotypes – of the two populations. Eventually, two new species are produced. The organisms can be regarded as a new species when they are no longer able to successfully breed with each other, or interbreed to produce fertile offspring.

Some animals and plants do sometimes interbreed – either naturally, or having been encouraged to by humans. But **hybrids**, if they do happen, are infertile. The key to our definition of a species is that *successful* **interbreeding** – that produces fertile offspring – is not possible between two different species.

Figure 4.4.4 A mule is a cross between a female horse and a male donkey

3 **Write down the definition of a species.**

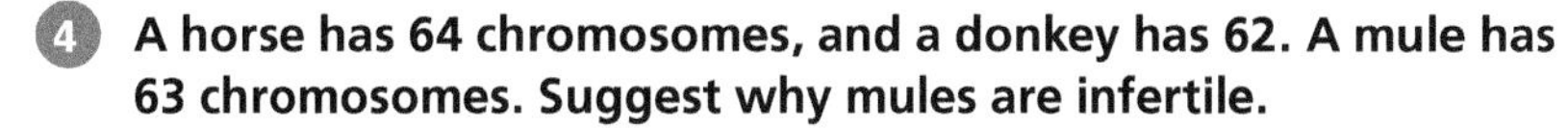

4 **A horse has 64 chromosomes, and a donkey has 62. A mule has 63 chromosomes. Suggest why mules are infertile.**

The Galápagos finches

Speciation is the formation of a new species over a long period of time, through separation of a population. We can see good evidence of this in the different species of finch on the Galápagos Islands. Birds from mainland South America would have been ground-dwelling seed eaters. They would have found different conditions when they arrived on the different islands, including variations in climate and food supplies.

Competition among the birds for these food supplies would have produced a struggle for existence. Slight variations in beak shape enabled some birds to exploit slightly different food supplies, for example, small seeds, nuts, cacti and insects. Birds also selected different habitats in which to live, so as to survive.

There are now 13 species of finch living on the Islands.

DID YOU KNOW?

You may see the Galápagos finches referred to as 'Darwin's finches', a description used by ornithologist David Lack in 1947. Darwin did not fully appreciate their evolutionary significance.

A tree finch.

A cactus finch feeding on a prickly pear cactus.

The woodpecker finch finds short twigs, breaks them to the right length, then uses them to spear and extract insects from trees.

Figure 4.4.5 Three species of Galápagos finch, all descended from the same ground-dwelling mainland finch

5 **Which factor(s) contributed to the struggle for existence amongst the finches on the Galápagos Islands?**

6 **How do the Galápagos finches provide evidence for natural selection?**

4.4c Evidence for evolution

Learning objectives:

- understand how, and the situations in which, fossils are formed
- understand how fossils are used as evidence for evolution of species from simpler life forms
- explain how antibiotic resistance in bacteria is evidence of evolution.

KEY WORDS

antibiotic resistance
fossils

Fossils are the remains of organisms that lived millions, or hundreds of thousands, of years ago. They are formed when an organism dies; if conditions are right, the hard parts of the organism become replaced by minerals. Fossils can provide evidence for evolution.

Figure 4.4.6 Fossils were well known in Darwin's time. *Ichthyosaurus* was found in 1811 by fossil hunter Mary Anning

The fossil record is incomplete

The fossil record shows us how organisms that lived in the past differ from those around us today. It also shows us how long a species existed for, by when it enters and leaves the fossil record. It shows us extinction, and the slow and successive appearance of new species.

There are **fossils** missing from the fossil record. Conditions must be just right for fossil formation, and the chances of this are very small. Many early life forms were microorganisms or they were soft-bodied, so fewer traces of them exist. However, some archaea and bacteria have left traces of the unique chemicals they produced. There is now clear evidence going back 3.5 billion years.

Despite this, the fossil record is incomplete. Darwin appreciated that, if life evolved, then there must be intermediate forms or 'missing links', between different organisms, as they evolved.

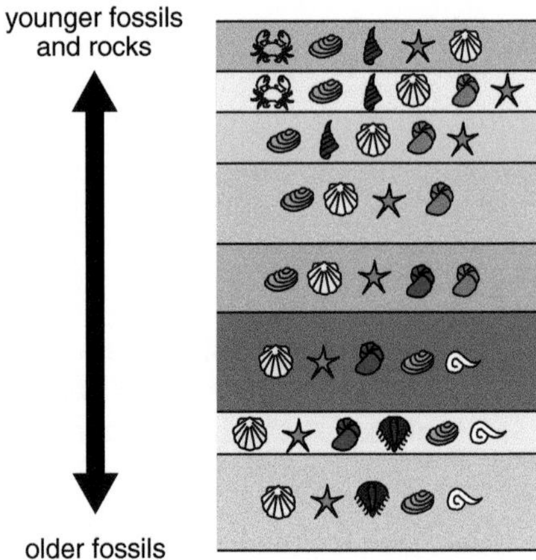

Figure 4.4.7 Geologists can work out the age of rocks that contain certain fossils. This shows them when different species existed on Earth

REMEMBER!

Fossils are rare, so any 'missing links' will be even rarer.

1. **How are the relative ages of fossil species worked out?**
2. **Why do we have few fossils of early life forms?**

Fossil horses

We can learn from the fossil record how life on Earth has changed. For example, there is an extensive fossil record of horse-like animals – and new ones are still being found. We can trace the evolution of the modern horse, *Equus*, from an animal the size of a small dog, called *Hyracotherium*, which had toes. These toes have been lost during evolution and hooves developed.

In contrast, certain bacteria have evolved very little in billions of years. It can be easy to distinguish fossil archaeans from fossil bacteria because of traces of the unique chemicals in their cell membranes.

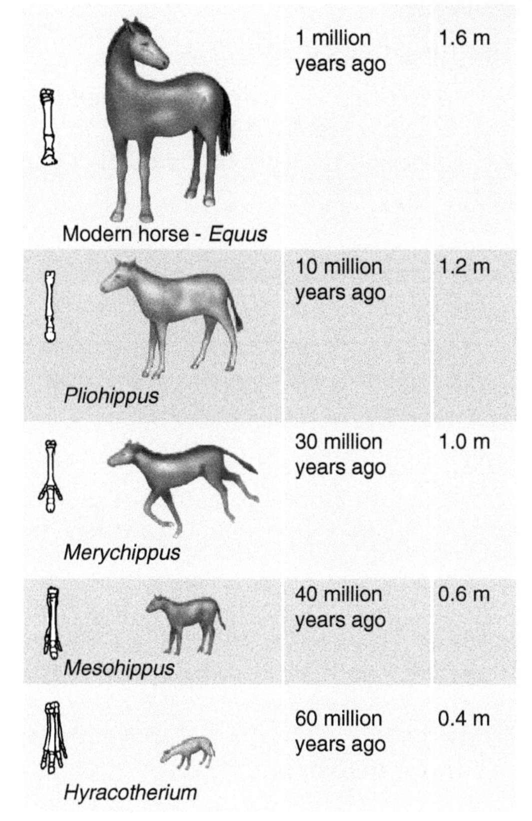

Figure 4.4.8 The evolution of the modern horse. We now know that this evolution is not a simple straight line – their evolutionary tree branches and has offshoots – with some intermediate stages actually being smaller

3 **Describe the trends in the evolution of the horse.**

4 **What is the evidence that tells us that some organisms have changed very little over time?**

Developing resistance

Bacteria and other microorganisms are becoming resistant to the drugs that have been designed to kill them. The mutation of genes in pathogenic bacteria produces new strains of the bacteria. Some of these strains may be resistant to an antibiotic used on a patient, and so the bacteria will not be killed.

As the bacteria reproduce, the genes for resistance spread through the population. **Antibiotic resistance** is an example of natural selection. Bacteria can evolve quickly because of their rapid reproductive rate.

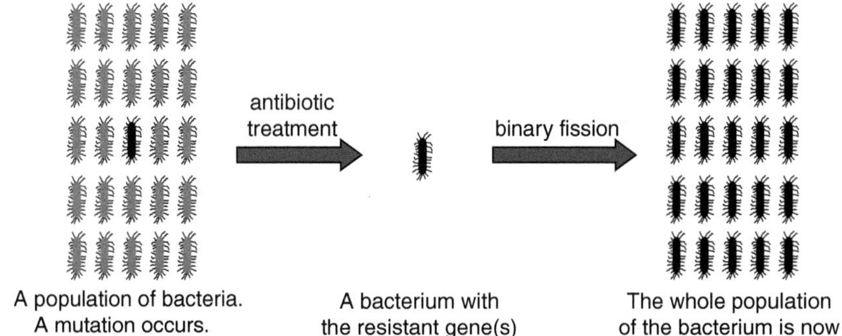

Figure 4.4.9 A simplified diagram of how resistance to antibiotics spreads throughout a population of bacteria. In reality, more than one mutation is required

Certain bacteria, such as *Staphylococcus aureus*, are building up resistance to antibiotics very quickly. *Staphylococcus* that are resistant to the antibiotic methicillin are called MRSA. The increased, and often inappropriate use, of antibiotics both by humans and farm animals, has led to an increase in antibiotic resistance.

DID YOU KNOW?

It has been 30 years since a new class of antibiotics was last introduced.

MAKING LINKS

Refer back to topic 3.3g, where you first learned about antibiotic resistance.

5 **Describe how antibiotic resistance develops.**

6 **Explain how this resistance is evidence for Darwin's theory of evolution.**

4.4d Identification and classification of living things

Learning objectives:

- describe how living things have been classified into groups using a system devised by Linnaeus
- describe how new models of classification have developed.

KEY WORDS

binomial system

In 2015, scientists discovered a new species of snake in a remote region of Western Australia. When a new species is discovered, scientists try to classify it.

Classifying organisms

All species are classified using the system developed by Swedish scientist Carl Linnaeus. Linnaeus divided all organisms into large groups called kingdoms. These were subdivided into smaller and smaller groups.

Linnaeus named species using the **binomial system**. Each organism has two names (hence binomial) – a genus name and a species name. For example, the dog and the wolf share many common characteristics, yet are different animals. They belong to the same genus, *Canis*, but their species is different. The dog is *Canis familiaris* and the wolf is *Canis lupus*.

REMEMBER!

You type a scientific name in *italics*, or if it's handwritten, underline it. The genus name is always written with a capital letter and the species name with a small letter.

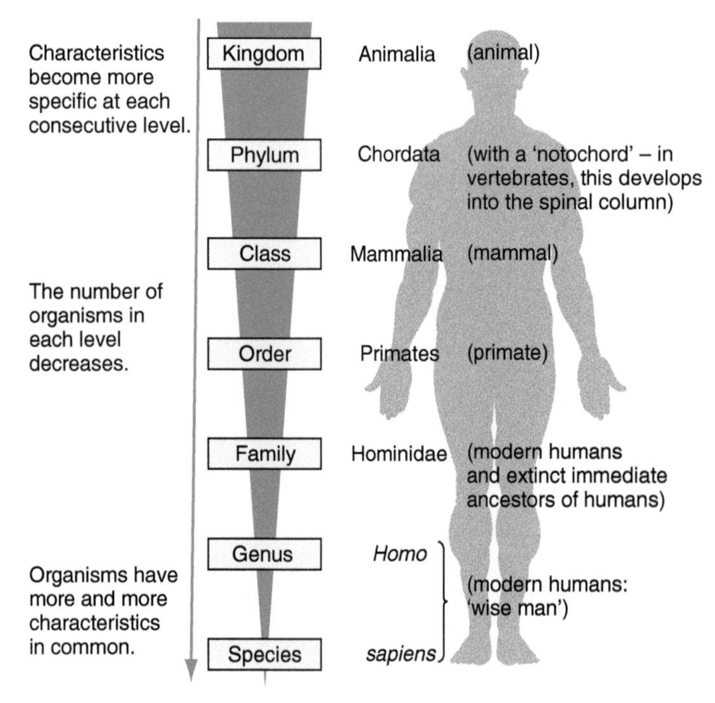

Figure 4.4.10 The classification of humans. Our scientific name is *Homo sapiens*. There are at least ten animals in our genus, but all are extinct except for humans

1. **Who devised the system for classifying organisms that we use today?**
2. **What system do we use to name an individual type of organism?**

Classification and evolution

Classification is important because it allows the clear identification of organisms. It helps to recognise and differentiate between different types of organisms, make scientific and biological predictions about organisms of the same type and classify how different types of organisms are related to each other.

Clear identification of organisms is very important in Biology. For example, if Darwin had not been able to distinguish between the 13 species of finch living on the Gálapagos Islands, he would not have been able to develop his theory of evolution by natural selection.

Scientists continue to rely on classification to explain relationships that exist between different organisms. This is helpful when attempting to trace back to the evolutionary ancestors of any species.

MAKING LINKS

Look back to topic 4.4b to remind yourself about Darwin's finches.

3 **Why is classification important?**

Development of classification

Early work in classification involved grouping organisms by comparing physical features and traits. Linnaeus used this method when he classified living things in the 18th century.

Improvements in tools such as microscopes made comparison of internal structures of organisms easier and increased understanding of biochemical processes led to the introduction of new models of classification.

Modern classification systems are based on theories about evolution developed from analysis of differences in DNA molecules. For example, the information for modern evolutionary relationships comes from organisms that are alive today and that are seen in fossil record. But scientists also look at DNA and the structures of proteins produced by the cells of the organism to group them.

Computer programs can generate vast amounts of information about genes to help classify organisms based on biochemical and molecular genetic traits of organisms as well as observed physical traits.

DID YOU KNOW?

The 'Barcode of Life' is an international initiative devoted to developing a 'DNA barcode' database of living species, which can be used in identification.

4 **What method of grouping did Linnaeus use to classify organisms?**

5 **Explain how modern classification models differ from early models.**

4.4e Selective breeding

Learning objectives:

- describe the process of selective breeding
- explain how selective breeding enables humans to choose desirable characteristics in plants and animals
- explain how selective breeding can lead to inbreeding.

KEY WORDS

breeds
environmental change
inbreeding
selective breeding

Humans have been domesticating animals and plants for thousands of years. DNA evidence suggests that dogs were the first animals to be domesticated, perhaps as long as 33 000 years ago.

Selective breeding

Among the first animals to be domesticated were goats, then sheep. They would have been kept and bred for their meat, milk and hides or skins, which were used for clothes and shelter.

Animals in a population show genetic variation. Humans would have selected those with the characteristics required – the ones that produced the most meat or milk, for instance – and allowed them to breed.

From the offspring of those animals, the humans would then have selected those animals producing the largest yields and bred those. This was repeated over many generations.

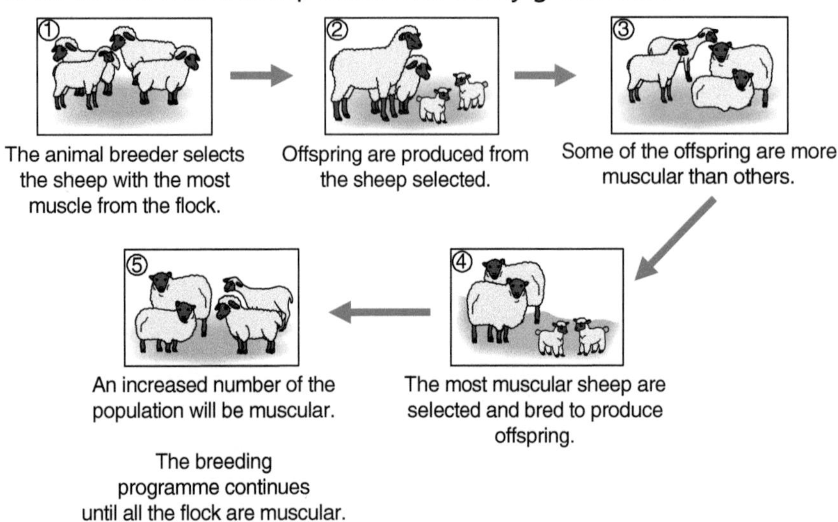

Figure 4.4.11 Sheep are bred for their lean meat and how quickly the meat is produced

This is called **selective breeding**. Darwin referred to it as artificial selection, in contrast to natural selection.

The outcomes of different selective breeding programmes are different **breeds**. The breed produced will depend on the characteristics required.

REMEMBER!

It is important to be able to describe the principles of selective breeding.

Figure 4.4.12 The Belgian Blue (top) has been bred for its meat, which is muscle. The Jersey (bottom) has been bred to produce rich, creamy milk

Related characteristics are often selected together, such as quality *and* volume of milk produced in cattle.

1. **Suggest what other characteristics sheep may be bred for.**
2. **Describe the technique of selective breeding.**

Producing new plant varieties

In modern agriculture, new crop varieties are bred for disease resistance. Increased yields and improved quality of food crops are also the main goals of plant breeders. Other beneficial characteristics are that crops:

- grow and mature quickly
- have a distinctive taste, aroma or colour, for example in strawberries
- have long shelf life, store well or can be frozen.

Selective breeding enables crop growers to plant large areas with identical plants, giving maximum yields. But this does mean that many crops are genetically uniform. The whole crop could be lost if there is **environmental change**.

3. **Suggest three characteristics that crop plants could be bred for.**
4. **What could happen to a crop made up of just one variety if the environment changed?**

Disadvantages of selective breeding

Selective breeding is a slow and imprecise method to produce new and beneficial combinations of genes. One of the disadvantages of selective breeding is that the whole set of genes is transferred. As well as the desirable genes, there may be genes that are not beneficial or may be harmful. Selective breeding repeated over a large number of generations can reduce the fitness of the new variety. To keep desirable features, breeders tend to **inbreed** organisms. Inbreeding is the production of offspring from mating individuals that are closely related genetically. Inbreeding may lead to some breeds being particularly prone to disease or inherited defects.

DID YOU KNOW?

Analysis of mitochondrial DNA suggests that modern dogs are related to the first dogs – which would have been kept for hunting – which originated in Europe between 19 000 and 32 000 years ago.

Figure 4.4.13 As a result of inbreeding, Dalmations are predisposed to deafness

5. **What are the disadvantages of selective breeding?**
6. **Explain the term 'inbreeding' using an example.**

4.4f Genetic engineering

Learning objectives:

- give examples of how plant crops have been genetically engineered to improve products and describe how fungus cells are engineered to produce human insulin.

KEY WORDS

genetic engineering
GM crops
vector

Genetic engineering involves taking specific genes from one organism and introducing them into the genome of another. Scientists can now transfer genes from more or less any organism, including plants, animals, bacteria and viruses.

REMEMBER!

The word 'vector' has two very different meanings in Biology and Physics. In Biology, vectors are bacterial plasmids or viruses, for example, that are used to transfer genes. In Physics, the term vector describes quantities that have both size and direction.

Genetic engineering

In genetic engineering, selected genes from one organism are transferred to another organism which may, or may not, belong to the same species. This process for genetic modification uses enzymes and carriers – called **vectors** – to transfer genes. It is much faster than selective breeding.

Genetic engineering involves cutting out the desired gene from the DNA of an organism and inserting it into the DNA of another species, (usually bacteria) so that it produces the protein coded for by that DNA. Bacterial DNA is contained in a circular plasmid and the cut piece of DNA is inserted into this plasmid. Bacteria grow and multiply rapidly; they are ideal organisms to produce large amounts of the protein made by the new DNA in a short time. The bacteria are cultured in special vessels called fermenters and the protein they produce can be extracted easily.

1. **What is a vector in the genetic engineering process?**
2. **Why do bacteria make effective vectors?**

Genetically engineered plants

Genetic engineering has transformed crop production. Genes from many organisms, often not even plants, are cut out of their chromosomes and inserted into the cells of crop plants. Such crop plants and other organisms are called genetically modified, **GM crops** or GM organisms (GMOs).

Plants have been engineered to be resistant to disease, and to increase yields, such as producing bigger, better fruit. Several types of crop plant have been produced that are resistant to diseases caused by viruses.

In the wet summer of 2012, potato plants became exposed to the potato blight fungus. In 2014 British scientists produced a

GM potato that is resistant to potato blight. Genes from two wild relatives of the potato were inserted into the Desiree potato variety. However, the cultivation of GM crops is highly regulated in Europe. The GM potato was withdrawn when countries opposed to GM were able to block Europe-wide approval. The situation is changing, but up to 2015 only one GM crop - a type of maize used for animal feed - was grown commercially in the EU.

3 **Give two reasons for the genetic modification of plant crops.**

4 **What types of organism cause disease in plants?**

Figure 4.4.14 Pesticides are sprayed over crops to protect them from diseases. Disease-resistant GM plants don't need the pesticides.

Producing human insulin

Patients with Type 1 diabetes need regular injections of the hormone insulin. Since the early 1920s, insulin was extracted from the pancreas of pigs or cattle. But these types of insulin differ slightly from human insulin in the amino acids they contain. They had some side effects.

Genetic engineering made it possible to genetically engineer the bacterium, *Escherichia coli*, and the fungus, yeast, to produce 'human' insulin. This is identical to the insulin produced by the human body.

Yeast produces a more complete version of the insulin molecule. Less processing is required, so this method is often preferred.

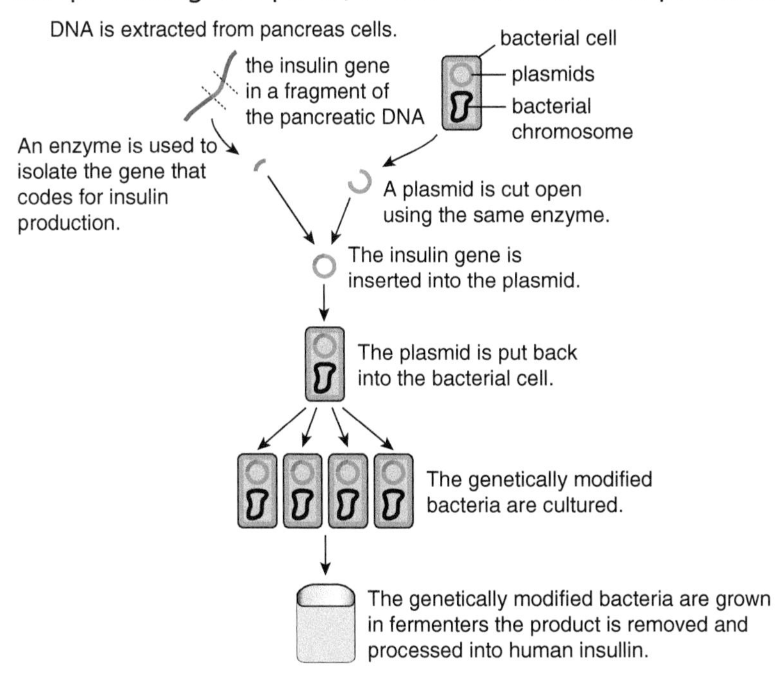

Figure 4.4.16 Insulin production using *Escherichia coli*

Figure 4.4.15 Human insulin production in India – this photograph shows the purification process

5 **What is genetic engineering?**

6 **Name two organisms that can be genetically engineered to produce insulin.**

4.4g Gene technology: benefits and risks

Learning objectives:

- explain the benefits of using gene technology in modern agriculture
- explain the risks of using gene technology in agriculture
- describe some of the practical and ethical considerations of using modern technology.

KEY WORDS

food security
genetically modified

Crops that have had their genes modified are called genetically modified (GM) crops. Producing GM crops with desired characteristics is faster than selective breeding. But are GM products safe?

Benefits of using gene technology

Food security means access by all people at all times to enough food for an active, healthy life.

Many people think that we have the ability, and a moral obligation, to achieve global food security by growing **genetically modified** (GM) organisms. GM crops have increased yields and will grow in poor soil and harsh environments. Scientists have also fortified plant crops with extra nutrients.

Weeds reduce crop-plant yields and encourage fungal disease. GM crops have had genes inserted that make them resistant to a particular herbicide. As the crop grows, the field is sprayed with herbicide. The crop is unaffected, but the weeds are killed.

Another key area is the development of GM crops that resist insect attack. *Bacillus thuringiensis* (Bt) is a soil bacterium that produces a natural insecticide. The gene for this has been inserted into crops. Bt insecticide is a protein that kills the caterpillar, or larva, that eats the crop plant. It only works on some groups of insect, such as butterflies and moths, which are the most serious pests.

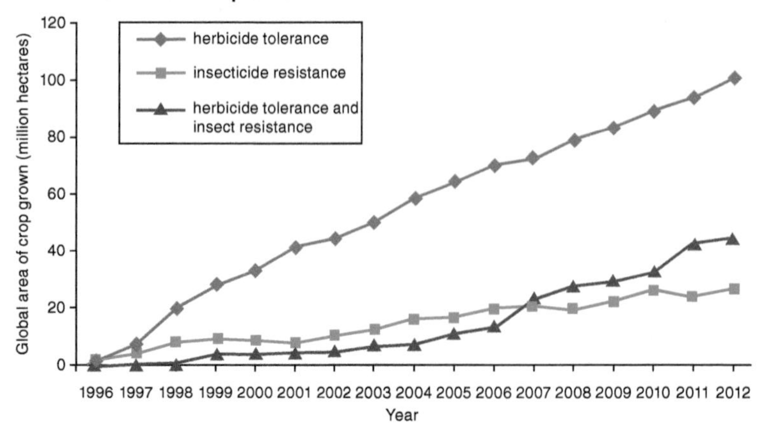

Figure 4.4.18 Global area of crop grown of herbicide-tolerant and insect-resistant plants

REMEMBER!

Be careful what you read about gene modification. Newspapers and TV tend to select stories that are either 'amazing breakthroughs' or 'disasters'. Most progress made in science isn't either of those.

Figure 4.4.17 After a few bites of the leaves of this Bt peanut, the caterpillar died

DID YOU KNOW?

Researchers aim to start growing a vitamin-enhanced 'super-banana' in Uganda. The bananas are genetically engineered to have increased levels of ⊠-carotene, which is converted to vitamin A by the body. A deficiency of vitamin A can be fatal.

 Link: 4.4g ⟶ 4.4h

1. **How might GM crops help to achieve global food security?**
2. **Look at the graph in Figure 4.4.19. Describe the trends shown by the graph.**

Risks of using gene technology

The risks of using GM crops can be divided into two main issues:

- **Damage to the environment:** Crops do not damage the environment simply because they are GM. Some farming practices, such as the overuse of herbicides have harmed the environment but these problems are similar for non-GM and GM crops. Studies of herbicide tolerant GM crops show that insect biodiversity is reduced. It did not matter whether or not the crop was GM; the important factor was the number of weeds that remained in the crop.
- **Crossbreeding:** GM crops may crossbreed with closely related plants and this includes non-GM varieties of the same crop. Crossbreeding between crops and their wild relatives could cause problems if this results in the wild relative acquiring characteristics that might make it more weedy and invasive. For example, herbicide resistant crops could be produced if a herbicide tolerant crop, GM or non-GM, were to breed with weedy relatives. Their offspring might be resistant to the herbicide if they inherit the tolerance gene from the crop. Other herbicides would then have to be used to control these weeds.

3. **What are the concerns about GM crops?**
4. **Do you consider GM products to be safe? Justify your answer.**

Figure 4.4.19 A monarch butterfly (top); and a monarch butterfly larva feeding on milkweed. One study in 1999 showed that the growth of monarch butterfly caterpillars was affected when their foodplants were dusted with Bt pollen

Practical and ethical considerations

Although genetic engineering presents an exciting range of possibilities, such as the potential to feed the hungry and to treat diseases, these promises also present potential problems. There are concerns about the effect of GM crops on populations of wild flowers and insects as a result of cross-pollination. There is also a concern that insects may evolve to become resistant so that GM crops are no longer protected.

An ethical question regarding GM technology is about using public money to fund GM. There have been cases where a GM crop has not delivered the intended improvements such as increased crop yields or virus resistance. A frequent criticism of GM is that it has failed to deliver more than herbicide tolerance and insect resistance. This is because these uses are based on genes available 20 years ago; with increasing knowledge of gene function, other characteristics are being developed.

REMEMBER!

Be prepared to analyse and discuss negative and positive aspects of genetic modification from information that you are given.

5. **What are the ethical concerns about GM crops?**
6. **Who owns GM technology?**

MATHS SKILLS

4.4h Using charts and graphs to display data

Learning objectives:

- understand when and how to use bar charts
- understand how to show sub-groups on bar charts
- understand how to plot histograms.

KEY WORDS

discrete
grouped bar chart
stacked bar chart

The International Union for Conservation of Nature and Natural Resources (IUCN) monitors species in danger of extinction. It must make this huge amount of data understandable to the public.

Bar charts

Scientists use graphs and charts to display, summarise and analyse data. They can be used to simplify the presentation of complex data and highlight patterns and trends.

Bar charts are used to display data collected for distinct groups. If the data points in the groups can be counted then this data is known as **discrete** data.

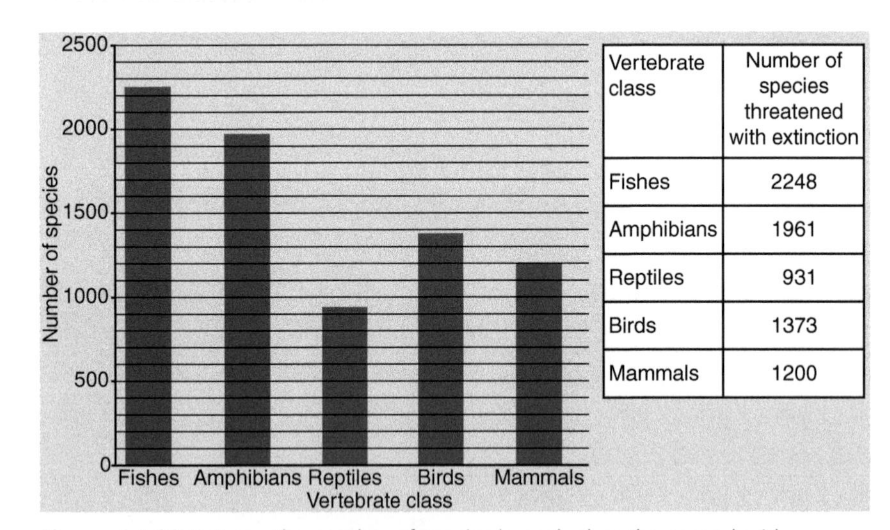

Vertebrate class	Number of species threatened with extinction
Fishes	2248
Amphibians	1961
Reptiles	931
Birds	1373
Mammals	1200

Figure 4.4.20 Data on the number of species in each class threatened with extinction are transferred from a table to a bar chart

When drawing a bar chart:

- the *height* of the bars is proportional to the measured number or frequency
- the *width* of the bars should be consistent
- the bars do not touch each other; they represent distinct groups
- the bars can be placed in *any* convenient order that makes sense.

1 **Draw a bar chart for the groups of organisms, shown below, found as fossils in a particular rock type.**

Group of organism	Fossils found in %
Algae	14
Arthropods	33
Molluscs	3
Sponges	22
Worms	3

2 **What percentage of fossil organisms were unclassified?**

DID YOU KNOW?

In his book of 1871, *The Descent of Man*, Darwin discussed the diversity of skull size and shape in different peoples.

More complex bar charts

Other types of bar chart could be used to provide further information about these endangered species.

From Figure 4.4.21, we don't know the number of species in each vertebrate class, so we don't have any indication of what proportion of each vertebrate class is endangered.

We can add this information to produce a dual bar chart (also known as a **grouped bar chart**).

A similar type of bar chart is the **stacked bar chart**. We often see these in the media. Here, different sub-groups are stacked on top of each other to produce a single bar for each group included.
Note that bar charts can be constructed horizontally as well as vertically.

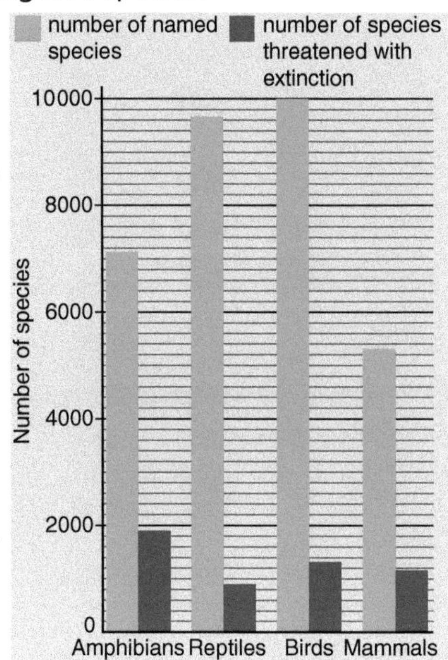

Figure 4.4.21 A dual bar chart for four of the vertebrate classes

3. **Construct a bar chart to illustrate the proportion of endangered species of fish:**

Number threatened with extinction	2248
Number of species	33100

> **REMEMBER!**
>
> Ensure that you choose sensible scales when drawing bar charts and histograms and draw them accurately.

Histograms

A histogram is a statistical diagram, similar to a bar chart. In a histogram, the area of each bar is proportional to the frequency of the class interval. The data are continuous instead of being discrete. They're commonly used to show the variation of a characteristic, such as height. Remember:

- the variable on the *x*-axis is continuous, so there are no gaps between bars
- the size of the categories is represented by the area of the bars, although for many of the histograms you will plot the ranges on the *x*-axis have been chosen to be equal.

Histograms can be used to show the range of variation across groups of organisms. One characteristic determined for both living humans and extinct human ancestors is the cephalic index (CI).

$$\text{Cephalic index} = \frac{\text{Maximum head width}}{\text{Maximum head length}} \times 100$$

The CI varies between human populations of different groups and over time.

Cephalic index (CI)	Boys in %	Girls in %
72.0≤CI<75.0	7.3	4.8
75.0≤CI<80.0	28.4	32.8
80.0≤CI<85.0	30.0	42.6
85.0≤CI<90.0	34.3	19.6

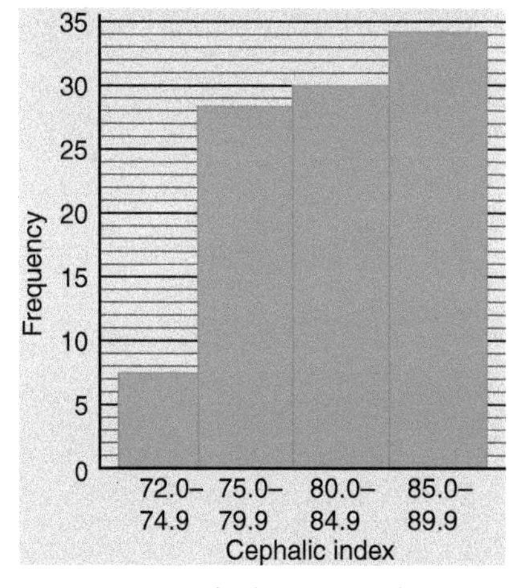

Figure 4.4.22 The histogram shows the range of CI for the boys in a group of school children. Note that the bar for the lowest category is narrower

4. **Plot a histogram for the range of CI of girls in the group.**
5. **Plot a histogram for the range of heights in your science class.**

Check your progress

You should be able to:

☐ distinguish between variation caused by genes and by the environment. →	☐ describe how variation contributes to an organism's survival. →	☐ describe how variation results in differences in phenotypes and genotypes in a population.
☐ describe evolution as a change in inherited characteristics of a population. →	☐ explain how evolution occurs through natural selection to give rise to phenotypes best suited to their environment. →	☐ explain the evidence for the occurrence of evolution and natural selection.
☐ describe the process of selective breeding. →	☐ describe the process of genetic engineering. →	☐ evaluate the risks and benefits of selective breeding and genetic engineering.

Worked example

Scientists in Germany have investigated the resistance of rats to two poisons – warfarin and bromadiolone. The rat populations originally had no resistance to the poisons. One poison started to be used before the other.

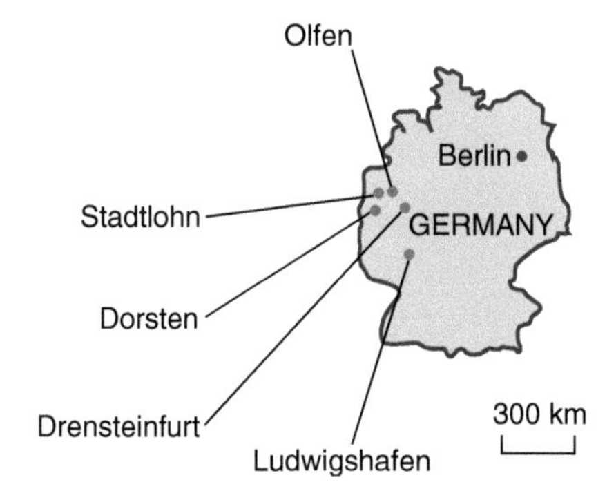

Some of their results are shown below.

Town	Not resistant to either poison (%)	Resistant to warfarin alone (%)	Resistant to both poisons (%)
Dorsten	44	56	0
Drensteinfurt	90	5	5
Ludwigshafen	100	0	0
Olfen	21	21	58
Stadtlohn	5	8	87

1 Use the data to explain how a population of rats with no resistance to the poisons might become resistant in time. **[3 Marks]**

In a population, some rats will be resistant to the rat poisons, and survive their use. These will pass on their resistant genes. These will spread through the population as the rats breed, until most of the rats become resistant.

2 Suggest reasons for the differences in the degree of resistance to the poisons. **[2 Marks]**

Resistance is greatest in Stadtlohn and Olfen. There is no resistance to the poisons in Ludwigshafen.

This could be because of the locations of the towns. If the towns are close together, populations can interbreed so that the resistant gene(s) spread. Ludwigshafen is further away.

This is a concise answer, but one that is not fully detailed. The student should have mentioned that some rats in a population will be resistant to the rat poison because of mutations. It is also important to say that these mutations will spread quickly through the population because the rats reproduce so rapidly

In this answer the student has overlooked the fact that mutations arise spontaneously in populations, as well as being spread by sexual reproduction.

These could also arise at a greater rate in Stadtlohn and Olfen, perhaps because of the numbers of rats in the different towns, which we do not know from the data given.

The data suggest that resistance arises first to one poison, followed by the other. You do not need to know the reason for this, but warfarin was a 'first generation' poison, while bromadiolone was developed because rats became resistant to warfarin.

End of chapter questions

Getting started

1. What is the name of the scientist who wrote *On the Origin of Species?* — 1 Mark
2. Why is our fossil record incomplete? — 2 Marks
3. What feature of bacteria and archaea help scientists to trace them back millions of years? — 1 Mark
4. What is a hybrid? — 1 Mark
5. Describe what is meant by genetic variation. — 1 Mark
6. Draw a bar chart for the groups of organisms shown below, found as fossils in a particular type of rock on the Dorset coastline. — 1 Mark

Group of organism	Fossils found %
Algae	11
Arthropods	30
Molluscs	4
Sponges	23
Worms	2

7. Of the fossils found at this location, calculate what percentage were unclassified (i.e. not listed in the table above). — 1 Mark

Going further

8. How can separated populations of organisms become new species? — 3 Marks
9. What is meant by the term 'natural selection?' — 3 Marks
10. The occurrence of a dark form of the two-spot ladybird was monitored by scientists between the 1960s and the 1980s.

   a Some of the scientists' results are shown in the table.

Year	Frequency of the dark form (%)
1960	47
1965	37
1970	27
1975	19
1980	12
1985	10

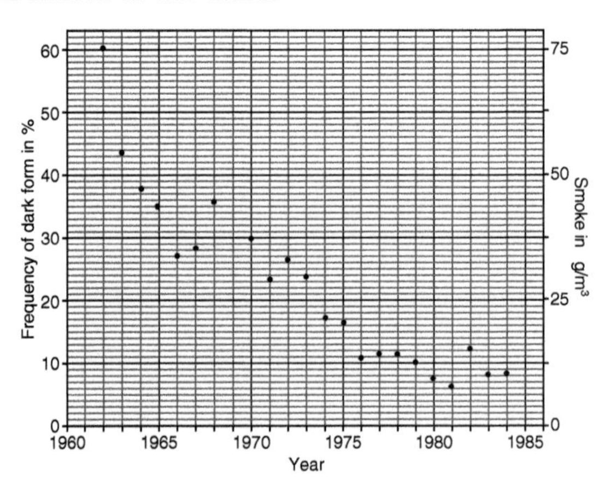

The scientists also recorded the yearly amounts of smoke in the air.
The readings for smoke concentration have already been plotted on the graph.

Plot the frequency of the dark form of the ladybird from 1960 to 1985 on a copy of the graph above.

Connect the points with a smooth curve.

4 Marks

b Suggest a reason for the change in frequency of the dark form of the ladybird. Predict what may have happened to the ladybird population after 1984.

2 Marks

More challenging

11 Describe how a human gene that codes for insulin production can be transferred to a bacterium.

2 Marks

12 Describe and explain two ethical objections to genetic engineering.

2 Marks

13 A plant breeder is trying to produce a new variety of wheat that will withstand drought. Describe how the plant breeder would do this.

6 Marks

Most demanding

14 Describe two separate examples of evidence that support the theory of evolution.

2 Marks

15 A scientist is investigating variation in bone mineral density measurements between human populations of different demographic groups over time.

The table below shows the data from a sample of people taken from a demographic.

Would it be more appropriate to draw a bar chart or a histogram to show the data below? Draw an appropriate chart or graph of the data.

Describe what the data show.

Bone mineral density [g/cm^2]	Number of people
81–100	7
101–120	18
121–140	31
141–160	35
161–170	27
171–180	11

4 Marks

16 The diagram shows the relationship between six different species (A, B, C, D, E and F), which have evolved from a common ancestor.

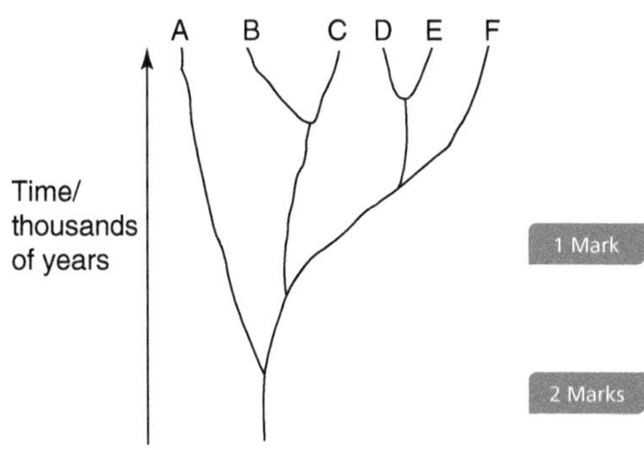

a From the diagram, state which two species are most closely related.

1 Mark

b Use the information in the diagram to explain why A and B are less closely related to each other than B and C.

2 Marks

c If specimens of species A, B and C had been provided instead of the diagram, how could you tell which two were most closely related?

1 Mark

Total: 40 Marks

Key

relative atomic mass **atomic symbol** name atomic (proton) number

1	2											3	4	5	6	7	0
							1 **H** hydrogen 1										4 **He** helium 2
7 **Li** lithium 3	9 **Be** beryllium 4											11 **B** boron 5	12 **C** carbon 6	14 **N** nitrogen 7	16 **O** oxygen 8	19 **F** fluorine 9	20 **Ne** neon 10
23 **Na** sodium 11	24 **Mg** magnesium 12											27 **Al** aluminum 13	28 **Si** silicon 14	31 **P** phosphorus 15	32 **S** sulfur 16	35.5 **Cl** chlorine 17	40 **Ar** argon 18
39 **K** potassium 19	40 **Ca** calcium 20	45 **Sc** scandium 21	48 **Ti** titanium 22	51 **V** vanadium 23	52 **Cr** chromium 24	55 **Mn** manganese 25	56 **Fe** iron 26	59 **Co** cobalt 27	59 **Ni** nickel 28	63.5 **Cu** copper 29	65 **Zn** zinc 30	70 **Ga** gallium 31	73 **Ge** germanium 32	75 **As** arsenic 33	79 **Se** selenium 34	80 **Br** bromine 35	84 **Kr** krypton 36
85 **Rb** rubidium 37	88 **Sr** strontium 38	89 **Y** yttrium 39	91 **Zr** zirconium 40	93 **Nb** niobium 41	96 **Mo** molybdenum 42	[98] **Tc** technetium 43	101 **Ru** ruthenium 44	103 **Rh** rhodium 45	106 **Pd** palladium 46	108 **Ag** silver 47	112 **Cd** cadmium 48	115 **In** indium 49	119 **Sn** tin 50	122 **Sb** antimony 51	128 **Te** tellurium 52	127 **I** iodine 53	131 **Xe** xenon 54
133 **Cs** cesium 55	137 **Ba** barium 56	139 **La*** lanthanum 57	178 **Hf** hafnium 72	181 **Ta** tantalum 73	184 **W** tungsten 74	186 **Re** rhenium 75	190 **Os** osmium 76	192 **Ir** iridium 77	195 **Pt** platinum 78	197 **Au** gold 79	201 **Hg** mercury 80	204 **Tl** thallium 81	207 **Pb** lead 82	209 **Bi** bismuth 83	[209] **Po** polonium 84	[210] **At** astatine 85	[222] **Rn** radon 86
[223] **Fr** francium 87	[226] **Ra** radium 88	[227] **Ac*** actinium 89	[261] **Rf** rutherfordium 104	[262] **Db** dubnium 105	[266] **Sg** seaborgium 106	[264] **Bh** bohrium 107	[277] **Hs** hassium 108	[268] **Mt** meitnerium 109	[271] **Ds** darmstadtium 110	[272] **Rg** roentgenium 111	[285] **Cn** copernicium 112	[286] **Uut** ununtrium 113	[289] **Fl** flerovium 114	[289] **Uup** ununpentium 115	[293] **Lv** livermorium 116	[294] **Uus** ununseptium 117	[294] **Uuo** ununoctium 118

* The Lanthanides (atomic numbers 58–71) and the Actinides (atomic numbers 90–103) have been omitted.
Relative atomic masses for **Cu** and **Cl** have not been rounded to the nearest whole number.

Glossary

abiotic factor physical or non-living conditions that affect the distribution of a population in an ecosystem, such as light, temperature, soil pH

absorbing in Biology, the process in which soluble products of digestion move into the blood from the small intestine; in Physics, the process in which matter takes in energy, e.g. when an atom takes in energy from an electromagnetic wave

abundance relative representation of a species in an ecosystem, usually measured as the number of individuals in a sample taken from that ecosystem

active transport in active transport, cells use energy to transport substances through cell membranes against a concentration gradient

activity the rate at which unstable nuclei decay in a sample of a radioactive material, measured in counts per second

adrenal gland one of two endocrine glands located on top of the kidneys that secrete the hormone adrenaline

adrenaline hormone released quickly from the adrenal glands during a 'flight or fight' situation

aerobic respiration respiration that involves oxygen

algae (singular: alga) large, diverse group of simple organisms that contain the pigment chlorophyll and photosynthesise

allele inherited characteristics are carried as pairs of alleles on pairs of chromosomes. Different forms of a gene are different alleles

alpha decay a type of radioactive decay in which an alpha particle is emitted by an atomic nucleus

alpha particle particle containing two neutrons and two protons (the same as a helium nucleus) emitted by an atomic nucleus during radioactive decay

alveoli (singular: alveolus) air sacs; the site of gaseous exchange in the lungs

amplitude maximum displacement of a wave or oscillating object from its rest position

anaerobic respiration respiration that does not involve oxygen

antibiotic medicine (e.g. penicillin) that works inside the body to kill bacterial pathogens

antibody protein normally present in the body or produced in response to an antigen, which it neutralises, thus producing an immune response

antibiotic resistance resistance to being affected by antibiotics that has developed in some bacteria; an example of evolution by natural selection

antiretroviral drug drug used to treat HIV infections; it interrupts the replication of the virus in infected cells

antitoxin chemical produced by white blood cells (lymphocytes) that neutralise toxins to make a safe chemical

aorta artery that carries oxygenated blood from the left ventricle to tissues around the body

approximately roughly, not accurately

aquifer underground layer of permeable rock or loose materials (gravel or silt) where groundwater is stored

artery blood vessel with thick elastic walls that carries blood away from the heart under high pressure

artificial pacemaker a device fitted under the skin that sends out electrical impulses to control the heartbeat

aspirin a type of painkiller

atom the basic 'building block' of an element, the smallest part of an element that can take part in a chemical reaction

atomic number the number of protons found in the nucleus of an atom

atomic radius distance between the nucleus at the centre of an atom and the electron in the outermost shell of the atom

atmosphere mass of gases that surround a planet such as the Earth

atria (singular: atrium) the upper chambers of the heart that receive blood from the body or lungs

average global temperature average surface temperature of the Earth; measurement often used when discussing global warming

basal metabolic rate the rate at which the body uses energy when at rest

Benedict's test a test for sugars such as glucose that turn the reagent from a blue solution to an orange-red precipitate

benign tumour slow-growing, harmless mass of cells that can usually be easily removed

beta decay a type of radioactive decay in which a beta particle is emitted by an atomic nucleus

beta particle fast-moving electron emitted by an atomic nucleus in some types of radioactive decay

bias when a data set is inaccurate or unrepresentative; can result from systematic errors

binomial system the scientific way of naming an organism whereby each organism has a genus name and a species name; e.g. *Homo sapiens*

biodiversity range of different plant and animal species living in an ecosystem

biotic factor any living organism or its effects that affect the distribution of a population in an ecosystem, such as predators and disease

body mass index measure of someone's weight in relation to their height

boiling point temperature at which the bulk of a liquid turns to vapour

breeds new forms of animals created with particular characteristics, created through selective breeding

bronchi (singular: bronchus) branches of the trachea – one going to each lung

calibrate the process of checking the accuracy of a piece of equipment using standard data

capillary small blood vessel that is one cell thick and permeable for diffusion of gases; capillaries join arteries to veins

carbohydrase enzyme that breaks down carbohydrates into simple sugars

carbohydrates one of the main groups of nutrients; e.g. glucose, starch

carbon capture and storage collecting carbon dioxide from the atmosphere and depositing it underground

carbon footprint amount of carbon dioxide released by a company, organisation or individual into the atmosphere as a result of their activities

carbon neutrality situation of removing as much carbon dioxide from the atmosphere as is put into it, to create a zero carbon footprint

carbon off-setting scheme to allow an individual or company to spend money on environmental projects around the world to balance their carbon dioxide emissions

carbon sink 'stores' of carbon dioxide in oceans or forests

carcinogen substance or agent that increases the risk of cancer, e.g. tobacco smoke

carrier someone who is heterozygous for a disease-causing gene so that they do not have the condition but 'carry' the defective allele, which they could pass on to their children

causal mechanism where a direct link has been made between a risk factor and a disease; e.g. smoking and lung cancer

cell cycle a series of three stages during which a cell divides

cell membrane layer around a cell that helps to control substances entering and leaving the cell

cellular respiration process in the cells of all organisms that produces energy from the breakdown of nutrients in food

cellulose polymer of glucose; strengthening component of plant cell walls

cell wall outer protective layer in the cells of plants, some algae, fungi and bacteria

central nervous system collectively, the brain and spinal cord

cervix the neck of the womb

chlorophyll pigment found in plants and algae that is used in photosynthesis (gives green plants their colour)

chloroplast cell structure found in green plants that contains chlorophyll

chromatogram results of chromatography, presented as a graphic record of the different components of a mixture

chromatography method for separating substances, used to identify compounds and check for purity

chromosome thread-like structure in the cell nucleus that contains DNA

cilia tiny hair-like structures found throughout the respiratory system that beat together to move mucus out of the lungs

classify to organise and present data in a logical order, e.g. in a table; in Biology, to organise living things into groups based on their characteristics or genetic relationships

climate general weather conditions in an area over a long period of time

clone in asexual reproduction, the offspring produced are identical to the parent (clones)

communicable disease infectious condition caused by a virus, protist, bacterium or fungus; e.g. measles, HIV

community all the plants and animals living in an ecosystem, e.g. a garden

compare to examine your data set next to a known set of results

compound measure measurement made up of two or more other measurements; e.g. density is a measurement of mass divided by volume

compression a region of a sound wave in which the particles are closer together

competition when different species in an ecosystem compete for the same resources

concentration gradient gradual change in concentration, from a region of high concentration to a region of low concentration, of particles in a solution or a gas

condom rubber sheath that covers the erect penis and prevents sperm reaching the egg; a barrier form of contraception

conservation in Biology, the protection of a species or ecosystem

conserved in Chemistry and Physics, unchanged (in relation to mass or energy, for example)

consumer organisms that eat other organisms in a food chain

contamination (radioactive) unwanted presence of materials containing radioactive atoms

contraceptive pill hormonal method of contraception, in tablet form

coronary artery vessel that provides the heart uscle with oxygen and glucose for respiration

coronary heart disease a condition caused by build-up of fatty deposits in the coronary artery, leading to reduced blood flow and less oxygen and glucose reaching the heart tissue, which can cause a heart attack

correlation relationship between two sets of results

curve of best fit curve on a graph that most closely matches all the data points to show a trend or pattern

cytoplasm part of a cell between the nucleus and the cell membrane

data measurements of quantities in an experiment; data can be recorded in a table and used to produce graphs

daughter cell cell produced during mitosis that is identical to the parent cell

decimal point full point used in a decimal fraction, placed between the units and the tenths

deforestation removal of large areas of trees to provide land for cattle or growing crops

density the density of a substance is found by dividing its mass by its volume

dependent variable quantity in an experiment that is measured for each change in the independent variable

desalination process for removing salt from seawater

diaphragm type of barrier method of contraception

diarrhoea symptom of food poisoning; frequent passing of liquid stools

differentiation when cells gain certain features needed for their function; they become specialised

diffuse reflection from a rough surface, scattered or widespread

diffusion net movement of molecules from a higher concentration to a lower concentration, down a concentration gradient, until they are spread equally

digestion process of breaking down food in the body

discrete describes distinct and separate data

displacement method method of measuring the density of an irregular solid using water

distillation the process of evaporation followed by condensation

distribution how living organisms are dispersed/spread out over an ecosystem

DNA (deoxyribonucleic acid) found as chromosomes in the nucleus – its sequence determines how our bodies are made, and gives each one of us a unique genetic code

dominant an allele that is expressed when one or two copies are present; represented by a capital letter

dose amount of a drug given to a patient; also the amount of ionising radiation absorbed by tissues

drug substance that is administered to cause a chemical change in the body

dual circulatory system a circulatory system with two circuits: one that transports oxygenated blood from the heart to the body tissues and deoxygenated blood back to the heart; and one that transports deoxygenated blood from the heart to the lungs and oxygenated blood back to the heart.

echo reflection of a sound or ultrasound wave

echo sounding process in which high-frequency sound waves are reflected off a surface to measure distances (e.g. a ship measuring the depth of the seabed)

ecosystem the interaction of a community of living organisms with the non-living parts of their surroundings

efficacy how effective a drug is at preventing or curing a disease

electromagnetic spectrum continous range of waves of different wavelengths and frequencies, from low-frequency radio waves to high-frequency gamma rays

electron small negatively charged particle within an atom that is outside the nucleus

electron shells energy levels around the nucleus of an atom, in which electrons orbit

electronic structure the number of electrons in sequence that occupy the shells, e.g. the 11 electrons of sodium are in sequence 2.8.1

emitting process in which energy or a particle is given out by an atom (e.g. in beta particle emission, a high-speed electron is given out)

emission spectrum particular pattern of coloured lines or bands representing the wavelengths of electromagnetic radiation emitted by an element

endocrine gland part of the body that releases chemical messengers, or hormones, directly into the bloodstream

endocrine system a control system in the body that communicates using chemical messengers, or hormones, to produce slow but long-lasting responses

endothermic describes a chemical reaction that takes in energy from its surroundings

energy energy is transferred between energy stores by processes such as heating, and by electromagnetic radiation; the total amount of energy stays and same before and after any change

energy level stable state of a physical system, e.g. electrons orbiting the nucleus of an atom can exist in only particular energy levels and be moved between them

energy store the energy associated with a fuel (chemical energy store), heated object (thermal energy store), moving object (kinetic energy store), a stretched spring (elastic potential energy store) and an object raised above ground level (gravitational potential energy store)

energy transfer process in which energy is moved from one store to another

environmental change a shift in the external conditions caused by human activity or a natural event

enzyme a biological catalyst that increases the speed of a chemical reaction

erosion wearing away of rock and soil by the action of wind, water or ice

estimate rough calculation, or judgement, of the value of a quantity such as mass, time or distance

eukaryotic describes cells containing a true nucleus in the cytoplasm; e.g. plant and animal cells

eugenics method of 'improving' a population by controlled breeding

evaporation when a liquid changes to a gas; the process by which the body cools through sweating

evolve to develop gradually over a long period of time, through the process of natural selection

exchange surface specialised area with large surface area : volume ratio for efficient diffusion

exothermic describes a chemical reaction that gives out energy to its surroundings

expression the 'switching on' of a gene to translate its code into a protein

fertility drug combination of hormones given to a woman with low FSH levels to stimulate egg production

field margins the edges of a field between the crop and the hedge or fence that forms the field boundary

field of view the area of a sample visible under a microscope

flaccid describes plant cells that are not full of water and so are floppy, without their walls pushing against neighbouring cells

food producing capacity ability of a region to produce food

food security when all people have access to a consistent supply of food to meet their needs

formulation a mixture that has been designed as a useful product

fossils the remains of organisms that lived millions of years ago

fraction number representing part of a whole, made up of a numerator (the number on top of the fraction) and a denominator (the number on the bottom of the fraction); e.g. $\frac{1}{5}$

frequency number of waves passing a set point in one second

FSH (follicle-stimulating hormone) a reproductive hormone that causes eggs to mature in the ovaries

gamete the male or female sex cells (sperm and egg)

gamma rays ionising electromagnetic radiation with shortest wavelengths in the electromagnetic spectrum

gas state of matter in which all the particles of a substance are separate and move about freely

gas exchange process of taking in oxygen and transferring out carbon dioxide, which takes place at a respiratory surface such as in the alveoli of the lungs

gene section of DNA that contains the instructions for a particular characteristic

genetically modified (GM) describes an organism that has had its genes changed to give the it particular desired characteristics

genetic engineering transfer of specific genes from one organism to another

genetic variation the product of meiosis, mutations and sexual reproduction, which all lead to changes in our genome

genome the entire genetic material of an organism

genotype the alleles present for a particular gene make up the organism's genotype

global warming the increase in the Earth's temperature due to increases in carbon dioxide levels

glucagon hormone released by the pancreas that, along with insulin, controls blood glucose concentrations

GM crops varieties of crops that have had their genomes modified by the insertion of genes from a plant or other organism

gonorrhoea sexually transmitted disease caused by bacteria, resulting in pain when urinating and a thick discharge from the vagina or penis

gradient the steepness of a line plotted on a graph, change in y / change in x

graticule piece of glass or plastic onto which a scale has been drawn that is used with a microscope

greenhouse gas any of the gases whose absorption of solar radiation is responsible for the greenhouse effect, e.g. carbon dioxide, methane

grouped bar chart bar graph in which two or more data sets are clustered next to each other for each group/bar; also called a dual bar chart

guard cells cells surrounding the stomata that open and close to control the exchange of gases and water loss

habitat place where an organism lives in an ecosystem; e.g. a worm's habitat is the soil

haemoglobin chemical found in red blood cells that binds to oxygen to transport it around the body

half-life the average time it takes for half of all the nuclei present in a sample of a radioactive element to decay, or the time it takes for the count rate (or activity) to halve

hazard anything that may cause injury (e.g. the risk of contamination by radioactive materials when used in a scientific experiment)

hertz (Hz) unit of wave frequency (1 Hz is equal to one complete wave per second)

heterozygous having two different alleles for a characteristic, e.g. someone with blond hair may also carry the allele for red hair

homeostasis regulation of internal conditions, such as temperature, in the body

homozygous having two alleles that are the same for a characteristic, e.g. a blue-eyed person will have two 'blue' alleles for eye colour

hormone chemical messenger that acts on target organs in the body

hybrid infertile organism created through interbreeding organisms from two different species

hydrocarbons compounds composed of the elements hydrogen and carbon only

hypothesis scientific question formed based on prior knowledge that can be tested in an experiment

immunity when the body is protected from a pathogen that it has already encountered and so can produce antibodies against it, rapidly

independent variable quantity in an experiment that is changed or selected by the experimenter

infrared radiation electromagnetic radiation with a range of wavelengths longer than visible light but shorter than microwaves; emitted in particular by heated objects

insulin hormone made by the pancreas that controls the concentration of glucose in the blood

interbreeding breeding between organisms of different species, which does not give fertile offspring

interdependence dependence on each other; e.g. plants need animals for carbon dioxide and seed dispersal, and animals need plants for food and oxygen (they depend on each other for survival)

in-vitro fertilisation a fertility treatment, whereby an egg is fertilised with sperm outside the body, in a laboratory

ion charged particle that can be positive or negative

irradiation exposure to ionising radiation

isotopes atoms with the same number of protons but different number of neutrons

IUD contraceptive device inserted into the uterus to prevent pregnancy

IVF a fertility treatment in which an egg is fertilised with sperm outside the body, in a laboratory

IVF cycle the process of IVF from stimulation of the ovaries to implantation of an embryo; if unsuccessful, another cycle can be attempted after about 2 months

joule unit of work done and energy; SI unit of energy, symbol J

lactic acid product of anaerobic respiration in muscles

latent heat energy needed for a substance to change state without a change in temperature (e.g. the latent heat of vaporisation is the energy needed to turn a sample of liquid water into gas)

LH (luteinising hormone) menstrual cycle hormone that stimulates an egg to be released from an ovary

lignin substance found in plant cell walls that gives them strength and rigidity

limiting factor factor such as light, temperature or carbon dioxide, which affects the rate of photosynthesis

linear describes something with a straight line; e.g. a linear graph

line of best fit line on a graph that most closely matches all the data points to show a trend or pattern

lipase enzyme that digests fats into fatty acids and glycerol

liquid state of matter in which the particles of a substance are close together and attract each other but have a limited amount of movement; a liquid has a definite volume but will spread out to fill its container

longitudinal wave wave motion in which the vibrations of the particles of the medium are parallel to the direction of energy transfer (e.g. sound waves)

lymphocyte white blood cell that produces antibodies and antitoxins to destroy pathogens

magnification factor by which an object is enlarged by a microscope; calculated as: size of image/size of real object

malignant tumour abnormal cancerous mass of cells that grows quickly and can spread to other parts of the body

mean average value calculated by adding up all the values in a data set then dividing by the number of values

meiosis cell division that results in gametes being produced, with half the number of chromosomes of the parent cell

melting point temperature at which a solid turns into a liquid

mental health a feeling of wellbeing, having a positive frame of mind

meristem regions at tips of roots and shoots where cell division and elongation take place

micrograph image captured using a microscope

microwaves electromagnetic radiation with a range of wavelengths longer than infrared but shorter than radio waves; used to cook food in

mitochondrion structure in a cell where respiration takes place

mitosis cell division that results in genetically identical diploid cells

mobile phase in chromatography, this is the phase that moves

model method of representing a system or situation to understand or explain it better, and to predict previously unknown features of it

monoculture where only one crop is grown in an area

mutation alteration in DNA (this happens in cancer)

myelin sheath fatty insulating layer that surrounds neurones and speeds up nerve transmission

nanometre units used to measure very small things (one billionth of a metre)

natural selection process by which advantageous characteristics that can be passed on in genes become more common in a population over many generations

negative feedback a regulatory process whereby changes in the body can be reversed once they have happened

net decline the ratio of the final value of the activity of a radioactive substance to the initial value in a given number of half-lives

neurone a nerve cell that is specialised to transmit electrical signals

neutral a neutral substance has a pH of 7

neutron small particle with no charge found in the nucleus of an atom

non-communicable disease condition caused by environmental or genetic factors that is not spread among people; e.g. cancer, cardiovascular disease

nuclear equation equation that uses symbols to show the elements involved in a nuclear decay, including the atomic numbers and mass numbers

nuclear model model of the atom with a small central nucleus, surrounded by orbiting electrons

nucleus central part of an atom that contains protons and neutrons

oestrogen main female reproductive hormone produced by the ovaries

opiates group of painkillers found in poppies

order of magnitude in microscopy, the difference between sizes of cells, calculated in factors of 10; description of a quantity in terms of powers of ten; e.g. a distance of 100 m (10^2 m) is two orders of magnitude larger than a distance of 1 m

organ group of tissues that carries out a specific function

organ system arrangement of organs in the body according to function; e.g. respiratory system

osmosis diffusion of water molecules through a partially permeable membrane, from a dilute solution to a concentrated solution

oxygen debt amount of oxygen that the body needs to breakdown lactic acid after muscles undergo anaerobic respiration

oxyhaemoglobin bright red substance formed when oxygen binds to haemoglobin in red blood cells; this is how oxygen is transported to tissues

pacemaker a group of cells in the right atrium that controls the heart rate

partially permeable membrane a membrane that allows some small molecules to pass through but not larger molecules

particle model model in which all substances contain large numbers of very small particles (atoms, ions or molecules); it is used to explain the different properties of solids, liquids and gases

particulates tiny particles of matter released into the atmosphere when fuels are burned and during other processes

pathogen harmful microorganism that invades the body and causes infectious disease

penicillin an antibiotic, isolated from *Penicillum* mould, which was discovered by Alexander Fleming

percentages numbers or amounts expressed per hundred

period time taken for one complete cycle of an oscillation

phagocyte type of white blood cell that enters tissues and engulfs pathogens then ingests them

phenotype characteristics of an organism that results from the expression of its genes

phloem specialised transporting cells that form tubules in plants to carry sugars from leaves to other parts of the plant

potometer device used for measuring the water uptake of a plant

photosynthesis process carried out by green plants in which sunlight, carbon dioxide and water are used to produce glucose and oxygen

physical change reversible change of state of a substance without changing its chemical composition; e.g. water freezing to become ice

pituitary gland known as the 'master gland' as it controls other endocrine glands, such as the thyroid gland

placebo treatment that does not contain a drug

plasma straw-coloured liquid part of blood

plasmids small rings of DNA found in prokaryotic cells

plasmolysis shrinking of a plant cell due to loss of water; the cell membrane pulls away from the cell wall

platelets cell fragments in blood that help in blood clotting

plum pudding model early model of the structure of an atom that suggests an atom is a solid sphere of positive electric charge with negatively charged electrons in it

population total number of one species in an ecosystem

potable water water that is safe to drink

precipitation water falling to the surface of the Earth in the form of rain, snow, sleet or hail; part of the water cycle

precision how closely grouped a set of repeated measurements are

predators animals that eat other animals (prey)

prey animals that are eaten by a predator

producer organism in a food chain that makes food using sunlight

progesterone reproductive hormone that causes the lining of the uterus to be maintained

prokaryotic describes single-celled organisms containing DNA in a loop not contained in a nucleus; e.g. bacteria

proportional if two quantities are proportional to each other, there is a constant relation between them, and a change in one of the quantities causes a change in the other quantity

protease digestive enzyme that breaks down proteins into amino acids

protist type of single-celled organism; an example is the pathogen that causes malaria

proton small positive particle found in the nucleus of an atom

pure in science, describes a single element or compound that is not mixed with any other substance

purity measure of how pure a substance is

quadrat a square or rectangular tool used when sampling slow-moving or stationary organisms in a habitat

qualitative reagent reagent to detect the type of substance in a sample, rather than to measure the amount of the substance

radiation energy given out in the form of electromagnetic waves or as moving particles; e.g. in radioactive beta decay an atomic nucleus emits high-speed electrons, and the Sun radiates electromagnetic waves including visible light

radioactive decay process in which an unstable atomic nucleus disintegrates and gives out ionising radiation, in the form of alpha particles, beta particles, gamma rays or neutrons

radioisotope atom with an unstable nucleus that undergoes radioactive decay

radio waves electromagnetic radiation with a range of wavelengths longer than microwaves; used for long-distance communication

random (radioactive decay) process in which the time of each particular event cannot be predicted, although a trend or average can be measured across many events; e.g. the decay of a radioactive element

random errors errors that cause repeat measurements to vary and scatter around a mean value

randomly describes something that occurs without any particular aim or pattern

range measure of spread; the difference between the biggest and smallest values in a set

range bar line drawn on a graph to show the range of a set of values

rarefaction region of a sound wave in which the particles are further apart

rate speed at which an event occurs over time; can be calculated using the gradient on a graph showing a measure of the change plotted against time

ratio relationship between two variables, expressed as 1 : 4, for example

ray diagram line diagram showing how light rays travel

reaction time time taken for a body to respond to an event

rearrange an equation to move parts of an equation around in order to calculate the value of a variable

receptor cell in the body that detects changes in the environment

recessive two copies of a recessive allele must be present for the characteristic to be expressed; represented by lowercase letters

red blood cells blood cells with a concave shape that are adapted to carry oxygen from the lungs to body cells

reflection process in which a surface does not absorb any energy, but instead bounces it back towards the source; e.g. light is reflected by polished surfaces

reflex action rapid automatic responses to a stimulus

reflex arc pathway taken by nerve impulses through the spinal cord during a reflex action

refraction bending of light rays as they travel from one medium to another, e.g. as they pass from air into a block of glass

regeneration restoration (of a habitat)

relative charge charge of a subatomic particle compared with the charge of a proton

relative mass mass of a subatomic particle compared with the mass of a proton

relay neurone neurone found in the spinal cord that transmits impulses from the sensory to the motor neurone

resolving power ability of a microscope to distinguish between two points; the resolving power of electron microscopes is higher than that of light microscopes

reverse osmosis method of desalination to provide drinkable water from seawater

R_f value in chromatography, the distance a substance moved divided by the distance the solvent moved

ribosome structure in a cell where protein synthesis takes place

risk probablility of harm from a hazard

risk factor lifestyle or genetic factor that increases the chance of developing a disease

root hair cell specialised cell in plant roots that is adapted for efficient uptake of water by osmosis and mineral ions by active transport

rose black spot fungal disease that affects plant growth; can be treated with fungicide or by destroying the leaves

run-off when fertilisers from a farmer's field are washed off into rivers or lakes (can lead to eutrophication)

Salmonella type of bacteria that causes food poisoning

scale on a graph, the range of values added to the axes

scanning electron microscope (SEM) works by bouncing electrons off the surface of a specimen that has had an ultrathin coating of a heavy metal, usually gold, applied; used to view surface shape of cells or small organisms

secondary sex characteristics features that develop during puberty, as a result of sex hormones being released, such as a deep voice and hair growth in boys, and breast development in girls

sedimentation process during water purification in which small solid particles are allowed to settle

sediments materials that settle to the bottom of a liquid, including on the ocean floor

segment part of a whole; e.g. a segment of a pie chart

selective breeding process of breeding organisms with desired characteristics (also known as artificial selection)

self-supporting describes an ecosystem; all ecosystems can support themselves; i.e. everything organisms need for growth and survival is present

sex determination whether a fertilised egg develops into a male (XY) or female (XX) depends on the 23rd pair of chromosomes

significant figures digits within a measured quantity that have meaning; e.g. a measurement made using a 30 cm long ruler with divisions marked in millimetres can only have three significant figures, such as 17.4 cm; it is meaningless to state 17.42 cm, because the ruler is not that precise (note: it may be possible on some rulers to estimate a measurement to the nearest 0.5 mm)

solid state of matter in which the particles are held together in a fixed structure by bonds

solute the part of a solution that is dissolved in a solvent

solvent liquid used to dissolve a solute

solvent front in paper chromatography, the furthest point reached by the solvent as it travels up the paper

specialised describes cells or tissues that are adapted to carry out their specific function

speciation formation of a new species over a long period of time, through separation of populations

specific heat capacity the energy needed to raise the temperature of 1 kg of a substance by 1°C; symbol *c*, unit J/kg °C

specific latent heat the energy needed to change 1 kg of a substance completely from one state to another state without any change in temperature; symbol *L*, unit J/kg

specific latent heat of fusion the energy needed to change 1 kg of a substance completely from solid to liquid without any change in temperature

specific latent heat of vaporisation the energy needed to change 1 kg of a substance completely from liquid to gas without any change in temperature

spectrum a continuous range of properties presented in decreasing or increasing order; e.g. the electromagnetic spectrum arranges the different types of electromagnetic wave in order of frequency and wavelength

spermicidal cream cream that is toxic to sperm; it is used alongside other contraceptive techniques, e.g. condoms

sphere shape that has the smallest surface area compared with its volume

stable community community in which population sizes remain more or less constant as the biotic and abiotic factors are balanced

stacked bar chart bar graph in which two or more data sets are stacked on top of each other for each group/bar

standard form way of writing very large or small numbers using powers of ten; e.g. 1.0×10^{-3}

statin drug that stops the liver producing so much cholesterol

stationary phase phase in chromatography that does not move; in paper chromatography it is the paper

stem cells unspecialised body cells (found in bone marrow) that can develop into other, specialised, cells that the body needs, e.g. blood cells

stent treatment for heart disease; a catheter with a balloon attached is inserted to open up a narrowed coronary artery

stomata small holes in the surface of leaves that allow gases in and out

subatomic particles particles that make up an atom, e.g. protons, neutrons and electrons

subject of an equation the unknown variable that you're trying to calculate in an equation; it is on its own, usually on the left-hand side of the equation

sublimate to change the state of a substance from a solid directly to a gas

substitute to replace a variable in an equation with a known quantity

surface area : volume (SA:V) ratio relationship between the surface area of an organism or structure and its volume; SA:V ratios are large in single-celled organisms for efficient diffusion

sustainable long-lasting; something that can be maintained for future generations

symbol one or two letters used to represent a chemical element; e.g. the symbol for calcium is Ca

synapse the gap between two neurones

systematic errors errors that affect all measurements in the same way, making them all higher or lower than the true value

tangent straight line that touches a curve on a graph, making equal angles to the curve on either side; used to find the gradient of the curve

target organ site where a hormone has its effect

testosterone male sex hormone produced by the testes that stimulates sperm production

thyroxine hormone produced by the thyroid gland that increases the body's metabolism

tissue group of cells that work together, with a particular function

tracer radioactive substance that is put into the body or fluid (such as in a pipe), so that the path of the substance can be followed by monitoring the radiation it emits

trachea in the human respiratory system, a large tube through which air travels to the lungs; in the insect respiratory system, small tubes containing water, through which gases diffuse into body cells

transect line across an area used when sampling organisms

translocation movement of sugars through a plant

transmission electron microscope (TEM) microscope that uses an electron beam to view thin sections of cells at high resolution

transpiration movement of water through a plant, in through the roots and out through the leaves

trend pattern that shows how the value of a property rises, or falls, when there are changes in a specified variable

tobacco mosaic virus virus that causes disease in a wide variety of plants, reducing photosynthesis and stunting their growth

transverse wave wave motion in which the vibrations of the particles of the medium are perpendicular to the direction of energy transfer (e.g. water waves or electromagnetic waves)

tumour abnormal growth of tissue, which may be benign or malignant (cancerous)

turgid describes plant cells that are full of water with their walls bowed out and pushing against neighbouring cells

type 1 diabetes condition in which the pancreas cannot produce enough, or any, insulin

type 2 diabetes condition in which the body cells no longer respond to insulin produced by the pancreas

ultraviolet electromagnetic radiation with a range of wavelengths shorter than visible light but longer than X-rays; emitted in particular by the Sun

unadulterated describes a substance or product that is not diluted or mixed with other ingredients

uncertainty measure of the range about the mean, calculated by the range divided by two;

uncertainty is reduced when accuracy and precision are increased

vaccination injection of a small quantity of inactive pathogen to protect us from developing the disease caused by the pathogen

vaccine preparation of an inactive or dead form of a pathogen given by injection or nasal spray

vacuole fluid-filled sac located in the cytoplasm of plant cells

valid describes results that are accurate, representative and repeatable

vector in Biology, a carrier, usually a plasmid, used to transfer a gene into an organism to be genetically modified; also an organism that transfers a disease-causing pathogen from one host to another but does not cause the disease itself; in Physics, a quantity that has both size and direction

vein blood vessel with a thin wall and valves to prevent backflow, that carries blood back to the heart at low pressure

vena cava vein that carries deoxygenated blood from the body to the right atrium of the heart

ventilation movement of air into and out of the lungs

ventricles lower chambers of the heart that pump blood around the body (left ventricle) or back to the lungs (right ventricle)

visible light electromagnetic radiation with a range of wavelengths shorter than infrared but longer than ultraviolet; detectable by the human eye

visible spectrum part of the electromagnetic spectrum that is visible to the human eye, between infrared and ultraviolet

volcanic from a volcano

wavefront line that joins all the points on a wave, at right angles to the direction the wave is travelling

wavelength distance between a point on one wave to the equivalent point on the adjacent wave

white blood cell type of blood cell that helps the immune system to fight infection

X-chromosome sex chromosome present in males (XY) and females (XX)

xylem tissue specialised for transporting water through a plant; xylem cells have thick walls, no cytoplasm and are dead, their end walls break down and they form a continuous tube

X-rays ionising electromagnetic waves used in X-ray photography (where X-rays are used to generate pictures of bones or teeth) and in CT scans; have a range of wavelengths shorter than ultraviolet and can have similar wavelengths to gamma rays

Y-chromosome sex chromosome found only in males (XY)

Index

ACKNOWLEDGEMENTS

The publishers gratefully acknowledge the permissions granted to reproduce copyright material in this book. Every effort has been made to contact the holders of copyright material, but if any have been inadvertently overlooked, the Publisher will be pleased to make the necessary arrangements at the first opportunity.

Topic 1

pp12–13 Ekaterina Koolaeva/Shutterstock; p14 Chaiwuth Wichitdho/Shutterstock, grey color/Shutterstock, Timur Kulgarin/Shutterstock; p15 Petr Malyshev/Shutterstock; p18 Andrey N Bannov/Shutterstock, GIPhotoStock/SCIENCE PHOTO LIBRARY; p19 SpaceKris/Shutterstock; p21 PRILL/Shutterstock, AlexanderZam/Shutterstock, Hurst Photo/Shutterstock; p24 Africa Studio/Shutterstock; p26 Waraphorn Aphai/Shutterstock; p27 practicalaction.org; p28 Douglas Freer/Shutterstock; p29 Kekyalyaynen/Shutterstock; p33 Tyler Olson/Shutterstock; p36 Lakeview Images/Shutterstock; p37 Anne Gilbert / Alamy Stock Photo, ANDREW LAMBERT PHOTOGRAPHY/SCIENCE PHOTO LIBRARY; p42 Smith1972/Shutterstock, Syda Productions/Shutterstock; p44 Ufuk ZIVANA/Shutterstock; p52 Syda Productions/Shutterstock; p62 Claudio Divizia/Shutterstock, STEVE GSCHMEISSNER/SCIENCE PHOTO LIBRARY; p63 royaltystockphoto.com/Shutterstock, eAlisa/Shutterstock, Giovanni Cancemi/Shutterstock, royaltystockphoto.com/Shutterstock; p64 Jubal Harshaw/Shutterstock, Pan Xunbin/Shutterstock; p65 STEVE GSCHMEISSNER/SCIENCE PHOTO LIBRARY; p66 Jose Luis Calvo/Shutterstock, Dimarion/Shutterstock; p67 CNRI/SCIENCE PHOTO LIBRARY, DR JEREMY BURGESS/SCIENCE PHOTO LIBRARY, AMMRF, UNIVERSITY OF SYDNEY/SCIENCE PHOTO LIBRARY; p68 TREVOR CLIFFORD PHOTOGRAPHY/SCIENCE PHOTO LIBRARY, Visuals Unlimited, Inc./Dr. Gladden Willis/Getty Images; p69 D. Kucharski K. Kucharska/Shutterstock, Power_J/Shutterstock, Lebendkulturen.de/Shutterstock, Lebendkulturen.de/Shutterstock; p70 StudyBlue inc; p71 StudyBlue inc; p83 Jose Luis Calvo/Shutterstock; p86 Leptospira/Shutterstock; p87 Lisa S/Shutterstock; p92 bogdan ionescu/Shutterstock; p93 Image Point Fr/Shutterstock, Suttha Burawonk/Shutterstock; p94 Joshua Rainey Photography/Shutterstock; p95 Aleksey Sagitov/Shutterstock, Andrey_Popov/Shutterstock; p96 SirinS/Shutterstock; p97 blackeagleEMJ/Shutterstock; p104 Anita Ponne/Shutterstock; p106 ZEPHYR/SCIENCE PHOTO LIBRARY; p107 Image Point Fr/Shutterstock; pp118–119 Nicola Renna/Shutterstock; p120 Valeriy Velikov/Shutterstock; p122 Ralf Herschbach/Shutterstock

Topic 2

pp118–119 Nicola Renna/Shutterstock; p120 Valeriy Velikov/Shutterstock; p122 Ralf Herschbach/Shutterstock; p123 BMJ/Shutterstock, Dave Massey/Shutterstock; p125 Maridav/Shutterstock; p128 JONATHAN PLEDGER/Shutterstock; p130 dreamerb/Shutterstock, Lebendkulturen.de/Shutterstock, Abel Tumik/Shutterstock, Monika Vosahlova/Shutterstock, Yann hubert/Shutterstock; p136 extender_01/Shutterstock; p142 Andril Vodolazhskyi/Shutterstock; p144 bikeriderlondon/Shutterstock; p146 India Picture/Shutterstock; p156 Heiti Paves/Shutterstock, Garsya/Shutterstock, Ethan Daniels/Shutterstock; p157 Jubal Harshaw/Shutterstock, MARCELODLT/Shutterstock, Brian Maudsley/Shutterstock; p159 ROSENFELD IMAGES LTD/SCIENCE PHOTO LIBRARY, M.I. WALKER/SCIENCE PHOTO LIBRARY; p163 Brandon Blinkenberg/Shutterstock, DR DAVID FURNESS, KEELE UNIVERSITY/SCIENCE PHOTO LIBRARY; p165 Jubal Harshaw/Shutterstock; p174 bjul/Shutterstock, Anton_Ivanov/Shutterstock; p175 Gemenacom/Shutterstock; p182 DR KEITH WHEELER/SCIENCE PHOTO LIBRARY; p183 Mariusz S. Jurgielewicz/Shutterstock; p185 NORM THOMAS/SCIENCE PHOTO LIBRARY, Anne Pilling, Nigel Cattlin/Visuals Unlimited/Corbis

Topic 3

pp190–191 NIBSC/SCIENCE PHOTO LIBRARY; p192 Stefano Carnevali/Shutterstock, fusebulb/Shutterstock, Pavel Chagochkin/Shutterstock; p193 WitthayaP/Shutterstock, Andrii Muzyka/Shutterstock, nevodka/Shutterstock; p194 Mita Stock Images/Shutterstock; p195 Antonio Guillem/Shutterstock; p196 DR TONY BRAIN/SCIENCE PHOTO LIBRARY; p197 Edyta Pawlowska/Shutterstock; p202 lzf/Shutterstock; p203 Andrii Muzyka/Shutterstock; p206 Santibhavank P/Shutterstock; p210 Roman Prishenko/Shutterstock, RAY ELLIS/SCIENCE PHOTO LIBRARY; p211 areeya_ann/Shutterstock, Image Point Fr/Shutterstock, Michael Kraus/Shutterstock, JPC-PROD/Shutterstock; p214 Keystone/Getty Images, originalpunkt/Shutterstock; p215 nevodka/Shutterstock; p216 ALAIN POL, ISM/SCIENCE PHOTO LIBRARY, annedde/istock; p222 CristinaMuraca/Shutterstock; p235 urbanbuzz/Shutterstock; p236 DIGITAL GLOBE/SCIENCE PHOTO LIBRARY; p237 Natasja Weitsz/Contributor/Getty Images; p240 Maksym Bondarchuk/Shutterstock, solkanar/Shutterstock, Australis Photography/Shutterstock; p241 CristinaMuraca/Shutterstock; p246 royaltystockphoto.com/Shutterstock, fusebulb/Shutterstock; p247 JPC-PROD/Shutterstock; p248 STEVE GSCHMEISSNER/SCIENCE PHOTO LIBRARY, Smith1972/Shutterstock; p249 Chaikom/Shutterstock, muuraa/Shutterstock; p250 LOWELL GEORGIA/SCIENCE PHOTO LIBRARY; p251 Jina K/Shutterstock; p252 MichaelTaylor3d/Shutterstock, royaltystockphoto.com/Shutterstock, Thailand Travel and Stock/Shutterstock; p256 Roberto Piras/Shutterstock; p258 JPC-PROD/Shutterstock; p260 EM Karuna/Shutterstock, P. FERGUSON, ISM/SCIENCE PHOTO LIBRARY; p261 Zaharia Bogdan Rares/Shutterstock; p262 Sarah Marchant/Shutterstock, Hellen Sergeyeva/Shutterstock, TwilightArtPictures/Shutterstock; p264 VOLKER STEGER/SCIENCE PHOTO LIBRARY; p265 MARTIN OEGGERLI/SCIENCE PHOTO LIBRARY, Dmitry Kalinovsky/Shutterstock; p266 royaltystockphoto.com/Shutterstock; p267 THOMAS DEERINCK/SCIENCE PHOTO LIBRARY; p269 James Cavallini/SCIENCE PHOTO LIBRARY, KWANGSHIN KIM/SCIENCE PHOTO LIBRARY, James Cavallini/SCIENCE PHOTO LIBRARY

Topic 4

pp276–277 kwest/Shutterstock; p278 Gualberto Becerra/Shutterstock, Jan Martin Will/Shutterstock, Hung Chung Chih/Shutterstock; p279 Khoroshunova Olga/Shutterstock, Unicus/Shutterstock, Filip Fuxa/Shutterstock; p280 Ammit Jack/Shutterstock, James Steidl/Shutterstock; p281 Khoroshunova Olga/Shutterstock; p282 Unicus/Shutterstock, attem/Shutterstock; p283 Lee Prince/Shutterstock, M Rutherford/Shutterstock, schankz/Shutterstock; p284 Carlos Caetano/Shutterstock; p285 Marek Velechovsky/Shutterstock; p288 Dudarev Mikhail/Shutterstock, Huguette Roe/Shutterstock; p290 Silken Photography/Shutterstock, Matty Symons/Shutterstock; p291 Pataporn Kuanui/Shutterstock, BOONCHUAY PROMJIAM/Shutterstock; p292 Filip Fuxa/Shutterstock, AnglianArt/Shutterstock,

Tomasz Darul/Shutterstock; p294 donikz/Shutterstock, Alexander Raths/Shutterstock, MikeDotta/Shutterstock; p295 Laurence Gough/Shutterstock, J. Helgason/Shutterstock; p296 kovgabor/Shutterstock; p300 ventdusud/Shutterstock, muratart/Shutterstock; p301 Kletr/Shutterstock; p308 apiguide/Shutterstock, patostudio/Shutterstock, Taiga/Shutterstock; 309 Peshkova/Shutterstock, Ethan Daniels/Shutterstock, Critterbiz/Shutterstock; p310 Tischenko Irina/Shutterstock; p312 Bjul/Shutterstock, Paula French/Shutterstock, Tom Roche/Shutterstock; p313 Polarpx/Shutterstock; p314 Vaclav Volrab/Shutterstock, Joanne Weston/Shutterstock, SJ Travel Photo and Video/Shutterstock; p316 richpav/Shutterstock; p317 Gajic Dragan/Shutterstock; p318 Zack Frank/Shutterstock, Martin Fowler/Shutterstock, Ron Zmiri/Shutterstock; p319 Rich Carey/Shutterstock; p320 Anton Gorlin/Shutterstock, Lukas Gojda/Shutterstock, ChiccoDodiFC/Shutterstock; p321 Richard Thornton/Shutterstock, Lodimup/Shutterstock, Gary Andrews/Shutterstock; p322 Stephen Lavery/Shutterstock, sunsetman/Shutterstock; p323 Ilona Ignatova/Shutterstock, Seaphotoart/Shutterstock; p325 Sarah Pettegree/Shutterstock; p328 A. BARRINGTON BROWN, GONVILLE AND CAIUS COLLEGE/SCIENCE PHOTO LIBRARY, STEVE GSCHMEISSNER/SCIENCE PHOTO LIBRARY, itsmejust/Shutterstock; p329 hxdbzxy/Shutterstock, Giovanni Cancemi/Shutterstock, Africa Studio/Shutterstock; p330 adike/Shutterstock; p331 alybaba/Shutterstock; p333 BIOPHOTO ASSOCIATE/SCIENCE PHOTO LIBRARY; p334 Puwadol Jaturawutthichai/Shutterstock; p336 Africa Studio/Shutterstock; p337 Gina Kelly/Alamy Stock Photo, Marion Bull/Alamy Stock Photo; p342 Sergey Novikov/Shutterstock, Kjersti Joergensen/Shutterstock, Esteban De Armas/Shutterstock; p343 Kenneth Keifer/Shutterstock, Everett Historical/Shutterstock, Natursports/Shutterstock; p344 MICHAEL W. TWEEDIE/SCIENCE PHOTO LIBRARY; p345 Zanna Holstova/Shutterstock; p 346 Ryan M. Bolton/Shutterstock, Claude Huot/Shutterstock, Guido Vermeulen-Perdaen/Shutterstock, Ben Queenborough/Shutterstock; p347 Chantelle Bosch/Shutterstock, Ben Queenborough/Shutterstock, Stubblefield Photography/Shutterstock, MIGUEL CASTRO/SCIENCE PHOTO LIBRARY; p348 sisqopote/Shutterstock; p352 dwphotos/Shutterstock, Nate Allred/Shutterstock; p353 Sergey Fatin/Shutterstock; p355 Denton Rumsey/Shutterstock, VOLKER STEGER/SCIENCE PHOTO LIBRARY; p356 sundetman/Shutterstock; p357 Lightspring/Shutterstock, Geoffrey Budesa/Shutterstock